Heterogeneous Catalysis and Advanced Oxidation Processes (AOP) for Environmental Protection (VOCs Oxidation, Air and Water Purification)

Heterogeneous Catalysis and Advanced Oxidation Processes (AOP) for Environmental Protection (VOCs Oxidation, Air and Water Purification)

Editor

Roberto Fiorenza

MDPI • Basel • Beijing • Wuhan • Barcelona • Belgrade • Manchester • Tokyo • Cluj • Tianjin

Editor
Roberto Fiorenza
Dip. Scienze Chimiche
Università di Catania
Catania
Italy

Editorial Office
MDPI
St. Alban-Anlage 66
4052 Basel, Switzerland

This is a reprint of articles from the Special Issue published online in the open access journal *Catalysts* (ISSN 2073-4344) (available at: www.mdpi.com/journal/catalysts/special_issues/aop_environ).

For citation purposes, cite each article independently as indicated on the article page online and as indicated below:

LastName, A.A.; LastName, B.B.; LastName, C.C. Article Title. *Journal Name* **Year**, *Volume Number*, Page Range.

ISBN 978-3-0365-3566-1 (Hbk)
ISBN 978-3-0365-3565-4 (PDF)

© 2022 by the authors. Articles in this book are Open Access and distributed under the Creative Commons Attribution (CC BY) license, which allows users to download, copy and build upon published articles, as long as the author and publisher are properly credited, which ensures maximum dissemination and a wider impact of our publications.

The book as a whole is distributed by MDPI under the terms and conditions of the Creative Commons license CC BY-NC-ND.

Contents

About the Editor vii

Preface to "Heterogeneous Catalysis and Advanced Oxidation Processes (AOP) for Environmental Protection (VOCs Oxidation, Air and Water Purification)" ix

Roberto Fiorenza
Heterogeneous Catalysis and Advanced Oxidation Processes (AOPs) for Environmental Protection (VOC Oxidation, Air and Water Purification)
Reprinted from: *Catalysts* 2022, 12, 317, doi:10.3390/catal12030317 1

Lilong Zhou, Chen Ma, Jonathan Horlyck, Runjing Liu and Jimmy Yun
Development of Pharmaceutical VOCs Elimination by Catalytic Processes in China
Reprinted from: *Catalysts* 2020, 10, 668, doi:10.3390/catal10060668 3

Roberto Fiorenza
Bimetallic Catalysts for Volatile Organic Compound Oxidation
Reprinted from: *Catalysts* 2020, 10, 661, doi:10.3390/catal10060661 53

Xin Huang, Luming Li, Rong Liu, Hongmei Li, Li Lan and Weiqi Zhou
Optimized Synthesis Routes of MnO_x-ZrO_2 Hybrid Catalysts for Improved Toluene Combustion
Reprinted from: *Catalysts* 2021, 11, 1037, doi:10.3390/catal11091037 79

Roberto Fiorenza, Roberta Agata Farina, Enrica Maria Malannata, Francesca Lo Presti and Stefano Andrea Balsamo
VOCs Photothermo-Catalytic Removal on MnO_x-ZrO_2 Catalysts
Reprinted from: *Catalysts* 2022, 12, 85, doi:10.3390/catal12010085 93

Abniel Machín, Kenneth Fontánez, José Duconge, María C. Cotto, Florian I. Petrescu and Carmen Morant et al.
Photocatalytic Degradation of Fluoroquinolone Antibiotics in Solution by Au@ZnO-rGO-gC_3N_4 Composites
Reprinted from: *Catalysts* 2022, 12, 166, doi:10.3390/catal12020166 111

William Vallejo, Alvaro Cantillo, Briggitte Salazar, Carlos Diaz-Uribe, Wilkendry Ramos and Eduard Romero et al.
Comparative Study of ZnO Thin Films Doped with Transition Metals (Cu and Co) for Methylene Blue Photodegradation under Visible Irradiation
Reprinted from: *Catalysts* 2020, 10, 528, doi:10.3390/catal10050528 129

Jan Bogacki, Piotr Marcinowski, Dominika Bury, Monika Krupa, Dominika Ścieżyńska and Prasanth Prabhu
Magnetite, Hematite and Zero-Valent Iron as Co-Catalysts in Advanced Oxidation Processes Application for Cosmetic Wastewater Treatment
Reprinted from: *Catalysts* 2020, 11, 9, doi:10.3390/catal11010009 143

Anne Heponiemi, Janne Pesonen, Tao Hu and Ulla Lassi
Alkali-Activated Materials as Catalysts for Water Purification
Reprinted from: *Catalysts* 2021, 11, 664, doi:10.3390/catal11060664 157

Kengo Hamada, Tsuyoshi Ochiai, Yasuyuki Tsuchida, Kyohei Miyano, Yosuke Ishikawa and Toshinari Nagura et al.
Eco-Friendly Cotton/Linen Fabric Treatment Using Aqueous Ozone and Ultraviolet Photolysis
Reprinted from: *Catalysts* **2020**, *10*, 1265, doi:10.3390/catal10111265 . . . . **175**

Roberto Fiorenza, Stefano Andrea Balsamo, Luisa D'Urso, Salvatore Sciré, Maria Violetta Brundo and Roberta Pecoraro et al.
CeO_2 for Water Remediation: Comparison of Various Advanced Oxidation Processes
Reprinted from: *Catalysts* **2020**, *10*, 446, doi:10.3390/catal10040446 . . . . **187**

Po-Yu Wen, Ting-Yu Lai, Tsunghsueh Wu and Yang-Wei Lin
Hydrothermal and Co-Precipitated Synthesis of Chalcopyrite for Fenton-like Degradation toward Rhodamine B
Reprinted from: *Catalysts* **2022**, *12*, 152, doi:10.3390/catal12020152 . . . . **203**

About the Editor

Roberto Fiorenza

Roberto Fiorenza (PhD) is a researcher in industrial chemistry at the department of Chemical Sciences of the University of Catania (Italy). He works on the synthesis and characterization of new materials (catalysts) for environmental and energy applications (purification of air, water, production and hydrogen, CO_2 valorization, etc.), focusing his research on semiconductors and on TiO_2-based materials. He is the author and co-author of various papers in international peer review journals and a reviewer of some international journals dealing with catalysis and material science. He teaches the courses Sustainable Industrial Chemistry (to Master's degree students in chemical science) and Sustainable Industrial Chemistry and Materials for the Environmental Protection (to PhD students).

In June 2019, he received the Adolfo Parmaliana 2019 Award from the catalysis group of the Italian chemical society for the best doctoral thesis on the theme catalysis for a sustainable development; the same work also earned him an award from the Gioenia Academy of Catania in the October 2018. In 2017, he received the Young Student Award at the international EMRS conference (European Materials Research Society) Spring Meeting 2017 in Strasbourg.

He carried out parts of his research work at the CNR-IMM of Catania (ESF SENTI project), at the University of Palermo with the research group of Prof. Leonardo Palmisano, at the University of Namur (Belgium) with the research group of Prof. Bao-Lian Su, at the Plataforma Solar de Almeria (Spain) with the research group of Prof. Sixto Malato and at the University of Duisburg-Essen (Germany) in the group of Prof. S. Reichenberger.

In February 2022, he achieved the National Scientific qualification as an associate professor in the Italian higher education system, for the disciplinary field of 03/C2 Industrial and Applied Chemistry.

Preface to "Heterogeneous Catalysis and Advanced Oxidation Processes (AOP) for Environmental Protection (VOCs Oxidation, Air and Water Purification)"

The quality of air and water is a key contemporary problem. The globalization economy and the current pandemic situation have given rise to new problems related to environmental protection. Catalysis science has always given smart, green, and scale-up eco-friendly solutions. In the last year, together with the traditional and efficient catalytic thermal treatments, new and emerging techniques such as photothermal treatments or advanced oxidation processes have provided good results both for air and water purification.

Based on the above considerations, this Special Issue on "Heterogeneous Catalysis and Advanced Oxidation Processes (AOP) for Environmental Protection (VOCs Oxidation, Air and Water Purification)"highlights the state of research on VOC oxidation (catalytic oxidation, photocatalytic oxidation or photothermal catalytic oxidation), air purification, and wastewater treatments (adsorption, membrane filtration, AOP, photocatalysis, Fenton and PhotoFenton, ozonation, etc.), as well as on the development of new catalysts for environmental protection.

From the high-quality papers presented in this book, researchers interested in these topics can find interesting and fascinating information, with contributions from some of the most important University and research centers in the world.

Roberto Fiorenza
Editor

Editorial

Heterogeneous Catalysis and Advanced Oxidation Processes (AOPs) for Environmental Protection (VOC Oxidation, Air and Water Purification)

Roberto Fiorenza

Department of Chemical Sciences, University of Catania, Viale A. Doria 6, 95125 Catania, Italy; rfiorenza@unict.it; Tel.: +39-0957385012

The quality of air and water is a crucial and critical contemporary problem. The more globalized economy, and the rapid growth of new economic powers, have given rise to new problems related to environmental protection. The science of catalysis has always given smart, green, and scalable, eco-friendly solutions. In the last year, together with the traditional and efficient catalytic thermal treatments, new and emerging techniques such as photothermal treatments or advanced oxidation processes (AOPs such as photocatalysis, Fenton and photo-Fenton, ozonation, etc.) have provided good results both in air and water purification [1,2]. All these technologies can also drive the transition from "classical" industrial chemistry to sustainable industrial chemistry, where not only are the processes involved in environmental purification versatile, green and environmentally friendly, but the employed catalysts are as well. This approach can help to overcome some relevant contemporary issues, some of which have been amplified by the COVID-19 pandemic situation. This has highlighted the necessity to ensure a high quality of air in both indoor and outdoor environments, or has drastically drawn attention to water polluted by the plastic waste (used face masks, for example). Plastics, together with other emerging water contaminants (such as pesticides, pharmaceutics, and antibiotics), are serious problems for water purity; conventional treatments do not work efficiently on these dangerous compounds.

All these aspects are well discussed and investigated in this Special Issue.

In particular, for air purification (including volatile organic compound (VOC) removal), the reviews of *Fiorenza* [3] and *Zhou and Yun* [4] examined two different aspects. The advantages and drawbacks on the utilization of bimetallic-based catalysts were analyzed in the first review, with particular attention on the bimetallic-gold-based samples, and considering both catalytic and the photocatalytic approaches [3]. In the second review, the attention focused on harmful pharmaceutical VOCs emitted in China. The developments in catalytic combustion, photocatalytic oxidation, non-thermal plasma, and electron beam treatments were discussed, together with the development of catalysts used in these processes [4].

The developments of eco-friendly catalysts and unconventional photocatalysts, not based on TiO_2, which can also represent possible solutions to the crisis of the raw material exportation [5] was also explored in two other papers of this Special Issue [6,7]; the good catalytic, photocatalytic, and phothermo-catalytic properties of MnO_x-ZrO_2 are presented, making these composites a promising future choice, as an example of an economical, not-critical, and high-performing catalyst applied for the removal of some dangerous VOCs such as toluene.

The innovative aspect that joins the water remediation from emerging contaminants with new materials was the core of this collection of papers. The degradation of fluoroquinolone antibiotics [8] and dyes [9], the treatment of cosmetic wastewaters [10], and the removal of bisphenol A [11] were originally investigated, employing Au@ZnO-rGO-gC_3N_4 [8], ZnO thin film [9], magnetite, hematite, and zero-valent iron [10], and alkaline active materials [11], respectively. These studies demonstrated synergisms between the photocatalytic properties of the new compounds, or that combinations of different AO can be the most attractive and valuable strategies to remove these recalcitrant pollutants from water. In this context, the use of aqueous ozone and UV photolysis represents sustainable solution for the bleaching of fabrics with a low environmental impact [12].

Finally, the Fenton and photo-Fenton-like processes were proposed to remove other water pollutants such as pesticides [13] and rhodamine B dye [14], using alternative catalysts such as reduced CeO_2 [13] and the chalcopyrite ($CuFeS_2$) [14].

In conclusion, as the Guest Editor of this Special Issue, I would like to extend appreciation to all the authors for their high-level articles, and I thank all the reviewers their comments on the manuscripts. I hope that readers will find the results in the article on this topic interesting and useful for their research. Thanks also to the editorial staff *Catalysts*, for their help and ensuring the success of this Special Issue.

Funding: This research received no external funding.

Conflicts of Interest: The author declares no conflict of interest.

References

1. Dong, G.; Chen, B.; Liu, B.; Hounjet, L.J.; Cao, Y.; Stoyanov, S.R.; Yang, M.; Zhang, B. Advanced oxidation processes microreactors for water and wastewater treatment: Development, challenges, and opportunities. *Water Res.* **2022**, *211*, 118. [CrossRef] [PubMed]
2. Zhao, W.; Adeel, M.; Zhang, P.; Zhou, P.; Huang, L.; Zhao, Y.; Ahmad, M.A.; Shakoor, N.; Lou, B.; Jiang, Y.; et al. A critical review on surface-modified nano-catalyst application for the photocatalytic degradation of volatile organic compounds. *Environ. Nano* **2022**, *9*, 61–80. [CrossRef]
3. Fiorenza, R. Bimetallic Catalysts for Volatile Organic Compound Oxidation. *Catalysts* **2020**, *10*, 661. [CrossRef]
4. Zhou, L.; Ma, C.; Horlyck, J.; Liu, R.; Yun, J. Development of Pharmaceutical VOCs Elimination by Catalytic Processes in China. *Catalysts* **2020**, *10*, 668. [CrossRef]
5. Cimprich, A.; Young, S.B.; Schrijvers, D.; Ku, A.Y.; Hagelüken, C.; Christmann, P.; Eggert, R.; Habib, K.; Hirohata, Hurd, A.J.; et al. The role of industrial actors in the circular economy for critical raw materials: A framework with case studies across a range of industries. *Miner. Econ.* **2022**. [CrossRef]
6. Huang, X.; Li, L.; Liu, R.; Li, H.; Lan, L.; Zhou, W. Optimized Synthesis Routes of MnO_x-ZrO_2 Hybrid Catalysts for Improved Toluene Combustion. *Catalysts* **2021**, *11*, 1037. [CrossRef]
7. Fiorenza, R.; Farina, R.A.; Malannata, E.M.; Lo Presti, F.; Balsamo, S.A. VOCs Photothermo-Catalytic Removal on MnO_x-Z Catalysts. *Catalysts* **2022**, *12*, 85. [CrossRef]
8. Machín, A.; Fontánez, K.; Duconge, J.; Cotto, M.C.; Petrescu, F.I.; Moran, C.; Márquez, F. Photocatalytic Degradation Fluoroquinolone Antibiotics in Solution by Au@ZnO-rGO-gC_3N_4 Composites. *Catalysts* **2022**, *12*, 166. [CrossRef]
9. Vallejo, W.; Cantillo, A.; Salazar, B.; Diaz-Uribe, C.; Ramos, W.; Romero, E.; Hurtado, M. Comparative Study of ZnO Thin Films Doped with Transition Metals (Cu and Co) for Methylene Blue Photodegradation under Visible Irradiation. *Catalysts* **2020**, *10*, [CrossRef]
10. Bogacki, J.; Marcinowski, P.; Bury, D.; Krupa, M.; Ścieżyńska, D.; Prabhu, P. Magnetite, Hematite and Zero-Valent Iron Co-Catalysts in Advanced Oxidation Processes Application for Cosmetic Wastewater Treatment. *Catalysts* **2020**, *11*, 9. [CrossRef]
11. Heponiemi, A.; Pesonen, J.; Hu, T.; Lassi, U. Alkali-Activated Materials as Catalysts for Water Purification. *Catalysts* **2021**, *11*, [CrossRef]
12. Hamada, K.; Ochiai, T.; Tsuchida, Y.; Miyano, K.; Ishikawa, Y.; Nagura, T.; Kimura, N. Eco-Friendly Cotton/Linen Fabric Treatment Using Aqueous Ozone and Ultraviolet Photolysis. *Catalysts* **2020**, *10*, 1265. [CrossRef]
13. Fiorenza, R.; Balsamo, S.A.; D'Urso, L.; Sciré, S.; Brundo, M.V.; Pecoraro, R.; Scalisi, E.M.; Privitera, V.; Impellizzeri, G. CeO_2 Water Remediation: Comparison of Various Advanced Oxidation Processes. *Catalysts* **2020**, *10*, 446. [CrossRef]
14. Wen, P.-Y.; Lai, T.-Y.; Wu, T.; Lin, Y.-W. Hydrothermal and Co-Precipitated Synthesis of Chalcopyrite for Fenton-like Degradation toward Rhodamine B. *Catalysts* **2022**, *12*, 152. [CrossRef]

Review

Development of Pharmaceutical VOCs Elimination by Catalytic Processes in China

Lilong Zhou [1,*], Chen Ma [1], Jonathan Horlyck [2], Runjing Liu [1] and Jimmy Yun [1,2,3,*]

[1] College of Chemical and Pharmaceutical Engineering, Hebei University of Science and Technology, Shijiazhuang 050018, China; machen33@126.com (C.M.); liurj@hebust.edu.cn (R.L.)
[2] School of Chemical Engineering, The University of New South Wales, Sydney NSW 2052, Australia; j.horlyck@outlook.com
[3] Qingdao International Academician Park Research Institute, Qingdao 266000, China
* Correspondence: llzhou@hebust.edu.cn (L.Z.); jimmy.yun@unsw.edu.au (J.Y.)

Received: 26 May 2020; Accepted: 9 June 2020; Published: 13 June 2020

Abstract: As a byproduct of emerging as one of the world's key producers of pharmaceuticals, China is now challenged by the emission of harmful pharmaceutical VOCs. In this review, the catalogue and volume of VOCs emitted by the pharmaceutical industry in China was introduced. The commonly used VOC removal processes and technologies was recommended by some typical examples. The progress of catalytic combustion, photocatalytic oxidation, non-thermal plasma, and electron beam treatment were presented, especially the development of catalysts. The advantages and shortages of these technologies in recent years were discussed and analyzed. Lastly, the development of VOCs elimination technologies and the most promising technology were discussed.

Keywords: VOCs; catalytic combustion; China; elimination technology; pharmaceutical industry

1. Introduction

Recently, the Chinese public has become increasingly concerned about the levels of chemical air pollution present in the form of haze. A main contributing factor to this pollution is the release of volatile organic compounds (VOCs) from the industry. VOCs are organic compounds with boiling points in the range of 50–260 °C at atmospheric pressure or with a Reid vapor pressure of over 10.3 Pa at room temperature (293.15 K) and atmosphere pressure (101.325 kPa) [1,2]. The pharmaceutical industry is a major source of these VOCs [3,4], which can have serious ramifications, such as toxicity, carcinogenesis, mutagenesis, photochemical pollution, haze, and fog [5–9]. The pharmaceutical industry has developed greatly in China, as the production of bulk drug intermediates and Chinese patent drugs increased from 205,070 to 340,830 kilotons, and 112,890 to 374,600 kilotons in the past 10 years (2007–2016), respectively [10]. Meanwhile, the amounts of released VOCs increased from about 174.8 to 393.2 kt in the pharmaceutical industry [11,12]. The guiding emission standards of VOCs went into effect on 1 January, 2018 in China, which led to new imputes to upgrade VOCs elimination equipment and technologies in factories across China [13].

2. Catalogue and Emission Amounts of VOCs in the China Pharmaceutical Industry

2.1. Catalogue of Pharmaceutical VOCs in China

Medicines in China are mainly produced in six ways, biological fermentation, chemical synthesis, extraction, coagulation preparation, bioengineering, and treatment of traditional Chinese medicine [4,14]. The different methods produce varying levels of pollution, with the order of emitted VOCs being biological fermentation > chemical synthesis > extraction > bioengineering > treatment of traditional Chinese medicine > coagulation preparation [15]. All these processes can be plagued by the need for high volumes of solvents, the consumption of large quantities of complex organic precursors, or the production of volatile byproducts. Some of the volatile organic compounds used or produced in these processes may be released into the atmosphere, which causes air pollution. The released VOCs primarily include alkanes, alcohols, ketones, aromatic hydrocarbons, halohydrocarbons, amines, esters, ether, aldehyde, carboxylic acid, and sulfur containing organic compounds [4,11,12,14,15].

Biological fermentation is often used to produce antibiotics, vitamins, and amino acids, via processes which include fermentation, separation, purification, and refinement [16,17]. The primary source of emitted VOCs are the solvents used in these processes, especially in the separation and purification steps. Additionally, H_2S is also produced as a byproduct in some fermentation processes.

The chemical synthesis technology is usually used for the production of medicine which can be used for prevention, cure, and diagnosis of the disease [18,19]. It contains the units to synthesize intermediates from raw materials, modify the structure of intermediates, purify the products, and dry the final products. The VOCs from this process comprise solvents and unreacted intermediates, which are more complex and often harder to be eliminated than those produced from the biological fermentation. These VOCs include heptane, toluene, xylene, methanol, n-propyl alcohol, isopropanol, phenol, aminomethane, dimethylamine, aminobenzene, cyclohexylamine, triethylamine, butyraldehyde, acetone, chloroform, chlorobenzene, etc.

Extraction involves the use of physical, chemical, and biological methods to separate a substance from a mixture. In the case of the pharmaceutical industry, extraction is carried out to separate organic compounds of interest from liquid solvents [16,20]. As a result, extraction processes require a large supply of solvents, including toluene, naphtha, methanol, ethanol, isopropanol, phenol, acetone, ether, acetic ether, diethylamine, dichloromethane, dichloloethane, chloroform, etc. VOCs are released during extraction mainly from the organic solvents used within the process.

The bioengineering method is a new way to produce some new medicines, via processes such as cloning antibodies, genetic engineering drugs, and genetic engineering vaccines [21]. The VOCs released from this process are similar to those of biological fermentation. These VOCs come from the solvents used within the process and as byproducts of certain reactions. They include n-hexane, methanol, ethanol, formaldehyde, acetaldehyde, formic acid, propanediol, acetone, aminoehtyl alcohol, acetonitrile, acetic acid, acetocaustin, N, N-dimethylformamide, phenol, butanone, 4-methyl-2-pentone, n-propyl alcohol, isopropanol, n-pentanol, isopropyl ether, isobutyl aldehyde, etc.

Another process of pharmaceutical production is the manufacture of traditional Chinese medicines or certain Chinese patent drugs [22]. As this is a traditional process which involves the use of only limited organic solvents to treat the natural animals and plants, the production of VOCs is limited. Hence, this process only emits a spot of VOCs, SO_2, and smoke.

Coagulation preparation involves the formation of larger particle agglomerates from fine particle suspensions. In the synthesis of pharmaceuticals, it is the process of mixing the active ingredients with helper constituents (often called coagulants) to produce a drug which has the desired particle size. The physical nature of the coagulation process means that the waste produced via this method is primarily a solid particle with little notable VOC production [4,14–16,22].

2.2. The Guiding Emission Standards of VOCs in China

The guiding emission standards of VOCs include six kinds of VOCs on the boundary of factories (Table 1) and four pollutant classifications with a total of 16 compounds in the areas surrounding workshops and installations (Table 2) [13]. These parameters are determined in 1 h by 3~4 samples for the average value. They are detected by the portable instruments or GC. These target VOCs are carcinogenic, odorous, and harmful organic compounds, which are mainly used as solvents and raw materials.

Table 1. The extreme emission value of volatile organic compounds (VOCs) on the boundary of factories and the highest concentration of selected VOCs allowed in 1 h for employees [13].

Pollutants	Limiting Value (mg·m^{-3})	The Highest Concentration of Selected VOCs Allowed in 1 h for Employees (mg·m^{-3})
Benzene	0.4	6
Formaldehyde	0.2	0.1
Trichloroethylene [a]	0.1	30
Dimethyl sulfate [a]	0.5	0.5
Dichloromethane [a]	4.0	0.5
Non-methane organic compounds	4.0	-
Ozone	20	0.26

a. They will be put into effect after the publication of national standards.

Table 2. The limiting value of specific pollutants in workshops and installations [13].

Classification	Pollutant	The Limiting Values (mg·m^{-3})	
		General Area	Key Area
Carcinogens	Trichloroethylene [a]	1	1
	Benzene	4	4
	Formaldehyde	5	5
Toxic substances	Phosgene	0.5	0.5
	HCN	1.9	1.9
	Acrolein	3	3
	Methyl sulfate [a]	5	5
	Cl_2	5	5
Photochemically active substances	Toluene	25	15
	xylene	40	20
	Dimethyl sulfoxide [a]	100	50
	Butylene oxide [a]	100	50
Other	NH_3	20	10
	HCl	20	10
	CH_3OH	50	30
	CH_2Cl_2 [a]	75	45

a. They will be put into effect after the publication of national standards.

2.3. VOCs Emissions in the Chinese Pharmaceutical Industry

China is the second largest producer of pharmaceutical products, only behind the United States of America. There are more than 1300 kinds of drug intermediates, 30 types of medicaments, and over 4500 pharmaceutical products made in China [23]. According to previous studies, the total VOCs emissions scaled linearly with the amount of final pharmaceutical products. [24]. The results showed that about 0.55 kg of VOCs were discharged to the atmosphere for the production of each 1 kg of final drug products.

The total VOCs emissions from the pharmaceutical industry increased by over 120% from about 174.8 kt in 2007 to 393.2 kt in 2016 (Figure 1) [12]. Although the VOCs emitted from the pharmaceutical industry only account for approximately 1.1% of China's total VOCs emissions, the absolute emission amount is very large. VOCs emitted from the pharmaceutical industry are potentially more harmful to human beings and ecosystems than the VOCs emitted from other sources. Compared with other emission sources, such as decoration, oil extraction and refining, catering, shoemaking and furniture manufacturing, the VOCs from pharmaceutical industries are more diverse with higher local concentrations and are harder to be eliminated.

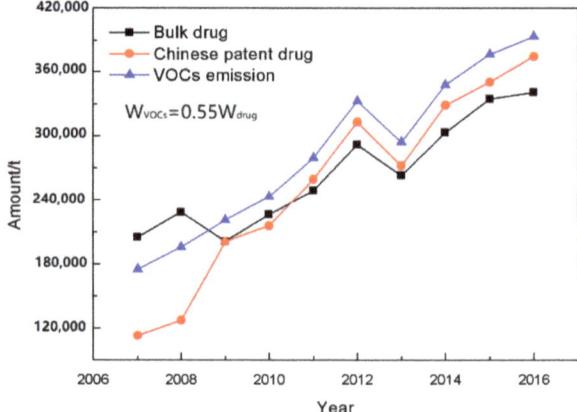

Figure 1. The production of bulk drugs and Chinese patent drugs and VOCs emissions from 2007 to 2016.

3. The Developing Technologies to Dispel VOCs

VOCs Elimination Technologies used in China's Pharmaceutical Industry

VOCs elimination technologies used in China were originally developed with the aim of recycling organic compounds to reduce cost, but recently the use and implementation of these technologies are targeted at minimizing the environmental impacts of VOCs. Technologies applied in the elimination of VOCs in China include condensation, absorption, adsorption, membrane purification, incineration, catalytic combustion, and the non-thermal plasma process (Figure 2) [15,25–27]. These methods are applied according to different working conditions, such as temperature and pressure, depending upon the VOCs targeted for removal. They also have distinct advantages and disadvantages. The development of these VOCs elimination technologies will be briefly introduced and discussed in this review. It is important to note that a wide range of VOCs which are often produced in a single waste stream and the various treatment technologies have different efficacies for the removal of certain VOCs. In fact, multiple technologies are often combined to eliminate the VOC mixtures from the waste streams.

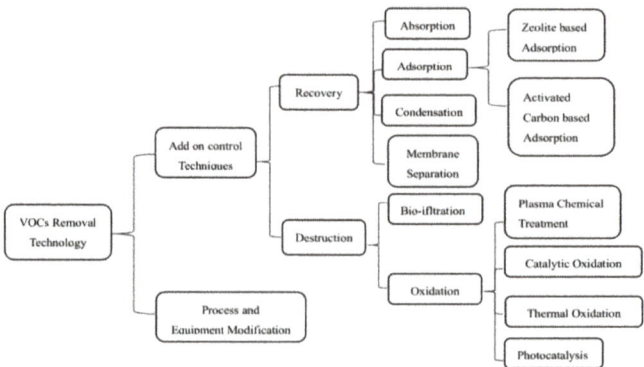

Figure 2. Classification of VOCs control techniques. Reprinted with permission from [26], 2000, Elsevier Ltd.

Condensation is the conversion of gas phase VOC mixtures with different vapor pressures to liquid via a decrease in temperature [28]. This technology is often used to recycle the solvents used in pharmaceutical production, with the key advantages that equipment requirements and operation of the condensation process are simple. It makes the cost of condensation lower than other technologies. Another advantage is that the gas produced from condensation is pure. Hence, condensation is often used as the first procedure to treat the waste gas from a pharmaceutical workshop before incineration and absorption to reduce the load on more complex and expensive downstream technologies. Water and air are the most commonly used cooling mediums for the condensation process, but ice, cold salt solutions and organic mediums have been used where cooling temperatures below 10 °C are required, such as the $CaCl_2$ solution, NaCl solution, and ethylene glycol aqueous solution [4]. The condensation efficiency is sensitive to the temperature and pressure and is suitable for the removal of high concentration VOCs which exist as liquefied at moderate temperatures [29,30].

The removal of VOCs via adsorption involves the use of porous materials, while absorption utilizes solvents. The porous materials used in adsorption need a high absorption capacity, large surface area, good pore structure, stable chemical properties, high physical strength, and tolerance of acidic/basic conditions. The adsorption materials commonly used in this technology include activated carbon, porous silica, zeolite, and porous resin [31–35]. The solvents commonly used in the absorption of VOCs are water, acid solution, alkali solution, and other organic compounds [36]. The components used in the adsorption process include spray columns, filled towers, columns of trays, and washing apparatuses.

Both the adsorption and absorption methods have a high VOC removal efficiency and can almost completely remove VOCs from waste gas with low energy consumption (Figure 3). They can be used to recycle organic solvents and valuable compounds while remaining economically feasible. This technology is often used for the treatment of a large flow of waste gas with low VOC concentration in processes such as fix bed adsorption, moving bed adsorption, fluid-bed adsorption, and pressure swing adsorption [37–39]. Disadvantages of the adsorption and absorption process are also noteworthy, such as huge equipment requirements, complex procedures, and the need for desorption and regeneration of saturated absorbents. Due to its high overall VOC removal efficiency, this technology is commonly used in many pharmaceutical factories.

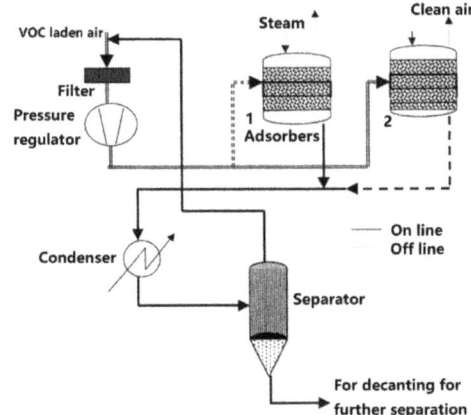

Figure 3. A typical activated carbon VOC removal (solvent recovery) plant. Reprinted with permission from [26], 2000, Elsevier Ltd.

Incineration is another widely used technology in pharmaceutical factories to eliminate VOCs (Figure 4). If VOC recycling is not technologically or financially feasible, incineration is a suitable method to completely eliminate VOCs [40]. Incineration is carried out by burning VOCs in a stove or kiln. Ideally, incineration results in the conversion of VOCs into CO_2 and H_2O in an efficient, simple, and safe manner. However, it has multiple shortcomings. If the VOCs concentration is too low to support the incineration, additional fuel is needed, which increases the running cost. Additionally, some VOCs are less suitable for incineration, because the incomplete combustion of halogenated and other harmful VOCs can result in the release of toxic chemicals such as dioxin, NO_x, and CO [40].

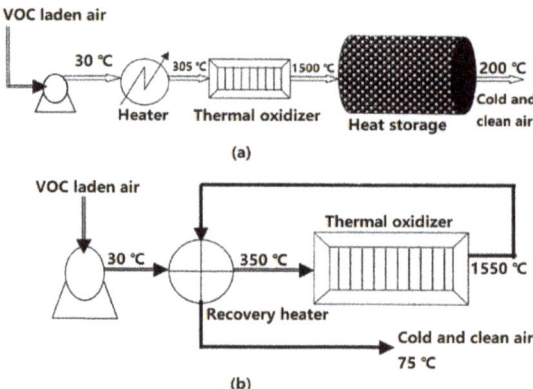

Figure 4. Schemes of thermal oxidation. (a) Regenerative thermal oxidation; (b) recuperative thermal oxidation. Reprinted with permission from [26], 2000, Elsevier Ltd.

To overcome the shortages of the incineration process for VOC destruction, the catalytic combustion technology was developed. In this process, VOCs are decomposed over catalysts at a low temperature (lower than 500 °C, Figures 5 and 6) [1,2,41–44]. The key factor which governs the catalytic combustion process is the activity of catalysts. Various kinds of catalysts have been used for the catalytic combustion of VOCs, such as noble metal catalysts, transition metal oxides, perovskite catalysts, and concentrated oxidation catalysts. The advantages for this technology are low operation temperature, decreased energy input requirements, high VOC removal efficiency, and minimal generation of toxic byproducts. Catalytic combustion is suitable for the treatment of waste streams containing VOCs across a wide range of concentrations. The main disadvantages of catalytic combustion are high investment requirements for equipment, short catalyst lifetime, and the need for process-specific designs, which are tailored to the waste stream. Nonetheless, the development of viable materials for the catalytic combustion process is still a hotspot for catalysis science.

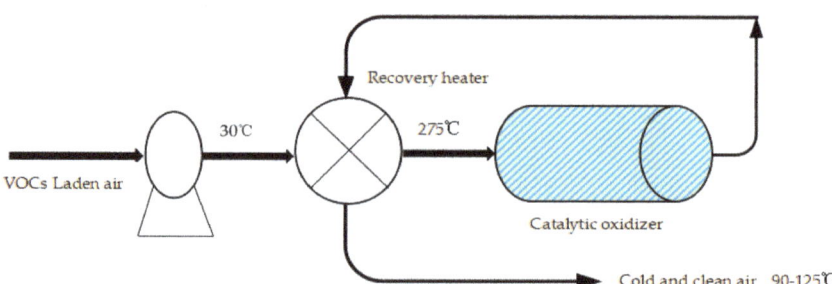

Figure 5. Scheme of catalytic oxidation. Reprinted with permission from [26], 2000, Elsevier Ltd.

The mechanism of catalytic combustion is considered by three types, with these being the Mars-Van Krevelen (MVK) model, Langmuir-Hinshelwood (LH) model, and Eley -Rideal (ER) model. In the Mars-Van Krevelen model, the VOCs molecules are initially adsorbed on the active sites, upon which they react with the oxygen species within the catalyst and are decomposed. Then, the reduced catalyst is re-oxidized by the supply of oxygen to the reactor. In the Langmuir-Hinshelwood model, the adsorbed VOCs molecules react directly with the adsorbed oxygen molecules, all occurring on the catalyst surface. In the Elay-Rideal mechanism, the adsorbed oxygen reacts with the VOCs molecules in the gas phase. The reaction pathway which follows depends on both the catalyst materials and the target VOCs in individual systems.

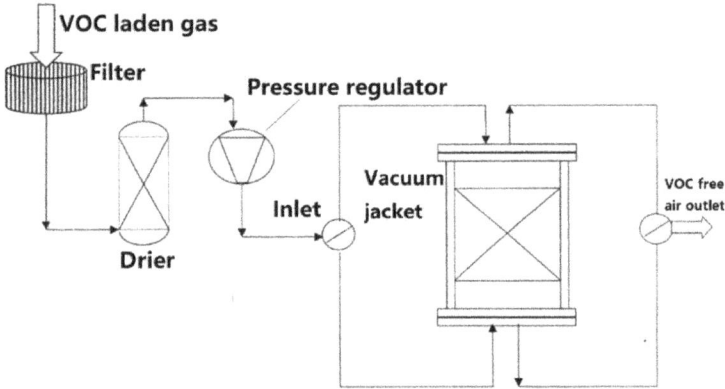

Figure 6. Schematic diagram of a reverse flow reactor. Reprinted with permission from [26], 2000, Elsevier Ltd.

Biodegradation is a widely used process for the treatment of pharmaceutical wastewater (Figure 7) [45,46]. It also can be applied to the treatment VOCs in the gas phase, especially for low concentration VOCs which are suitable for the growth of microorganisms. This process works with using the VOCs as a feedstock for the microorganisms, where they are converted to cytoplasm, CO_2, and H_2O. The sulfur and nitrogen elements in VOCs can be transformed to H_2S, nitrate, or N_2 at moderate temperatures. However, VOCs emitted from the pharmaceutical process often contain aromatics or halogens, which would poison the microorganism, rendering this method largely unviable in the abatement of pharmaceutical VOCs.

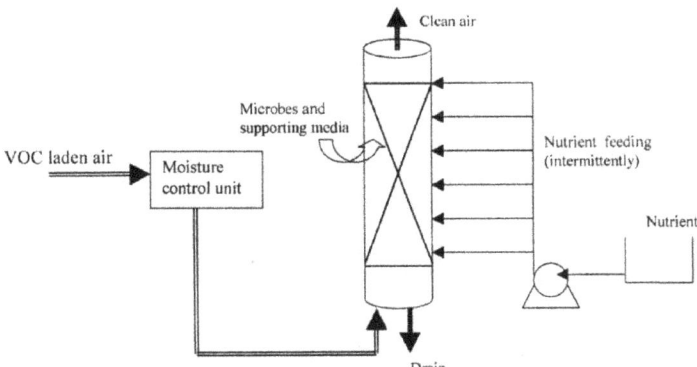

Figure 7. A simple schematic sketch of a bio-filtration system. Reprinted with permission from [26], 2000, Elsevier Ltd.

Non-thermal plasma is also a commonly used technology for the elimination of VOCs [47–49]. Free electrons and radicals formed during the plasma process react with VOCs and lead to the degradation of VOCs to CO_2 and H_2O. There are some advantages of the non-thermal plasma technology, such as a low press drop across the reactor, compact size, and simple equipment structure. The process can be started immediately without warm up and can treat VOCs with solid particles and liquid drops.

The non-thermal plasma technology can be combined with catalytic combustion to achieve a superior abatement of pharmaceutical VOCs.

The application of different VOCs elimination technologies depends on the factors such as the range and concentration of VOCs present in the waste stream, the volume of waste stream, and the funds available for installation and running costs. Jie Hao et al. summarized the scope of application for different VOCs elimination technologies (Table 3) [50]. Condensation and adsorption recycling are suitable for recycling VOCs. Catalytic combustion and incineration can remove the VOCs with a moderate concentration (3000–1/4 LEL) at high temperatures, while biodegradation and non-thermal plasma are suitable for abatement of low concentration VOCs at moderate temperatures. Hence, when selecting the efficient and economic VOCs abatement technologies, the scope of application for different technologies needs to be considered.

Table 3. VOCs elimination techniques and their operating conditions [50].

Abatement Technologies	VOC Concentration (mg·m^{-3})	Discharge Rate (m^3·h^{-1})	Temperature (°C)
Adsorption recycling	100–1.5×10^4	<6 × 10^4	<45
Preheated catalytic combustion	3000–1/4 LEL *	<4 × 10^4	<500
Thermal storage catalytic combustion	1000–1/4 LEL	<4 × 10^4	<500
Preheated incineration	3000–1/4 LEL	<4 × 10^4	>700
Thermal storage incineration	1000–1/4 LEL	<4 × 10^4	>700
Adsorption concentration	<1500	10^4–1.2 × 10^5	<45
Biodegradation	<1000	<1.2 × 10^4	<45
Condensation	10^4–10^5	<10^4	<150
Non-thermal	<500	<3 × 10^4	<80

* Lower explosive limit (LEL).

In a survey of 771 industrial applications of VOCs elimination processes, including 330 cases in China and 441 cases in various other countries (Figure 8), Jinying Xi et al. examined how often the various technologies were used [51]. The data showed that the most commonly used technology in China was adsorption (38%), followed by catalytic combustion (22%) and biodegradation (15%), while in other countries, the most often utilized were biodegradation (29%) and catalytic combustion (29%), followed by adsorption (16%). Due to its simple operation, low capital cost, and ability to recycle VOCs across a wide range of concentrations, adsorption was the most widely used technology in China. However, in some instances, the adsorption equipment was not well maintained and used correctly for recycling of VOCs. Adsorption, membrane purification, and condensation were the most effective methods, therefore most commonly applied, in instances where VOCs were present in concentrations above 10,000 mg·m^{-3}. Catalytic combustion and incineration were used for the destruction of VOCs in the concentration range of 2000~10,000 mg·m^{-3}, where recycling is not financially viable. Biodegradation and non-thermal plasma were applied for the treatment of VOCs in a lower concentration than 2000 mg·m^{-3}. Some examples are introduced in the following section to illustrate the application of these technologies for the elimination of pharmaceutical VOCs in China [3,4,14,51,52].

In the production of cefuroxime axetil, cefuroxime sodium, and cefotaxime sodium, the emitted VOCs include methanol, acetone, dichloromethane, DMF, acetic ether, and cyclohexane. One reported setup for the removal of these VOCs via a combination of condensation and adsorption technologies is described in Figure 9. Firstly, a portion of the various solvents was removed across a three-stage condensing unit comprised of a single stage of recycled water condensation, followed by two stages of 7 °C water condensation. The VOCs which were unable to be removed via the condensation process were removed by a two-stage activated carbon adsorption tower. The VOC-rich waste gas feedstock had a total VOC concentration of 2400 mg·m^{-3}, of which 800 mg·m^{-3} was attributed to methanol. It entered the first condensation unit at a rate of 2000 m³·h⁻¹. More than 95% of the total VOC content was removed after the combined condensation and adsorption process. As a result, the emitted concentration and discharge rate of methanol was reduced to 18.9–29.3 mg·m⁻³ and 0.08 kg·h⁻¹, respectively, while the concentration and discharge rate of the other VOCs present was 55–63 mg·m⁻³ and 0.16 kg·h⁻¹, respectively. Both of these levels were in accordance with the emission standards of Hebei province where the factory was located.

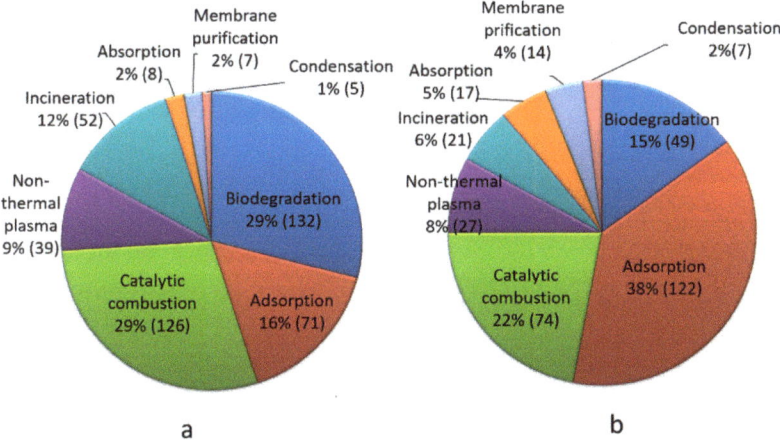

Figure 8. The ratios of VOCs elimination technologies used in China (a) and other countries (b) (the numbers in the brackets are the numbers of companies using the related technologies). Reprinted with permission from [51], copyright 2012, CNKI.

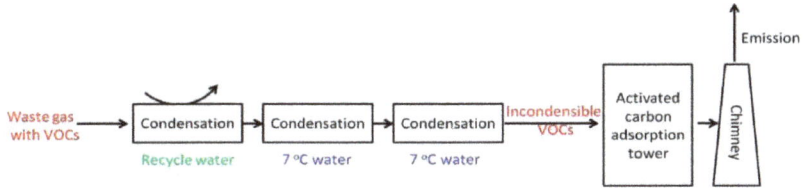

Figure 9. Representation of the VOC abatement process in a factory which produced cefuroxime axetil, cefuroxime sodium, and cefotaxime sodium. Reprinted with permission from [4], 2016, CNKI.

Yan Li has investigated the VOC abatement process in four separate companies in Taizhou city, Zhejiang province [4]. The main products in these four companies were clindamycin, clindamycin phosphate; losartan potassium, valsartan, nevirapine; meropenem, imipenem; and ciprofloxacin, spirolactone, respectively. There were three different procedures used in these four companies. The various procedures included adsorption, catalytic combustion, and non-thermal plasma technologies. The author assessed the efficiencies of the processes by measuring the concentration of a range of VOCs including benzene, toluene, xylene, methanol, formaldehyde, dichloromethane, chloroform, acetic ether, butylene oxide, acetonitrile, dimethylformamide, dysodia, and isopropanol before and after treatment. The results showed that after these elimination processes, the concentrations of VOCs in emitted waste gas were lower than the required concentration in standards for emissions of atmospheric pollutants from the pharmaceutical industry in Zhejiang province [53].

An example of a catalytic VOC treatment process is shown in Figure 10 below, which uses a regenerative catalytic oxidizer. The waste gas from the workshop was collected and combined with the waste gas from the sewage station, storehouse, and solid waste pile. The gas mixture was first washed with a water and alkali solution, then dehydrated and defogged. The dry gas was filtered, then heated with a preheater with an attached regenerative heat transfer. The preheated gas mixture entered the catalytic reactor and was combusted. After catalytic combustion, the gas was washed again with a water and alkali solution and emitted to high altitude atmosphere by a fan. The removal efficiencies of dysodia and non-methane hydrocarbons were 68.07% and 94.33%, respectively. The running fee of the whole VOCs elimination system was about RMB 1 million per year.

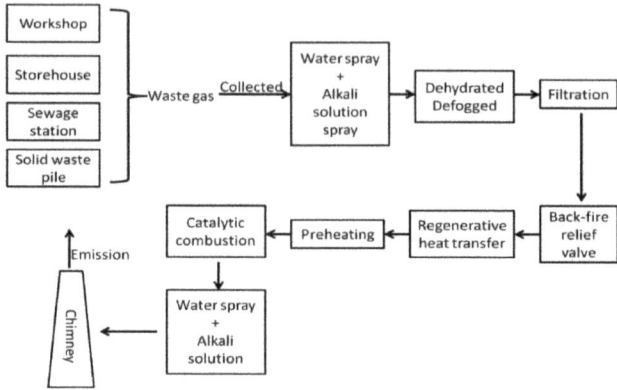

Figure 10. VOCs abatement process in the A factory. Reprinted with permission from [4], 2016, CNKI.

Two companies used a process which combines a regenerative thermal oxidizer (RTO) with three condensation stages connected in a series (Figure 11). The waste gas was pretreated in the workshop below determining the lower explosion limit (LEL) of the mixture. Then, fresh air was added in accordance with the determined LEL, such that the proper fuel, oxygen ratio, would be present in the RTO. After that, the gas mixture was oxidized in the RTO at a temperature of 850 °C with 98% of the thermal energy recycled. The high temperature gas exiting the RTO was cooled in a cooling tower, then further cooled across a three-stage condensation setup before it was emitted to the atmosphere. The dysodia concentration in the total vent was 300 mg·m^{-3}, while the concentration of non-methane hydrocarbons was lower than 85 mg·m^{-3}. These concentrations correspond to removal rates of 89% and 92% for dysodia and non-methane hydrocarbons, respectively. The main drawback of this procedure was that the condensation step was not effective for dichloromethane recycling and a lot of HCl generated from the incineration of chlorinated organic compounds, which in turn led to the corrosion of the equipment.

A process which combines the non-thermal plasma and catalytic oxidation techniques (oxidation of VOCs by H_2O_2 in a low pH) to treat VOCs in waste gas is outlined in Figure 12. The waste gas, with a high VOC concentration, was pretreated, combined with exhaust gas, and washed with an alkali solution. Then, the washed gas entered the catalytic combustion/oxidation tower, where a portion of the VOCs content was oxidized to CO_2 and H_2O. The oxidized waste gas was washed with water and entered a dehydrator to remove humidity. After that, the waste gas from the sewage station was added to the treated gas, upon which the gas mixture was treated with a non-thermal plasma, to remove additional VOCs. The gas was dehydrated again and entered into the second catalytic oxidation tower. The gas was washed by the alkali solution again and emitted to the atmosphere. The dysodia concentration was reduced by 84% to no more than 250 mg·m^{-3}. The removal efficiency of non-methane hydrocarbons content was approximately 92%, with a concentration of less than 85 mg·m^{-3} in the emitted gas.

These procedures (Figures 9–12 Figure 9 Figure 10 Figure 11 Figure 12) are representative of the range of VOCs elimination technologies currently in use in China. Almost all kinds of VOCs abatement technologies have been applied in the treatment of waste from the Chinese pharmaceutical industry. While the application of these technologies has successfully decreased the emission of VOCs, the high cost of investment and low efficiency are still the main factors which hinder the application of these technologies in the industry. The adsorption, absorption, and biodegradation technologies face challenges concerning the production of secondary pollutants, the desorption of adsorbed VOCs, and the production of waste water and sludge. The incineration process is effective in removing the issue of secondary waste production, but it requires a large energy input and has safety risks posed by the high temperature and use of a flame within a factory. The catalytic combustion and non-thermal plasma technologies partially circumvent the issues of waste and high temperature, but they are currently costly techniques with a short equipment lifetime. Thus, the continuing improvement of these technologies and development of new technologies is needed. With the new emission standards/law coming into effect, VOC elimination processes need to be upgraded in many Chinese pharmaceutical factories. As such, there is a desire for novel, effective, and energy efficient technologies in the near future.

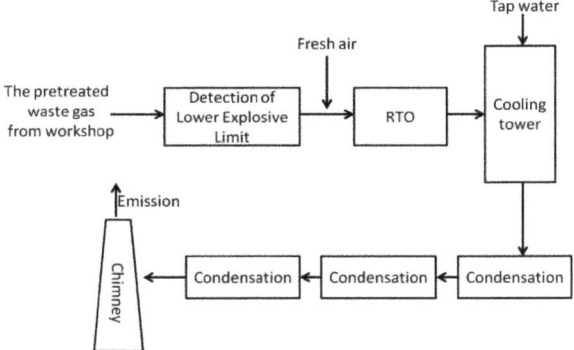

Figure 11. VOCs abatement process in the B and C factory, utilizing a regenerative thermal oxidizer (RTO) and condensation. Reprinted with permission from [4], 2016, CNKI.

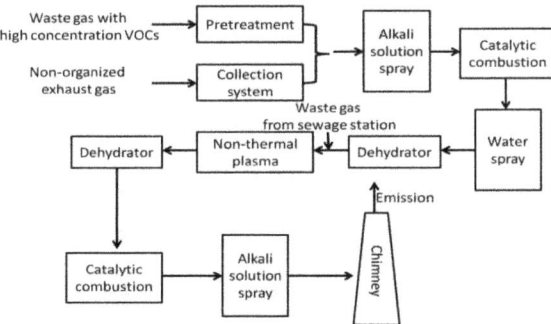

Figure 12. VOCs abatement process in the D factory. Reprinted with permission from [4], 2016, CNKI.

4. The Developing Technologies for VOCs Elimination

As outlined in the previous section, the VOC abatement technologies currently in use in China are hindered by some key limitations, such as high construction and running costs, low removal efficiencies for complex VOCs, and high temperature or energy requirements. To address these shortcomings, significant research is being carried out to improve the traditional methods of VOC removal, such as adsorption, catalytic combustion, and non-thermal plasma [1,2,33,41,43,44,47–49,54]. In addition, new technologies which can avoid the disadvantages of the traditional technologies have emerged, such as photocatalytic oxidation, condensation-oxidation, and electron beam treatment [2,42,48,55]. A summary of the various catalysts and conditions that have been investigated as active materials for the catalytic oxidation of VOCs is contained in Table 4 below.

Table 4. Reported catalysts for VOCs elimination.

VOCs	Refer	Catalyst	Catalytic Performance	Remarks	Catalytic Mechanism
	[56]	0.3% Pt/γ-Al$_2$O$_3$	T$_{95}$: 220 °C	-	-
	[57]	Pd-AuFeCeIM	T$_{100}$: 200 °C	-	-
	[58]	Layered copper manganese oxide (LCMO)	T$_{50}$: 218 °C T$_{90}$: 240 °C	The introduction of water vapor had no significant effect on the catalyst. The catalyst activity decreased about 30% after SO$_2$ was added. The addition of CO$_2$ has no effect on the catalytic activity of the catalyst.	The larger surface area and the formation of Cu^{2+}-O^{2-}-Mn^{4+} entities at the interface between CuO and layered MnO$_2$ promoted the CO and VOCs oxidation.
	[59]	LaCoO$_3$	T$_{50}$: 330 °C	Water vapor has no significant effect on the catalytic performance of catalyst.	Factors improve activity: (i) The larger exposed surface, (ii) the composition of the support provides increased oxygen mobility.
	[60]	LaMnO$_3$ LaCoO$_3$	T$_{50}$: 301 °C T$_{50}$: 323 °C	-	-
Benzene	[61]	γ-MnO$_2$/SmMnO$_3$	T$_{50}$: 213 °C T$_{90}$: 226 °C	-	Toluene-Mn ions+ surface reactive oxygen → benzaldehyde → benzoic acid → chain carboxylic acids → CO$_2$. Benzene → (splitting of benzene, hydrogenation, H abstraction by OH radicals) six-member ring cyclitols+aldehydes ketones etc. → a class of fulvene (isomerization) →CO$_2$ + H$_2$O.
	[62]	Mn/meso-TiO$_2$	-	-	Benzene is mainly degraded by photo-generated electron–hole pairs and hydroxyl radicals.
	[63]	Mesoporous TiO$_2$	-	-	
	[64]	Graphite-SiO$_2$-TiO$_2$	-	The graphite-SiO$_2$-TiO$_2$ composites exhibited higher photocatalytic activity for degradation of benzene gas under visible light irradiation than that of pure TiO$_2$.	Optimum concentration of graphite facilitates the separation of photogenerated electron–hole pairs for graphite-SiO$_2$-TiO$_2$ composites by visible light.
	[65]	Pt/Al$_2$O$_3$	-	-	-

Table 4. *Cont.*

VOCs	Refer	Catalyst	Catalytic Performance	Remarks	Catalytic Mechanism
	[66]	2.3 wt% Pt/3DOM-Mn$_2$O$_3$	T_{50}: 165 °C T_{90}: 194 °C	-	A first-order reaction mechanism. Toluene → cat. Surface → benzyl alcohol → benzoic acid and benzaldehyde → (rise temperature) maleic anhydride → CO$_2$ and H$_2$O.
	[67]	Pd(shell)-Au(core)/TiO$_2$	T_{50}: 220 °C T_{90}: 230 °C	-	Langmuir-Hinshelwood mechanism.
	[68]	Au-Pd/CeO$_2$	T_{50}: 120 °C T_{90}: 150 °C	-	Due to the synergistic effect between Au and Pd nanoclusters, Au-Pd/CeO$_2$ bimetallic catalysts are much better than Au and Pd single metal catalysts.
	[69]	Mn$_2$O$_3$	T_{50}: 231 °C T_{90}: 239 °C	-	Toluene +cat. → benzyl species + active oxygen species → aromatic alkoxide → enzaldehyde → benzoate species+active oxygen specie → maleic anhydrides→CO$_2$.
Toluene	[58]	Layered copper manganese oxide (LCMO)	T_{50}: 187 °C T_{90}: 207 °C	-	-
	[59]	LaCoO$_3$	T_{50}: 244 °C	-	-
	[70]	La$_{0.8}$Sr$_{0.2}$CoO$_3$	T_{50}: < 160 °C T_{99}: 300 °C	-	Partial substitution of strontium (Sr) for lanthanum (La) greatly increased the oxygen vacancies in the surface regions to enhance catalytic activity.
	[71]	SrTi$_{1-x}$Cu$_x$O$_3$ SrTi$_{1-x}$Mn$_x$O$_3$	T_{50}: 343 °C T_{90}: 398 °C T_{50}: 302 °C T_{90}: 335 °C T_{50}: 250 °C	-	Incorporation of Mn attributed a higher amount of oxygen vacancies in the perovskite surface to promote the toluene conversion. Supra facial mechanism.
	[72]	LaMn$_{1-x}$B$_x$O$_3$ (B = Co, Ni, Cu, Al)	234 °C 226 °C 242 °C T_{90}: none 278 °C 271 °C 318 °C	-	Addition of nickel into LaMnO$_3$ can improve the catalytic oxidation of toluene by the generation of more Mn^{4+} species and oxygen vacancies and the enhancing reducibility at a low temperature.

Table 4. Cont.

VOCs	Refer	Catalyst	Catalytic Performance	Remarks	Catalytic Mechanism
	[61]	γ-MnO2/SmMnO$_3$	T_{50}: 187 °C T_{90}: 208 °C	-	-
	[73]	LaMnO3/δ-MnO2	T_{90}: 258 °C	-	-
	[74]	LaMnO$_3$	T_{50}: 258 °C T_{90}: 275 °C	-	-
	[75]	MnO$_x$/H-Beta-SDS MnO$_x$/K-Beta-SDS MnO$_x$/Si-Beta	T_{50}: 253, 262, 280 °C T_{90}: 285, 295, 312 °C	-	Mars-Van Krevelen mechanism. Organic molecules + lattice oxygen→oxygen vacancy + CO_2 + H_2O, oxygen vacancy + O_2→lattice oxygen.
	[76]	Mn/R-SBA-15	T_{98}: 240 °C	-	MVK mechanism. Mn_2O_3-MnO_2/R-SBA-15 supply more lattice oxygen species.
	[77]	LaFeO3/black-TiO$_2$	The removal efficiency of black TiO$_2$ and LaFeO$_3$ for toluene was 89% and 98%, respectively, and the removal efficiency for IPA was 90% and 94%, respectively.	-	Cat.+photon (the wavelength shorter than 440 nm)→ electrons + O_2 → O_2^{2-}, O^{2-} +toluene and IPA → CO_2 + $H_2O_{(g)}$.
	[78]	Pd/CeO$_2$/γ-Al$_2$O$_3$	T_{98}: 205 °C	-	-
	[79]	CuO/ZnO Nanocomposite photocatalysts	-	-	Photooxidative activity and stability over ZnO are improved by loading CuO. (1) Direct removal caused by the collision of electrons or oxidation caused by the gas-phase radicals ($O\bullet$, $OH\bullet$, $N_2\bullet$, $NO\bullet$, $NO_2\bullet$) in gas phase; (2) the reaction between adsorbed toluene or other intermediates and the active species ($O\bullet$, $OH\bullet$) on the catalyst surface. O_2 might be fixed on the catalyst surface via facile interconversion between Fe^{2+} and Fe^{3+} states and then be transported to the toluene or intermediates leading to CO_2 formation.
	[80]	FeO$_x$/SBA-15	-	Under the condition of 3% Fe loading, the oxidation of toluene is the best one.	

Table 4. Cont.

VOCs	Refer	Catalyst	Catalytic Performance	Remarks	Catalytic Mechanism
Butanol	[56]	0.3% Pt/γ-Al$_2$O$_3$	T$_{95}$: 200 °C	-	-
CH$_3$SSCH$_3$	[81]	Pt-Au/Ce-Al	T$_{50}$: 425 °C T$_{90}$: 480 °C		The addition of Au improved the selectivity of ceria-containing catalysts by decreasing the formation of byproducts. This may have had a connection to a lower amount of reactive oxygen after the Au addition.
	[81]	Cu-Au/Ce-Al	T$_{50}$: 275 °C T$_{90}$: 370 °C	-	
Chlorobenzene	[82]	Pt/CrOOH	T$_{50}$: 340 °C T$_{90}$: 378 °C	-	Platinum accelerates the hydrolysis of Cr-Cl bonds formed at the CrOOH surface and determines the catalyst stability.
Dichloroethane	[82]	Pt/CrOOH	T$_{50}$: 283 °C T$_{90}$: 317 °C	-	
Dichloromethane	[83]	CeO$_2$	T$_{90}$: 260 °C T$_{90}$: 160 °C	The addition of 3% (v/v) water can obviously inhibit the catalytic decomposition of VOCs on CeO$_2$.	Trichloroethylene + CeO$_2$ →C$_2$HCl→HCl, Cl$_2$, CO$_2$ and trace CO.
Dichlorobenzene	[84]	H-ZSM-5 and Na-ZSM-5	-	-	-
Trichloro benzene	[84]	H-ZSM-5 and Na-ZSM-5	-	-	-
DCE	[85]	CeO$_2$@SiO$_2$-400	T$_{50}$: 219 °C T$_{90}$: 275 °C	The conversion of 1 vol% H$_2$O and 3 vol% H$_2$O decreased by 6% and 19%, respectively.	DCE+acid sites→HCl+ VC. (1) VC→reactive carbonium ion→adsorbed alcohol species→acetate species. Or (2) VC→1,1,2-trichloroethane→dichloroethylene formed (subsequent chlorination reactions) →H$_2$O, CO$_x$ and HCl.
Ethylbenzene	[61]	γ-MnO$_2$/SmMnO$_3$	T$_{50}$: 201 °C T$_{90}$: 217 °C		

Table 4. Cont.

VOCs	Refer	Catalyst	Catalytic Performance	Remarks	Catalytic Mechanism
Formaldehyde	[86]	TiO_2	TiO_2 degraded almost 100% of formaldehyde or acetaldehyde at a starting concentration of 400–500 ppb with a relative humidity of 40%.	-	The rate-determining step is the adsorption (external diffusion) on the catalysts active sites, thus the higher surface area, the higher the degradation.
Acetaldehyde	[86]	TiO_2	-	-	-
Acetone	[78]	$Pd/CeO_2/\gamma\text{-}Al_2O_3$	T_{98}:220 °C	-	-
	[87]	$CuO/g\text{-}Al_2O_3$	-	5.0 wt% $CuO/g\text{-}Al_2O_3$ catalyst has the highest removal rate of acetone, reaching 67.9%.	Both short-lived radicals and acetone/intermediates can be adsorbed on the catalyst surfaces to initiate a series of surface oxidation reactions, forming CO, CO_2, H_2O, and byproducts.
Ethyl acetate	[78]	$Pd/CeO_2/\gamma\text{-}Al_2O_3$	T_{98}:275 °C	-	-
Benzoquinone	[88]	$g\text{-}C_3N_4$	-	-	OH radicals+phenol→dihydroxycyclohexadienyl radical adducts→phenoxy radicals→H_2O(a very slow process), adducts+dissolved O_2→dihydroxy intermediates (-HO_2) →CO_2+H_2O.
Hydroquinone Catechol	[88]	$g\text{-}C_3N_4$	-	-	-

Table 4. Cont.

VOCs	Refer	Catalyst	Catalytic Performance	Remarks	Catalytic Mechanism
O-xylene	[89]	Ag/NiO$_x$-MnO$_2$	T$_{50}$: 145 °C T$_{90}$: 190 °C	The catalytic activity decreased with the addition of water vapor, but recovered after the removal of water vapor.	Mars-Van Krevelen mechanism. (i) Electrophilic O$_l$ +aromatic ring→maleate; (ii) O$_2$+ Cat.→electrophilic oxygen species (O$_2^{2-}$, O^{2-}, O$^-$), electrophilic oxygen species+aromatic ring→maleate, carbonate+(O$_2$: O$_2^{2-}$, O^{2-}, O$^-$ or O$_2$) +Ag→nucleophilic oxygen, nucleophilic oxygen + maleate and carboxylates→CO$_2$ and H$_2$O.
	[58]	Layered copper manganese oxide (LCMO)	T$_{50}$: 213 °C T$_{90}$: 227 °C	-	-
	[61]	γ-MnO2/SmMnO$_3$	T$_{50}$: 232 °C T$_{90}$: 250 °C	-	-
O-dichlorobenzene	[90]	CeMn30	GHSV = 15,000 mL/(g·h) T$_{50}$: 291 °C T$_{90}$: 347 °C GHSV = 7500 mL/(g·h) T$_{90}$: 224 °C GHSV = 30,000 mL/(g·h) T$_{90}$: 360 °C	The addition of low concentration of water vapor is beneficial to the early reaction, but will weaken the catalytic activity of catalyst.	O-dichlorobenzene molecules+basic oxygen (Mn-O-Ce-Vö, nucleophilic substitution)→HCl-H-O-Ce-Vö+phenyl. Mn-O-Ce weakened HCl-H-O-Ce-Vö and then prevented Deacon reaction and chlorine poisoning. Phenyl+ active lattice oxygen→organic intermediates →CO$_2$+H$_2$O. (Mars-Van Krevelen mechanism).
CO	[58]	Layered copper manganese oxide (LCMO)	T$_{50}$: 62 °C T$_{90}$: 76 °C	-	-
Cyclohexane	[70]	La$_{0.8}$Sr$_{0.2}$CoO$_3$	T$_{50}$: 180 °C T$_{99}$: 260 °C	-	-
CH$_3$SH	[91]	La(13)/HZSM-5	T$_{100}$: 40 °C	-	The La-modified HZSM-5 increased basic sites and displayed better adsorption ability to CH$_3$SH, decreased in strong acid sites and suppressed the formation of coke deposit.

Table 4. Cont.

VOCs	Refer	Catalyst	Catalytic Performance	Remarks	Catalytic Mechanism
C_2H_3CN	[92]	Cu-ZSM-5(SiO_2/Al_2O_3 = 26)	T_{90}: 325 °C	The catalyst has good resistance to steam poisoning.	Without H_2O: C_2H_3CN-SCC+O_2→NCO→oxidation products; with H_2O: C_2H_3CN-SCC+O_2 + H_2O→NH_3→products.
Cumene	[84]	H-ZSM-5 and Na-ZSM-5	-	-	-
Isopropanol	[60]	$LaMnO_3$ $LaCoO_3$	T_{50}: 216 °C 237 °C	-	-
	[79]	CuO/ZnO Nanocomposite photocatalysts	-	-	Photooxidative activity and stability over ZnO are improved by loading CuO.
	[93]	Au-Ag/CeO_2	T_{50}: 105 °C T_{90}: 158 °C	-	MVK redox mechanism.
Propyl alcohol	[60]	$LaMnO_3$ $LaCoO_3$	T_{50}: 203 °C 222 °C	-	-
	[70]	$La_{0.8}Sr_{0.2}CoO_3$	T_{90}: 160 °C	-	-
Propane	[94]	$LaCoO_3$	T_{50}: 208 °C T_{90}: 238 °C	-	MVK mechanism.
Methane	[95]	Nanocubic MnO_2	T_{50}: 293 °C T_{90}: 350 °C	-	CH_4+lattice oxygen or surface oxygen vacancies (MnO_2-C)→carboxylate species (Langmuir-Hinshelwood route), + active oxygen species→CO_2 and H_2O.

Table 4. Cont.

VOCs	Refer	Catalyst	Catalytic Performance	Remarks	Catalytic Mechanism
VOCs	[96]	MnO_2	T_{50}: 233 °C T_{90}: 256 °C	-	Toluene + Mn cations→CO_2 and H_2O.
	[97]	$LaMnO_3$ and $LaCoO_3$	-	-	-
	[98]	La^{3+}-TiO_2 and Nd^{3+}-TiO_2	1.2% La^{3+}-TiO_2 had the highest photocatalytic activity.	-	-
	[99]	TiO_2/Pd	The conversion rate of VOCs reached 90% when the residence time was 27 s.	-	-
	[100]	MesoTiO$_2$/hydro-CF	-	-	Promotion effects on degradation of gaseous polar acetone come from well crystallized anatase nanocrystals, hydro-CF skeleton for adsorption, and fast mass transportation within the hierarchical frameworks.
	[101]	CsX, NaX, and HY	T_{100}: 200 °C	-	Mars-Van Krevelen mechanism, involving several redox steps.
O_3	[102]	Mn/ZSM-5	O_3 can be efficiently decomposed by the Mn/ZSM-5 and used for benzene degradation through the OZCO.	-	Benzene+•OH→phenol→benzoquinone→CO_2 + H_2O.
	[103]	$CoMnO_x$/TiO_2	-	When the temperature is 320 °C, the decomposition efficiency of O_3 is 98%.	-
Vinyl chloride	[104]	HCl modified $La_{0.5}Sr_{0.5}MnO_3$	T_{100}: 300 °C	-	Doping and acid treatment obviously promote the active Mn^{4+} species amount and oxygen activation ability, and affect the chloric by-product distribution. Lower temperature inhibits the Deacon reaction and chlorination.

4.1. Catalytic Combustion

Catalytic combustion has received attention recently, as it shows great potential to address the shortcomings of the incineration method. Catalytic combustion is suitable for waste gas streams with low VOC concentrations and a moderate flow rate. Compared with the incineration method, catalytic combustion has been shown to efficiently remove VOCs from waste streams with a wide range of VOC concentrations. In addition, it operates at a lower temperature (293 versus 673 K) than incineration and resistant to the production of undesirable byproducts, such as dioxins and NO_x [105]. The diversity of VOC species necessitates the development of different kinds of catalysts for the combustion method, with commonly used catalysts including noble metals, non-noble metal oxides concentrated oxidation catalysts [1,2,41–44].

Previous reviews covering certain aspects of the catalytic combustion of VOCs have been published. For example, K. Everaert et al. reviewed, analyzed, and discussed the reaction kinetics, reactors, and reaction conditions of catalytic combustion research prior to 2004 [41]. Muhammad Shahzad Kamal et al. and Zhixiang Zhang et al. covered the recent progress in the development of combustion catalysts [2,44]. L.F. Liotta and W.B. Li et al. reviewed the mechanism of VOCs catalytic combustion over noble metal catalysts and non-noble metal catalysts, respectively [1,42]. In this section, we will briefly introduce the more recent development of new catalysts materials which have been used in this process, analyze the advantages and disadvantages they provide, and strategies for their successful implementation.

4.1.1. Noble Metal Catalysts

The general consensus of previous studies is that noble metal catalysts show the best catalytic performance in the combustion of non-halogenated VOCs. The noble catalysts which have been investigated include platinum, palladium, ruthenium, iridium, gold, and silver (Table 5) [2,41–44]. Due to their size-dependence catalytic properties and high price, noble metal catalysts are often supported on porous supports, such as γ-Al_2O_3, SiO_2, zeolite, and other non-metal oxides to increase the dispersion of noble metal nanoparticles and surface area, which can improve the catalytic efficiency of noble metal catalysts [106–111]. Catalysts are often supported on the substrate, which is in the form of a monolith or honeycomb material, such as cordierite, aluminum, and stainless steel [112]. Noble catalysts display high efficiencies for VOC removal at a lower temperature than other kinds of catalysts. Pt shows the best catalytic combustion of VOCs, exhibiting equal removal efficiencies at operating temperatures of up to 100 K lower than those used for other noble metals [42,44]. Two key factors were found to influence the catalytic performance of alumina supported Pt, dispersion and loading amount, with an increase in either of these properties associated with an obvious improvement of catalytic activity [106,113,114]. The metal particle size is also an important factor that influences the catalytic activity. Changes to the active metal particle size can also enhance catalytic performance. For instance, when the crystalline size of alumina supported Pt increased from 1.0 to 15.5 nm, the oxidation rate was found to increase by a factor of 10 [115]. The same phenomenon was also found in the oxidation of propylene over alumina supported Pt and Pd catalysts.

An important factor in the performance of noble metal catalysts for VOC combustion is specificity; in other words, the target VOCs molecules. For instance, P. Papaefthimiou et al. found that Pt and Pd supported on alumina showed good performance for the oxidation of benzene and butanol, but not ethyl acetate, with Pd generally outperforming Pt [57]. M. J. Patterson et al. discovered that alumina supported rhodium is the most active noble metal for 1-hexene, but not for aromatics, while benzene can be decomposed the most easily on platinum, and palladium showed the best catalytic performance in abatement of toluene [116]. The impact of certain non-VOC species in the waste stream can also impact performance, where these species can potentially 'poison' the catalyst or cause side reactions to occur. One example is the presence of CO, which was found to have little effect on the performance of Pd catalysts, but significantly inhibit the activity over Pt [117].

Table 5. Noble metal catalysts for catalytic combustion of VOCs.

Catalyst	VOCs	Catalytic Performance	Remarks	Catalytic Mechanism
0.3% Pt/γ-Al$_2$O$_3$	Butanol	T$_{95}$: 200 °C	-	-
Pd-AuFeCeIM	Benzene	220 °C	-	-
	Benzene	T$_{100}$: 200 °C	-	Langmuir-Hinshelwood mechanism.
Pd(shell)-Au(core)/TiO$_2$	Toluene	T$_{50}$: 220 °C T$_{90}$: 230 °C	-	
Au-Pd/CeO$_2$	Toluene	T$_{50}$: 120 °C T$_{90}$: 150 °C	-	Due to the synergistic effect between Au and Pd nanoclusters, Au-Pd/CeO$_2$ bimetallic catalysts are much better than Au and Pd single metal catalysts.
Au-Ag/CeO$_2$	Isopropanol	T$_{50}$: 105 °C T$_{90}$: 158 °C	-	MVK redox mechanism.
Pt-Au/Ce-Al	CH$_3$SSCH$_3$	T$_{50}$: 425 °C T$_{90}$: 480 °C	-	The addition of Au improved the selectivity of ceria-containing catalysts by decreasing the formation of byproducts due to lower amount of reactive oxygen after the Au addition.
Cu-Au/Ce-Al	CH$_3$SSCH$_3$	T$_{50}$: 275 °C T$_{90}$: 370 °C	-	-
Ag/NiOx-MnO$_2$	O-xylene	T$_{50}$: 145 °C T$_{90}$: 190 °C	The catalytic activity decreased with the addition of water vapor, but recovered after the removal of water vapor.	Mars-Van Krevelen mechanism. (i) Electrophilic O$_1$ +aromatic ring→maleate; (ii) O$_2$+ Cat.→electrophilic oxygen species (O$_2^{2-}$, O^{2-}, O$^-$), electrophilic oxygen species + aromatic ring→maleate, carbonate+(O$_2$, O$_2^{2-}$, O^{2-}, O$^-$ or O$_2$) +Ag→nucleophilic oxygen, nucleophilic oxygen + maleate and carboxylates→CO$_2$ and H$_2$O.
2.3 wt% Pt/3DOM-Mn$_2$O$_3$	Toluene	T$_{50}$: 165 °C T$_{90}$: 194 °C	-	A first-order reaction mechanism. Toluene →cat. surface→benzyl alcohol→benzoic acid and benzaldehyde→(rise temperature) maleic anhydride→CO$_2$ and H$_2$O.

Transition metal oxides can be utilized as both supports and promoters for noble metal catalysts. Pt, Pd, Ru, Au supported on MgO, SnO_2, Co_3O_4, NiO, TiO_2, CeO_2, La_2O_3, ZrO_2 or PrO_2 have been explored for the oxidation of toluene, benzene, xylene, propene, light alkane, ethanol, propanol, butanol, formaldehyde, acetone, and acetic acid, respectively [42]. The role of the transition metal oxides is not only to supply a large surface area to disperse the noble metal particles, but in some cases, it can also improve the catalytic performance of noble metal particles by enhancing the mobility of lattice oxygen species. Previous studies showed that active oxygen species formed on cobalt oxide spinel-type crystallites can enhance the catalytic oxidation over PdO supported on alumina [118,119].

The reduction properties also influence the oxidation ability of supported noble metal catalysts. T. Mitsui et al. prepared SnO_2, CeO_2, and ZrO_2 supported Pt and Pd catalysts for the abatement of acetaldehyde [120]. These prepared catalysts were treated in an H_2/N_2 flow and calcined in the atmosphere. The results showed that the SnO_2 supported Pt and Pd showed the best catalytic performance among the calcined catalysts in the atmosphere, while after treatment in an H_2/N_2 flow, the catalytic activity of SnO_2 supported Pt and Pd decreased due to the formation of inactive inter-metallic phases (PtSn and Pd_3Sn_2). In contrast, CeO_2 and ZrO_2 supported catalysts showed the improved catalytic activity after reduction. In the elimination of formaldehyde over a TiO_2-supported catalyst, Pt/TiO_2 showed a superior catalytic performance to Rh/TiO_2, Pd/TiO_2, Au/TiO_2, and neat TiO_2 [121]. Other research showed that a series of supported Pt, Pd, and Au catalysts can even partially eliminate formaldehyde at room temperature [122].

Recently, the use of MnO_x based materials as catalyst supports has gained attention. In one instance, supporting Ag on NiO-doped MnO_2 showed a high activity towards the combustion of o-xylene [89]. The improvement in xylene oxidation was attributed to the enhanced oxygen activation and mobility afforded to the catalyst support via the addition of NiO and Ag. Wenbo Pei et al. explored the use of ordered, mesoporous Mn_2O_3 supports with embedded Pt particles for the catalytic combustion of toluene [66]. They found that the strong interaction between Pt and Mn_2O_3 in the ordered structure improves the activity and stability of the catalyst.

The single noble metal catalysts cannot satisfy the requirements of VOCs combustion. Therefore, some mixed noble metal catalysts have been developed to combine the advantages of different noble metal catalysts, such as Pt-Au, Cu-Au, and Pd-Au. T. Tabakova et al. found that the Pd deposition on the deposited gold showed the best catalytic performance for benzene combustion, which was totally eliminated at 200 °C. It also showed good stability [57]. M. Hosseini et al. showed that the deposition of palladium on aurum supported on TiO_2 (Pd(shell)-Au(core)/TiO_2) can significantly improve the catalytic activity for oxidation of toluene and propylene [67]. Der Shing Lee et al. deposited Au-Pd bimetallic nanoparticles on CeO_2 for toluene degradation, which showed a much better catalytic performance than Au/CeO_2 and Pd/CeO_2 catalysts due to the synergistic effect of gold and palladium [68]. The addition of non-noble metals also can improve the catalytic activity of noble metal catalysts. Roberto Fiorenza et al. prepared $Au-Ag/CeO_2$ and $Au-Cu/CeO_2$ bimetallic catalysts for alcohol oxidation and CO oxidation. These two catalysts showed higher selectivity for intermediate products higher CO conversion at a low temperature (100 °C) than Au/CeO_2 [93]. The addition of Au also can improve the performance of Al_2O_3 supported Cu-Pt catalysts in DMDS oxidation [81].

Noble metal catalysts showed a high catalytic activity and remarkable thermal stability in catalytic elimination of VOCs. However, the use of noble metal catalysts is also associated with distinct disadvantages. Firstly, the high cost of noble metal limits their application in the industrial abatement of VOCs. Secondly, the presence of chlorine, sulfur, CO, and water can suppress the catalytic performance significantly [42,123–125]. The regeneration and recycling of noble metal catalysts poisoned by Cl and S is difficult, so they are not suited to the treatment chlorine and sulfur-containing VOCs. In fact, the release of chlorine and sulfur containing VOCs is particularly common in pharmaceutical production processes, so chlorine- and sulfur-containing VOCs need to be removed prior to the treatment by noble metal catalysts, which will further increase the cost of waste gas purification.

4.1.2. Non-Noble Metal Catalysts

To address the cost of noble metals, non-noble metal oxides catalysts were developed for the abatement of VOCs. The materials which have been studied as non-noble metals include the derivatives of transition metals and rare earth elements, such as Ti, Cu, Mn, Al, Ce, Co, Fe, Cr, and V (Table 6) [1,126–129]. Although transition metal oxides catalysts generally showed a lower catalytic activity than noble metal catalysts for the oxidation of VOCs, they have many advantages, such as a resistance to chlorine and sulfur poisoning, tunable material properties, low cost, long on-stream lifetime, easy regeneration, and low environmental impact. The non-noble metal oxides catalysts applied and studied in VOCs abatement include CuO_x, MnO_2, FeO_x, NiO_x, CrO_x, and CoO_x.

The non-noble metal oxides catalyst systems which we will discuss include ones in which the metal oxide is both supported or unsupported. Due to the presence of mobile oxide species in lattice, Co_2O_3 displays excellent reduction and oxidation abilities. As such, studies have shown Co_2O_3 to be one of the best catalysts used for the combustion of benzene, toluene, propane, 1,2-dichloroethane, and 1,2-dichlorobenzene [130–133]. The catalytic activity of Co_2O_3 is determined by the method of preparation, treatment conditions, and surface area.

MnO_2 is another commonly used metal oxide catalyst, which has been applied and studied in the abatement of n-hexane, acetone, benzene, ethanol, toluene, propane, trichloroethene, ethyl acetate, and NO_x [133–139]. The catalytic activity can be tuned by the preparation method and depends on the structure, surface area, support materials, and oxidation states of catalysts. In the catalytic combustion of ethyl acetate and hexane, MnO_2 even achieved a better activity than Pt/TiO_2 [140]. Yonghui Wei et al. removed the La atoms from the $LaMnO_3$ perovskite to prepare MnO_2 with a high surface area (>150 m^2/g), upon which it showed excellent catalytic activity in the oxidation of toluene [96]. Zhang Kai et al. synthesized the nano-cubic MnO_x which has a large specific surface area, many oxygen vacancies, and good low temperature reducibility. The conversion of toluene via combustion was more than 90% at 350 °C [95]. Xueqin Yang et al. found that the acid treatment did not change the morphology of the catalyst, but could improve the oxidation ability of the catalyst by increasing the number of Mn^{4+} species and structural defects on the surface of the catalyst [69]. A common theme throughout the implementation of Mn-based metal oxide catalysts for VOC oxidation is the availability and mobility of oxygen within the MnO_x, which is attributed to the oxidation and reduction ability of the Mn afforded by the multiple oxidation states in which it can exist.

Copper oxides are another kind of efficient catalysts used in total oxidation of methane, methanol, ethanol, and acetaldehyde [141,142]. The main factors which influence the catalytic activity are the Cu oxidation state and the availability of lattice oxygen. The addition of other metal oxides, such as CeO_2, can enhance the catalytic ability noticeably [133].

Chromium oxides are also promising oxidation catalysts, especially for the combustion of halogenated VOCs [124,133,143]. For chromium oxides, highly crystalline samples showed a better catalytic activity than amorphous ones [144]. Rotter et al.'s research showed that, when using TiO_2 as a support material, chromium oxides achieved higher catalytic oxidation of trichloroethylene than manganese oxide, cobalt oxide, and iron oxide [82]. Chromium has also been successfully supported on silica, alumina, porous carbon, and clay to eliminate pollutants such as carbon tetrachloride, chloromethane, trichloroethylene, ethyl chloride, chlorobenzene, and perchloroethylene [142]. However, chromium oxides also suffer deactivation due to the reaction between chromium and chlorine to form Cr_2Cl_2 [145,146].

CeO_2 is a widely used catalyst in oxidation reactions due to its strong interactions with other metals, high oxygen storage capacity, and ready shuttling between the Ce^{3+} and Ce^{4+} states [147–149]. Dai et al. compared the removal of chlorinated alkanes and alkylenes over CeO_2 [83]. The results showed that CeO_2 is more efficient when it comes to oxidizing chlorinated alkanes than chlorinated alkylenes. CeO_2 also faced the deactivation problem due to the absorption of Cl_2 and HCl on the surface [150], so the design of chlorine resistance metal oxide catalysts is still a challenge which must be overcome.

Vanadium oxides were also developed to decompose chlorinated VOCs, such as polychlorinated pollutants and dichlorobenzene due to its tolerance of chlorine and sulfur compounds [151]. The presence of water can enhance and suppress the catalytic activity of V_2O_5 via the removal of surface absorbed chlorine and reduction of active sites, respectively [152]. Other non-noble metal oxides were also investigated for abatement of VOCs, such as NiO and FeO_x, which require further improvement of catalytic efficiency [153].

The above discussion suggests that the use of a single metal oxide as VOC oxidation catalysts is too often plagued by either a low catalytic activity or catalyst poisoning. Thus, focus has shifted to the development of mixed metal oxide catalysts such as Mn-Ce, Mn-Cu, Co-Ce, Sn-Ce, Mn-Co, and Ce-Cu oxides [119,154–157]. The logic here is that combining two metal oxides with different materials and catalytic properties allows for a synergistic enhancement in performance. The previous studies showed that the rate determining step of VOC catalytic combustion was the oxygen removal from the catalysts lattice [119], so the goal of mix-metal oxides catalysts design was the enhancement of the lattice oxygen species availability.

The addition of copper into CeO_2 can promote the catalytic efficiency due to a synergistic effect, leading to a highly efficient decomposition of ethyl acetate, ethanol, propane, benzene, and toluene [158–160]. MnO_x-CeO_2 has been applied for the destruction of ethanol, formaldehyde, hexane, phenol, ethyl acetate, and toluene [155,161–164]. Mn-Co oxides catalysts also showed improved catalytic activity relative to either MnO_x or Co_2O_3 in the combustion of ethyl acetate and n-hexane [100]. CeO_2-CrO_x showed excellent catalytic activity for the decomposition of chlorinated VOCs [165], while the removal of chlorobenzene over MnO_x-TiO_2 and MnO_x-TiO_2-SnO_x showed much better catalytic performance than not only the individual oxides, but also achieved removal efficiencies on par with noble metal catalysts [166]. A three-dimensional ordered mesoporous material of mixed cerium-manganese oxide was prepared for the efficient catalytic combustion of chlorine-containing VOCs due to its large specific surface, enriched Ce^{3+} content, oxygen vacancies, active oxygen species, and acidic sites. It showed good water resistance and high airspeed applicability. However, catalyst deactivation caused by inorganic chlorine adsorption still occurred [90]. Layered copper manganese oxide has been prepared for the catalytic combustion of CO and VOCs, which showed efficient activity due to the interfacial structure of mixed phases and the formation of the Cu^{2+}-O^{2-}-Mn^{4+} entity [58]. Acidic sites can be provided by support to prevent the decrease of catalytic activity. CeO_2@SiO_2 was prepared to catalyze the combustion of 1,2-dichloroethane. SiO_2 can provide weak acid sites, as well as promote the adsorption and activation of 1,2-dichloroethane and the desorption of generated HCl [85].

A vast number of preparation methods exist for the synthesis of mixed metal oxide catalysts, including thermal decomposition, impregnation, co-precipitation, and the sol-gel method [167–169]. The selection of preparation methods depends on the properties of catalysts and the application situation. Furthermore, as discussed above, complex ordered microporous, multilayer or core-shell structures have recently been applied to catalytic oxidation processes to access properties which come from having a highly controlled particle composition and morphology. The chance to alter not only the metal centers present within mixed metal oxide systems, but also the relative metal ratios.

Table 6. Non-noble metal catalysts for catalytic combustion of VOCs.

Refer	Catalyst	VOCs	Catalytic Performance	Remarks
[69]	Mn_2O_3	Toluene	T_{50}: 231 °C T_{90}: 239 °C	-
[82]	Pt/CrOOH	Dichloroethane Chlorobenzene	T_{50}: 283 °C T_{90}: 317 °C T_{50}: 340 °C T_{90}: 378 °C	-
[83]	CeO_2	Tetrachloroethylene dichloromethane	T_{90}: 260 °C T_{90}: 160 °C GHSV = 15,000 mL/(g·h)	The addition of 3% (v/v) water can obviously inhibit the catalytic decomposition of VOCs on CeO_2.
[90]	CeMn30	O-dichlorobenzene	T_{50}: 291 °C T_{90}: 347 °C GHSV= 7500 mL/(g·h) T_{90}: 224 °C GHSV = 30,000 mL/(g·h) T_{90}: 360 °C	The addition of low concentration of water vapor is beneficial to the early reaction, but will weaken the catalytic activity of the catalyst.
[96]	MnO_2	VOCs	T_{50}: 233 °C T_{90}: 256 °C	-
[95]	Nanocubic MnO_2	Methane	T_{50}: 293 °C T_{90}: 350 °C	-

4.1.3. Perovskite Catalysts

Perovskite-type oxides are a kind of composite oxides which have a similar structure with $CaTiO_3$, and can be expressed by ABO_3. The common way to modify the perovskite catalysts is replacement of the cation B by B′ to tune the redox ability or enhance the stability [1]. With the replacement of the cation B, the crystal lattice would be distortion which leads to the enhancement of redox ability and improvement of stability. The most commonly used perovskite for catalytic combustion of VOCs is $LaBO_3$, in which B can be Co, Fe, Ni, Mn, and Sr (Table 7) [60,97,170,171]. Huang et al. used Sr partially replaced La in $LaCoO_3$ for catalytic combustion of propyl alcohol, toluene, and cyclohexane [70]. The results showed that the doped $LaCoO_3$ showed a better catalytic performance than the undoped one, and the modified catalysts were stable in the reaction. R. Spinicci et al. compared the catalytic activity of $LaMnO_3$ and $LaCoO_3$ for catalytic combustion of acetone, isopropanol, and benzene [60]. They suggested that $LaMnO_3$ showed a better performance than $LaCoO_3$. In oxidation of isopropanol, acetone was the intermediate product. The surface oxygen species played a key role in this process. The increase of oxygen pressure is positive for the catalytic combustion of VOCs over these perovskite catalysts. G. Sinquin et al. applied $LaMnO_3$ and $LaCoO_3$ for the catalytic combustion of chlorinated VOCs, such as CH_2Cl_2 and CCl_4. $LaMnO_3$ showed a better chlorine resistance than $LaCoO_3$ [97]. Mihai Alifanti et al. supported $LaCoO_3$ on cerium-zirconium oxides ($Ce_{1-x}Zr_xO_2$, x = 0–0.3) for the catalytic combustion of benzene and toluene [172]. The results showed that all the supported catalysts showed a better performance than $Ce_{1-x}Zr_xO_2$ and 20% loaded $LaCoO_3$ showed about 10 times higher catalytic activity than $LaCoO_3$ for toluene oxidation due to its large surface area and good oxygen mobility. S. I. Suárez-Vázquez et al. synthesized $SrTi_{1-x}B_xO_3$ (B = Mn, Cu) for toluene destruction [71]. Mn could replace Ti and enter the perovskite structure, while Cu could not. The Mn doped catalysts showed the highest catalytic activity and can completely decompose toluene to CO_2 at a temperature lower than 350 °C. Perovskite also can be prepared from solid waste such as the obsoleting lithium battery. Mingming Guo et al. prepared manganese-based perovskite catalyst from the waste lithium battery for catalytic combustion of toluene, which showed a better catalytic activity than pure manganese perovskite catalyst due to more Mn^{4+} ions and lattice oxygen species, as well as high specific surface area [72]. In order to increase the amounts of active sites, Junxuan Yao et al. removed the La ions from $LaCoO_3$ to obtain the disordered Co_3O_4. It showed a better catalytic activity for propane combustion than the one prepared by other methods [94]. To further improve the catalytic activity, γ-MnO_2 was calcined on the surface of $SmMnO_3$ which had a large specific surface area, high Mn^{4+}/Mn^{3+} and O_{latt}/O_{ads}. Compared with $SmMnO_3$, it showed a better catalytic activity and stability (10 vol% water) in the process of catalytic reaction [61]. Jingsi Yang et al. assembled the $LaMnO_3$ perovskite in MnO_2 and adjusted La/Mn to 15. The redox ability of the catalyst was improved by enhancing the interaction between the active phase and the support [73]. The ratio of citric acid and metal ion ($La^{3+}Mn^{2+}$) was also tested to find out the best composite of perovskite catalysts. Zakaria Sihaib et al. prepared $LaMnO_3$ with different ratios. The results show that the catalyst with the ratio of 0.5 to 1.5 has the best catalytic performance and the amount of citric acid affects the specific surface area of perovskite catalyst [74]. Li Wang et al. added Sr into the $LaMnO_3$ to prepare $La_{0.5}Sr_{0.5}MnO_3$. The amount of Mn^{4+} and the oxidation ability of vinyl chloride has been improved after HCl modification [104]. The perovskite catalysts showed a good catalytic activity for combustion of VOCs at low temperature due to their tunable redox property by replacing the B atom. However, they also have some disadvantages, such as low thermal stability. The catalytic activity and stability of perovskite catalysts need to be further improved.

Table 7. Perovskite catalysts for catalytic combustion of VOCs.

Refer	Catalyst	VOCs	Catalytic Performance	Remarks	Catalytic Mechanism
[58]	Layered copper manganese oxide (LCMO)	CO Benzene Toluene O-xylene	T_{50}: 62 °C T_{90}: 76 °C T_{50}: 218 °C T_{90}: 240 °C T_{50}: 187 °C T_{90}: 207 °C T_{50}: 213 °C T_{90}: 227 °C	The introduction of water vapor had no significant effect on the catalyst. The catalyst activity decreased about 30% after SO_2 was added. The addition of CO_2 has no effect on the catalytic activity of the catalyst.	The larger surface area and the formation of Cu^{2+}-O^{2-}-Mn^{4+} entities at the interface between CuO and layered MnO_2 promoted the CO and VOCs oxidation.
[59]	$LaCoO_3$	Benzene Toluene	T_{50}: 330 °C T_{50}: 244 °C	Water vapor has no significant effect on the catalytic performance of catalyst.	Factors improve activity: (i) The larger exposed surface, (ii) the composition of the support provides increased oxygen mobility.
[60]	$LaMnO_3$	Isopropanol Benzene Propyl alcohol	T_{50}: 216 °C 301 °C 203 °C	-	-
	$LaCoO_3$	Isopropanol Benzene Propyl alcohol	T_{90}: 237 °C 323 °C 222 °C		
[61]	γ-MnO_2/$SmMnO_3$	Toluene Benzene O-xylene Ethylbenzene	T_{50}: 187 °C T_{90}: 208 °C T_{50}: 213 °C T_{90}: 226 °C T_{50}: 232 °C T_{90}: 250 °C T_{50}: 201 °C T_{90}: 217 °C	-	Toluene-Mn ions+ surface reactive oxygen → benzaldehyde → benzoic acid → chain carboxylic acids → CO_2.

Table 7. Cont.

Refer	Catalyst	VOCs	Catalytic Performance	Remarks	Catalytic Mechanism
[70]	$La_{0.8}Sr_{0.2}CoO_3$	toluene Cyclohexane Propyl alcohol	T_{50}: <160 °C T_{99}: 300 °C T_{50}: 180 °C T_{99}: 260 °C T_{90}: 160 °C	-	Partial substitution of strontium (Sr) for lanthanum (La) greatly increased the oxygen vacancies in the surface regions to enhance the catalytic activity.
[71]	$SrTi_{1-x}Cu_xO_3$ $SrTi_{1-x}Mn_xO_3$	Toluene	T_{50}: 343 °C T_{90}: 398 °C T_{50}: 302 °C T_{90}: 335 °C	-	Incorporation of Mn attributed a higher amount of oxygen vacancies in the perovskite surface to promote the toluene conversion. Supra facial mechanism.
[72]	$LaMn_{1-x}B_xO_3$ (B = Co, Ni, Cu, Al)	Toluene LMLi LMNi LMCo LMAl LMLi LMNi LMCo LMAl	T_{50}: 250 °C 234 °C 226 °C 242 °C T_{90}: none 278 °C 271 °C 318 °C	-	Addition of nickel into $LaMnO_3$ can improve the catalytic oxidation of toluene by generation of more Mn^{4+} species and oxygen vacancies and the enhancing reducibility at low temperature.
[73]	$LaMnO3/\delta$-$MnO2$	Toluene	T_{90}: 258 °C	-	-

Table 7. *Cont.*

Refer	Catalyst	VOCs	Catalytic Performance	Remarks	Catalytic Mechanism
[74]	LaMnO$_3$	Toluene	T$_{50}$: 258 °C T$_{90}$: 275 °C	-	-
[85]	CeO$_2$@SiO$_2$-400	DCE	T$_{50}$: 219 °C T$_{90}$: 275 °C	The conversion of 1 vol% H$_2$O and 3 vol% H$_2$O decreased by 6% and 19%, respectively.	DCE+acid sites→HCl+ VC. (1) VC→reactive carbonium ion→adsorbed alcohol species→acetate species. Or (2) VC→1,1,2-trichloroethane→ dichloroethylene formed (subsequent chlorination reactions) →H$_2$O, CO$_x$ and HCl.
[94]	LaCoO$_3$	Propane	T$_{50}$: 208 °C T$_{90}$: 238 °C	-	MVK mechanism.
[97]	LaMnO$_3$ and LaCoO$_3$	Chlorinated VOCs	-	-	-
[104]	HCl modified La$_{0.5}$Sr$_{0.5}$MnO$_3$	Vinyl chloride	T$_{100}$: 300 °C	-	Doping and acid treatment obviously promote the active Mn^{4+} species amount and oxygen activation ability, and affect the chloric byproduct distribution. The lower temperature inhibits the Deacon reaction and chlorination.

4.1.4. Concentrated Oxidation Catalysts

The low concentration of VOCs can limit the catalytic efficiency of catalysts. Some porous materials were investigated to the concentration and decomposition of VOCs, such as zeolite, γ-Al_2O_3 (Table 8, Figure 13) [84,101,173–175]. R. Beauchet et al. tried to decompose the isopropanol and o-xylene mixture over the CsX, NaX, and HY zeolite [173]. O-xylene and isopropanol were totally decomposed at 250 °C over the NaX zeolite. The addition of Pt on the zeolite can significantly increase the catalytic activity. The main obstacle for the application of zeolite was the coke that formed during the reaction, which led to the short lifetime. Amir Ikhlaq et al. studied the mechanism and kinetics of decomposition of chlorinated VOCs by ozonation over the γ-Al_2O_3 and ZSM-5 zeolite [174]. The results suggest that ozone reacts with the absorbed VOCs on the surface of catalysts. However, the ozonation process will increase the cost in most of the industrial processes. Yuexin Peng et al. supported MnO_2 on the Al-rich β-zeolite to degrade toluene [75]. The T_{90} is 285 °C, which is much lower than the MnO_2 supported on γ-Al_2O_3 due to the lattice oxygen species in MnO_2 and absorbed oxygen species on the zeolite. Cu and Co were also used to modify the β-zeolite for the destruction of toluene and trichloroethylene [76]. The modified zeolite showed good stability in the reaction. The catalytic activity is mainly from the supported metal oxides. The suitable acidity and strong oxidation stability can improve the CO_2 selectivity. Dedong He et al. modified the HZSM-5 zeolite with a series of rare earth elements, including La, Ce, Pr, Nd, Sm, Y, and Er, for the catalytic decomposition of CH_3SH [91]. Cu was also used to modify ZSM-5 for the combustion of acrylonitrile. The isolated Cu is the active center. The SiO_2/Al_2O_3 ratio can affect the ion exchange capacity and the catalytic performance of the catalyst. When the ratio is 26, the catalyst shows the best catalytic activity [92]. SBA-15 was used to support MnO_x for the combustion of toluene and showed good catalytic activity [176]. The La modified HZSM-5 zeolite showed much better activity and stability than HZSM-5 due to the tunable acidity, which can promote the adsorption and activation of the CH_3SH molecule and inhibit the formation of coke deposit. The previous researches showed the potential application in the industry. However, to fulfill the requirements of the industry, the concentration-catalysis process to remove VOCs needs more research on lifetime, catalytic efficiency.

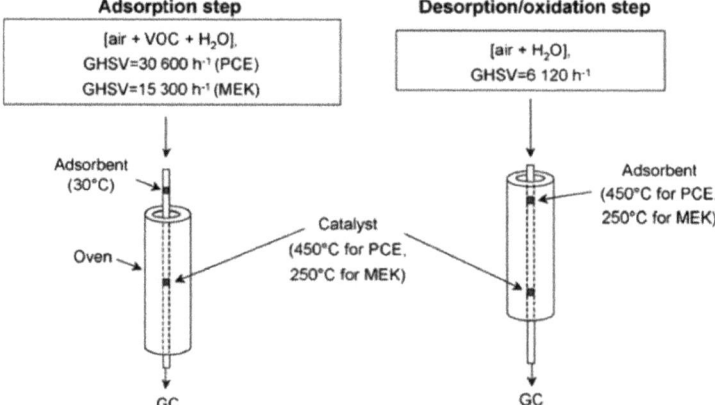

Figure 13. Scheme of adsorption/oxidation apparatus. Reprinted with permission from [1], 2009, Elsevier Ltd.

Table 8. Concentrated oxidation catalysts for catalytic combustion of VOCs.

Refer	Catalyst	VOCs	Catalytic Performance	Remarks	Catalytic Mechanism
[75]	MnO_x/H-Beta-SDS MnO_x/K-Beta-SDS MnO_x/Si-Beta	Toluene	T_{50}: 253, 262, 280 °C T_{90}: 285, 295, 312 °C	-	Mars-Van Krevelen mechanism. Organic molecules+lattice oxygen→oxygen vacancy+CO_2+H_2O, oxygen vacancy+O_2→lattice oxygen.
[84]	H-ZSM-5 and Na-ZSM-5	Cumene Dichlorobenzene Trichloro benzene	-	-	
[91]	La(13)/HZSM-5	CH_3SH	T_{100}: 40 °C	-	The La-modified HZSM-5 increased basic sites and displayed better adsorption ability to CH_3SH, decreased in strong acid sites and suppressed the formation of coke deposit.
[92]	Cu-ZSM-5(SiO_2/Al_2O_3 = 26)	C_2H_3CN	T_{90}: 325 °C	The catalyst has good resistance to steam poisoning.	Without H_2O: C_2H_3CN-SCC+O_2→NCO→oxidation products; with H_2O: C_2H_3CN-SCC+O_2+H_2O→NH_3→products.
[101]	CsX, NaX, and HY	VOCs mixture	T_{100}: 200 °C	-	Mars-Van Krevelen mechanism, involving several redox steps.
[176]	Mn/R-SBA-15	Toluene	T_{98}: 240 °C	-	MVK mechanism. Mn_2O_3-MnO_2/R-SBA-15 supply more lattice oxygen species.

4.1.5. The Influence Factors on Catalytic Performance

In the real industrial process, a lot of factors influence the catalytic efficiency. Firstly, the kind of VOCs determines the selection of catalysts. For example, noble metal catalysts show the best VOCs eliminating efficiency, but they are not suitable for the destruction of chlorine and sulfur containing VOCs due to the poison of the catalysts. Secondly, the surface area of the catalysts is the main factor that influences the catalytic activity. Research shows that MnO_2 with a higher surface area showed much better catalytic performance than the one with a lower surface area [133]. Thirdly, the crystal type of catalysts also influences the catalytic performance of the catalysts with the same content. For instance, the catalytic performance of TiO_2 with different crystal types, namely rutile and anatase, showed a different catalytic activity [121,167]. The humidity is the common content in the industrial waste gas. In most of the reports, water molecules can suppress the catalytic activity due to the complete adsorption on active sites and destruction of catalysts [177]. In other researches, humidity plays a positive role in the oxidation. In the catalytic combustion of chlorobenzene over VO_x/TiO_2, VO_x-WO_x/TiO_2, and VO_x-MoO_x/TiO_2, water can remove the adsorbed Cl^- from the catalysts surface and react with chlorine to produce HCl [134]. The water vapor can show a different effect on the same VOCs over different catalysts. Kullavanijayam et al. reported that water enhanced the catalytic oxidation of cyclohexene over the ceria–alumina supported Pt and Rh, but it had a negative effect on the oxidation of cyclohexane over the ceria–alumina supported Pd catalyst [178]. CO is another poison for precious metal catalysts, so during the catalytic combustion, enough air is needed to avoid the generation of CO. The life of catalysts is also important for catalytic combustion. The main obstacle for the development of catalysts for the catalytic combustion of halogen and sulfur containing VOCs is the short life of common catalysts. Although V oxides show less catalytic efficiency than other catalysts, it is still used in the elimination of halogen and sulfur containing VOCs due to its long catalytic life [152]. Since after the installation of catalytic combustion equipment, it will be in operation for a long time, therefore, the life of the catalysts determines the cost of catalytic combustion.

The catalytic combustion process has been well developed in recent years. The diversity of the catalysts has been investigated. However, there are still some obstacles on the way to the industrial application, such as short lifetime, high cost, and no universality to different kinds of VOCs. More work should be done on these problems.

4.2. Photocatalytic Oxidation

Photocatalytic reactions have drawn a lot of attention and have been well developed in recent years since Fujishima found the splitting water to H_2 and O_2 over TiO_2 [55,179,180]. Different kinds of photocatalysts have been developed to treat VOCs containing waste water, such as TiO_2, WO_3, ZnO_2, CdS, g-C_3N_4, and BiOBr (Table 9) [56,180–185]. The mechanism of photocatalysis is that when the light with a suitable wavelength radio on catalysts (semiconductors), the electrons and holes were separated and generated on the surface of catalysts, then the radicals of •OH and O^{2-} was formed on the surface of catalysts, the VOCs reacted with these radicals and decomposed to CO_2 and H_2O at last [186,187]. In the water solution, water can react with the catalysts and form •OH, which is positive for the decomposition of VOCs in water. The photocatalytic elimination of VOCs in the gas phase follows a similar mechanism, but the radicals are main O^{2-} due to the shortage of humidity. In this section, the progress in the elimination of VOCs in the gas phase by photocatalytic methods was mainly discussed.

The most studied photocatalyst for the elimination of VOCs is TiO_2 [44]. Wilson F. Jariam et al. degraded 17 kinds of VOCs with the concentration range of 400–600 ppmv over TiO_2 under the radiation of ultraviolet light [98]. The results showed that trichloroethylene (99.9%), isooctane (98.9%), acetone (98.5%), methanol (97.9%), methyl ethyl ketone (97.1%), t-butyl methyl ether (96.1%), dimethoxymethane (93.9%), methylene chloride (90.4%), methyl isopropyl ketone (88.5%), isopropanol (79.7%), chloroform (69.5%), and tetrachloroethylene (66.6%) were decomposed efficiently over TiO_2. The photodegradation of isopropylbenlene (30.3%), methyl chloroform (20.5%), and pyridine (15.8%)

on TiO_2 was not as efficient as other VOCs. The catalytic lifetime was also tested by toluene. The conversion of toluene decreased to 20.9% after a 150 min test, but the deactivated catalysts can be easily regenerated by washing with H_2O_2 and illumination. F. B. Li et al. prepared La ion doped TiO_2 by the sol-gel method for photodegradation of benzene, toluene, ethylbenzene, and o-xylene in the gas phase [188]. The results showed that the La ion doped TiO_2 performed much better than the pure TiO_2. This was due to the improved adsorption ability and the enhanced electron–hole pairs separation by the presence of Ti^{3+} and the electron transfer between the conduction band/defect level and lanthanide crystal field state. Tânia M. Fujimoto et al. supported the palladium on TiO_2 for the photocatalytic decomposition of octane, isooctane, n-hexane, and cyclohexane in a low concentration (100~120 ppmv) [99]. The modified catalysts showed excellent catalytic activity in the decomposition of VOCs rapidly. V. Héquet et al. used a closed-loop reactor to study the mixture effect over the P_{25} TiO_2/SiO_2 mixture (Figure 14) [189]. They have developed the accurate analytical methods to identify and quantify the majority of the potential formed intermediates, which provide an efficient way to study the reaction mechanism. Yajie Shu et al. used Mn doped TiO_2 to degrade benzene by O_3 under vacuum ultraviolet (VUV) irradiation [62]. The doped TiO_2 showed better performance than the undoped one and P_{25} due to the formation of highly reactive oxidizing species. Jian Ji et al. showed that compared with the one without UV radiation, the UV radiation can improve the removal efficiency of benzene by about 10% (Figure 15) [63]. Marta Stucchi et al. developed a simultaneous photodegradation system for the VOC mixture elimination by TiO_2 powders, which showed a good efficiency [86]. Huiling Huang et al. developed Mn modified ZSM-5 as catalysts for VUV photolysis combined with ozone-assisted catalytic oxidation and studied the mechanism [102]. The catalysts showed good efficiency. Although most of the catalysts showed excellent activity in the elimination of VOCs under UV light, the light utilization efficiency was still low due to the low percent of the UV light in nature light. Xufang Qian et al. designed mesoporous TiO_2 films coated with carbon foam for photodegradation of acetone and toluene, which can converse more than 90% of VOCs to CO_2 under visible light due to the plausible carbon doping and the strong interaction between the TiO_2 precursor and the hydro-carbon foams [100]. The graphite-SiO_2-TiO_2 composite and BiOBr@SiO_2 flower-like nanospheres were also used for photodegradation of VOCs under visible light and showed good catalytic activity [64]. The low solar utilization ratio was the main obstacle to improve photocatalytic efficiency. In order to improve the efficiency, materials with a good light adsorption ability was applied. Yun-En Lee et al. prepared black-TiO_2 and LFO/black-TiO_2 and they showed excellent photo catalytic activity for the removal of toluene and IPA due to their good light adsorption ability [77]. The modification of g-C_3N_4 by hydroxyl groups can enhance visible light-driven photocatalytic properties of g-C_3N_4 obviously which can improve the adsorption energy of g-C_3N_4 for water and phenol [88]. The structure of catalysts was also carefully tuned to improve the catalytic activity [190]. Bettini S et al. insulated a layer of SiO_2 between zinc oxide and nano silver and controlled the thickness of the insulated layer, which enhanced the photocatalytic oxidation ability of the catalyst significantly [78]. The photocatalytic activity of ZnO can also be improved by doping CuO [191]. The photocatalytic elimination of VOCs is one of the most promising methods. However, the industrial application of this method is still a problem.

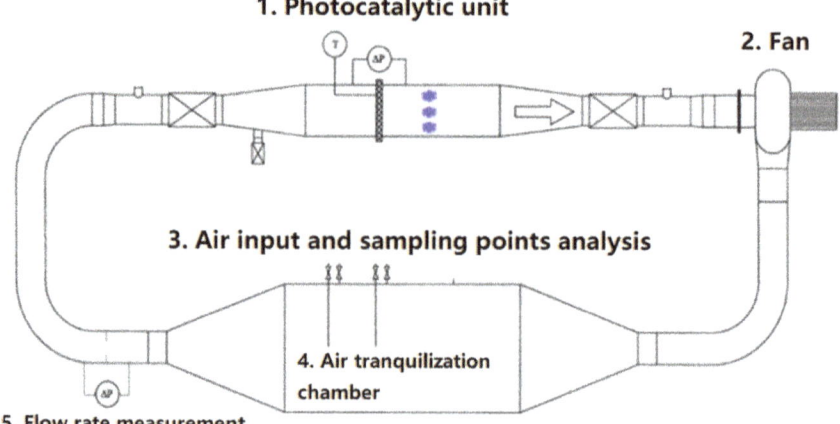

Figure 14. A 420-L of continuous closed-loop photocatalytic reactor: (**1**) Photocatalytic unit containing the TiO$_2$ photocatalytic medium and the ultraviolet (UV) lamps, (**2**) fan, (**3**) air input and sampling points for analysis, (**4**) air tranquilization chamber, (**5**) flow rate measurement. Reprinted with permission from [176], 2018, Elsevier Ltd..

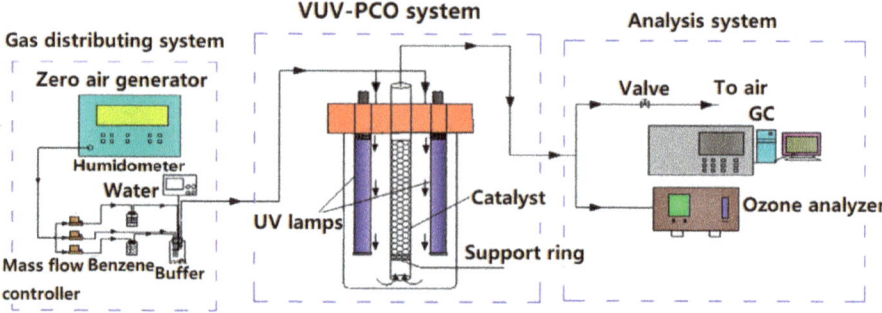

Figure 15. The schematic diagram of the VUV-PCO system. Reprinted with permission from [63], 2017, Elsevier Ltd.

There is a large amount of reports on the photocatalytic elimination of VOCs. The advantages of this process are green, energy saving, and clean. However, the reported catalysts are not efficient enough for an industrial application at the present stage. Firstly, the reaction needs more time than other methods. Secondly, most of the reported catalysts could not use visible light as energy to start the reactions, which limited the improvement of quantum efficiency, so it still needs to develop efficient photocatalysts and design effective reaction systems.

Table 9. Photocatalysts for elimination of VOCs.

Refer	Catalyst	VOCs	Catalytic Performance	Remarks	Catalytic Mechanism
[62]	Mn/meso-TiO$_2$	Benzene	-	-	Benzene→(splitting of benzene, hydrogenation, H abstraction by OH radicals) six-member ring cyclitols+aldehydes ketones, etc. →a class of fulvene (isomerization)→ $CO_2 + H_2O$.
[63]	Mesoporous TiO$_2$	Benzene	-	-	Benzene is mainly degraded by photo-generated electron-hole pairs and hydroxyl radicals.
[64]	Graphite-SiO$_2$-TiO$_2$	Benzene	The graphite-SiO$_2$-TiO$_2$ composites exhibited higher photocatalytic activity for degradation of benzene gas under visible light irradiation than that of pure TiO$_2$.		Optimum concentration of graphite facilitates the separation of photogenerated electron-hole pairs for graphite-SiO$_2$-TiO$_2$ composites by visible light.
[77]	LaFeO$_3$/black-TiO$_2$	Toluene	The removal efficiency of black-TiO$_2$ and LaFeO$_3$ for toluene was 89% and 98%, respectively, and the removal efficiency for IPA was 90% and 94%, respectively.	-	Cat.+photon (the wavelength shorter than 440 nm)→ electrons $+O_2 \to O_2^{2-}$, O^{2-} +toluene and IPA →$CO_2+H_2O_{(g)}$.
[78]	Pd/CeO$_2$/γ-Al$_2$O$_3$	Toluene Acetone Ethyl acetate	T_{98}: 205 °C 220 °C 275 °C	-	-
[79]	CuO/ZnO nanocomposite photocatalysts	Toluene Iopropanol	-	-	Photooxidative activity and stability over ZnO are improved by loading CuO.
[86]	TiO$_2$	Formaldehyde acetaldehyde	TiO$_2$ degraded almost 100% of formaldehyde or acetaldehyde at a starting concentration of 400–500 ppb with a relative humidity of 40%.	-	The rate-determining step is the adsorption (external diffusion) on the catalysts active sites, thus the higher the surface area, the higher the degradation.
[88]	g-C$_3$N$_4$	Benzoquinone Hydroquinone catechol	-	-	OH radicals+phenol→dihydroxycyclohexadienyl radical adducts→phenoxy radicals→H_2O (a very slow process), adducts+dissolved O_2→dihydroxy intermediates (-HO$_2$) →CO_2+H_2O.

Table 9. *Cont.*

Refer	Catalyst	VOCs	Catalytic Performance	Remarks	Catalytic Mechanism
[98]	La^{3+}-TiO_2 and Nd^{3+}-TiO_2	VOCs mixture	1.2% La^{3+}-TiO_2 had the highest photocatalytic activity.	-	-
[99]	TiO_2/Pd	VOCs	The conversion rate of VOCs reached 90% when the residence time was 27 s.	-	-
[102]	Mn/ZSM-5	O_3	O_3 can be efficiently decomposed by the Mn/ZSM-5 and used for benzene degradation through the OZCO.	-	Benzene+•OH→phenol→benzoquinone →CO_2 + H_2O.
[100]	Meso-TiO_2/hydro-CF	VOCs	-	-	Promotion effects on degradation of gaseous polar acetone come from well crystallized anatase nanocrystals, hydro-CF skeleton for adsorption, and fast mass transportation within the hierarchical frameworks.

4.3. Non-Thermal Plasma Process

In the non-thermal plasma process, electrons and their surroundings are not in a thermal equilibrium, so the electrons are heated by electric discharges instead of the gas itself, and produce the electrons with high energy, active radicals and ions which can promote numerous chemical reactions in the ionized zones. It can be used to treat the high flow for both low (<100 ppmv) and high (>1000 ppmv) concentrations of VOCs, including toluene, benzene, acetone, trichloroethylene, etc. (Table 10) [47]. The main bottlenecks for the commercialization of the technology are the formation of poison byproducts and high energy consumption. The discharge methods are important for the VOCs removal efficiency, which includes corona discharge, surface discharge, microwave discharge, dielectric barrier discharge, and packed bed dielectric barrier discharge [190–194]. Among these discharge types, the packed bed dielectric barrier discharge shows the most potential in the industry application [195]. Savita K. P. Veerapandian et al. reviewed the packed bed DBD [47]. The influences of dielectric constant, packing materials size, shape, surface properties, and the byproducts formation were discussed.

Using porous and catalytic materials as the packed bed can increase the resistant time of VOCs and decrease the unwanted byproducts, such as O_3, NO_x, and CO (Figure 16) [196]. The packed bed materials can be non-catalytic porous materials, such as activated carbon, porous Al_2O_3, glass, zeolite, graphene oxide, and the catalysts, such as metal oxides, noble metal loaded metal oxides, and catalytic porous materials [197]. For example, using Al_2O_3 as the packed bed can concentrate VOCs molecules on its surface and weaken the bond energy of VOCs, which can enhance the dissociation when these adsorbed molecules encounter the active species in the plasma and increase the collision probability and deep oxidation of VOCs [198]. Lee et al. used porous γ-Al_2O_3 as the packed bed to oxidize toluene to CO_2. It showed a 100% conversion of toluene and high CO_2 selectivity [199]. Gandhi's research results showed that a large surface area and pore volume of alumina can not only increase the conversion of ethylene to CO_2 or CO, but also decrease the selectivity of unwanted byproducts, such as acetaldehyde, acetylene, methane, N_2O, and O_3 [200]. Other researches also showed the same phenomenon in the abatement of benzene, acetone, formaldehyde, TCE, chlorobenzene in non-thermal plasma with the packed bed which has a large surface area and pore volume, such as porous alumina, TiO_2, zeolite, and porous metal oxides [199,201–203]. Due to the large surface area, the retention time and concentration of VOCs molecules increased which led to the increase of collisions between VOCs and active species [200], the adsorption effect can weaken the chemical bond of VOCs [204], and form micro-discharges in the micro-pores in addition to the micro-discharges in the gas phase.

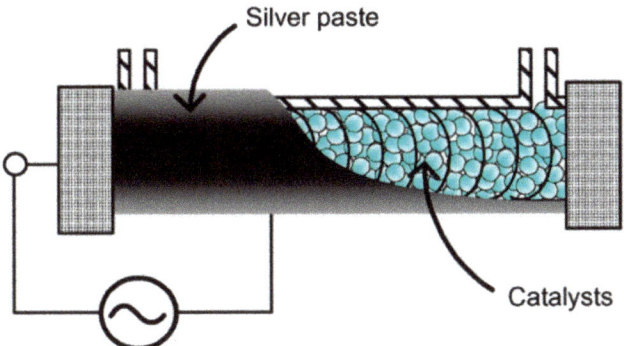

Figure 16. Schematic diagram of the plasma-driven catalyst (PDC) reactor. Reprinted with permission from [196], 2007, Elsevier Ltd.

The addition of catalysts in the non-thermal plasma process as the packed bed can enhance the VOCs removal efficiency significantly due to the plasma assisted reactions on the surface of the catalyst.

These plasma-assisted reactions can generate some active radical species, such as intermediates with electrons, O·, OH·, N_2·, NO·, N_2O· in the gas phase and O·, OH· on the surface of catalysts, which can improve the VOCs removal efficiency and increase the CO_x selectivity [80]. The catalysts include metal oxides, such as CuO, MnO_2, CeO_2, NiO, CoO_2, and Fe_2O_3, and noble metals, such as Pt, Au, and Ag [65,196,203,205–209]. The different kinds of oxygen species in catalysts, namely lattice oxygen, surface oxygen, and absorbed oxygen, are very helpful for the oxidation of VOCs, which can activate the oxygen molecules in the gas phase. P.J. Asilevi et al. established a laboratory scale DBD reactor for the removal of toluene by ·O and ·OH which were generated from the reaction between O_2 and H_2O and the removal efficiency of VOCs can be improved by increasing the oxygen concentration and relative humidity [210]. Lu et al. used FeO_x/SBA-15 as the packed bed to eliminate toluene in the non-thermal plasma process. The results showed that the presence of Fe^{2+} can increase the toluene removal efficiency and CO_x selectivity obviously and reduce the formation of unwanted and toxic byproducts [80]. Zhu et al. prepared 5 wt% CuO/-Al_2O_3 as the packed bed to the abatement acetone. It showed better performance than the one without CuO due to the better reducibility and abundant active oxygen species [87]. Li et al. prepared Pt/Al_2O_3 and it showed higher decomposition efficiency than the unloaded Al_2O_3 due to the presence of Pt, which increases the number of active sites and reduces the activation energy of the decomposition reaction and suppresses harmful NO_x formation [65]. The research by Hua Song et al. showed that $CoMnO_x$/TiO_2 can obviously improve the plasma removal of VOCs [103].

There are other factors that influence the VOCs removal efficiency, such as humidity and the plasma structure. In industrial conditions, the humidity in the VOCs stream is the factor which must be considered. The humidity has both a positive and negative effect on the removal of VOCs in the non-thermal plasma process. The water molecules can modify the surface of the packed bed, especially the catalytic packed bed, quench the free electrons and active species, and produce OH radicals during the discharge process [103,202,207,211,212]. A lot of studies showed that the presence of humidity can reduce the VOCs removal efficiency by quenching the high energy electrons and reactive species, and modify the catalysts surface ($BaTiO_3$) in the decomposition of toluene, benzene, xylene, and C_2F_2 [103,207,212]. On the other hand, the humidity can suppress the formation of toxic byproducts, such as CO and NOx by deactivation of some oxygen species in the catalysts [213,214]. The optimization of the plasma structure also can enhance the VOCs decomposition efficiency. Muhammad Farooq Mustafa et al. designed a double dielectric barrier discharge reactor, in which the conversion of tetrachloroethylene, toluene, trichloroethylene, benzene, ethyl acetate, and carbon disulfide can be 100% with $BaTiO_3$ and HZSM-5 as catalysts, respectively [215].

There are also some disadvantages in the non-thermal plasma process, such as high energy consumption, low VOCs conversion (mostly about 30% to 70%), low CO_x selectivity, and produced unwanted byproducts. The other barrier for the application of non-thermal plasma in the industry is the high cost and the high energy consumption. The way to solve these problems is the development of a highly efficient catalytic packed bed and the optimization of an electrode structure.

Table 10. Catalysts for VOCs elimination through the non-thermal plasma process.

Refer	Catalyst	VOCs	Catalytic Performance	Remarks	Catalytic Mechanism
[65]	Pt/Al$_2$O$_3$	Benzene byproducts	-	-	-
[80]	FeO$_x$/SBA-15	Toluene	-	Under the condition of 3% Fe loading, the oxidation of toluene is the best one.	(1) Direct removal caused by the collision of electrons or oxidation caused by the gas-phase radicals (O•, OH•, N$_2$•, NO•, NO$_2$•) in the gas phase; (2) the reaction between adsorbed toluene or other intermediates and the active species (O•, OH•) on the catalyst surface. O$_2$ might be fixed on the catalyst surface via a facile interconversion between Fe^{2+} and Fe^{3+} states and then be transported to the toluene or intermediates leading to CO$_2$ formation.
[87]	CuO/g-Al$_2$O$_3$	Acetone	-	5.0 wt% CuO/g-Al$_2$O$_3$ catalyst has the highest removal rate of acetone, reaching 67.9%.	Both short-lived radicals and acetone/intermediates can be adsorbed on the catalyst surfaces to initiate a series of surface oxidation reactions, forming CO, CO$_2$, H$_2$O, and byproducts.
[103]	CoMnO$_x$/TiO$_2$	O$_3$	-	When the temperature is 320 °C, the decomposition efficiency of O$_3$ is 98%.	-

4.4. Electron Beam Treatment

In the electron beam technology, an electron beam generated from an EB accelerator and absorbed by the background gas to ionize the air and form some active oxidizing radicals, such as ·OH and ·HO$_2$ and excited ions as O (^{3}P), which is critical for the destruction of VOCs [216–218]. The oxidizing radicals reacted with VOCs and decomposed them to inorganic compounds. There are three types of accelerators used for VOCs decomposition, including transformer accelerator, ultrahigh frequency accelerator, and linear microwave accelerator [217]. This technology has been used for the elimination of acetaldehyde, benzene, dodecane, ethylbenzene, hexadecane, pentane, styrene, tetradecane, toluene, trimethylamine, turpentine, xylene, dimethyl sulfide, dimethyl disulfide, methyl mercaptan, and chlorinated VOCs in a lab or pilot scale [218–228]. The electron beam can treat the waste gas with a low concentration of VOCs (10 to hundreds ppm). The initial electrons from an electron beam can hardly decompose the VOCs molecules, while it can react with background gases and form active radicals. The way to improve the abatement rate of aromatic VOCs includes the addition of humidity (formation of OH·), ammonia, chlorinated VOCs, ozone, and using an absorbed dose [220,229]. The reaction temperature, VOCs molecular structure, and background gas also influence the decomposition efficiency. However, there are also some drawbacks of the electron beam technology, the generation of active oxidizing radicals, e.g., ·OH and ·HO$_2$ can lead to the creation of toxic intermediates, especially in the case of the VOCs emission with an unknown composition, such as the production of toxic byproducts (aerosol, ozone, CO, and other trace organic compounds). The equipment is also too complex. It still needs more research for the industrial application.

5. Outlook of the Different Kinds of Technologies

Different kinds of VOCs were emitted from different parts of pharmaceutical production processes. With the new VOCs discharge standards coming into force, it is critical for the pharmaceutical companies to eliminate VOCs exhaustively in China. According to the production processes and conditions, the suitable VOCs elimination technologies must be selected with the highest efficiency and at the lowest cost. Adsorption, absorption, incineration, catalytic combustion, biodegradation, and non-thermal plasma technologies have been applied in factories for the abatement of VOCs. Adsorption and absorption are commonly the first used methods to recycle useful VOCs. Then, the rest of the VOCs were decomposed by other technologies, such as incineration, catalytic combustion, non-thermal plasma, biodegradation, and non-thermal plasma. The key point for the improvement of catalytic combustion and non-thermal plasma is the development of efficient catalysts or packed bed materials. Some new technologies are emerging for the abatement of VOCs, such as photocatalysis and electron beam radiation. These new technologies have shown good potential for the elimination of VOCs with a high efficiency and low energy cost. However, they still have some engineering problems for the industrial application. It is important to develop VOCs elimination technologies to decrease the amounts of VOCs in the discharged waste gas. The final solution for the VOCs is avoiding the emission of VOCs from the whole pharmaceutical production processes by developing new chemical and engineering technologies for the production of medicine.

Author Contributions: L.Z. has written the main content of this paper. C.M. has added some new research papers of this topic and summarized the catalysts used in different researches. J.H. has improved the English writing of the whole paper and written parts of photocatalysis parts. R.L. has written the parts of VOCs elimination policies in China and improve the structure of the paper. J.Y. has designed the whole structure of the paper and improved the manuscript a lot. All authors have read and agreed to the published version of the manuscript.

Funding: This research was funded by the Scientific Technology Research Program for the University in Hebei Province Youth Project (No. QN2020126).

Acknowledgments: The authors gratefully acknowledge the help of Jing Wang and Ping Chen from Tianjushi Engineering&Technology Group Co., Ltd. who gave us a lot of suggestions about the VOCs treatment technologies and policies in China. The authors gratefully acknowledge the help of Rose Amal for the help of improvement the parts of catalytic processes for the treatment of VOCs.

Conflicts of Interest: The authors declare no conflict of interest.

References

1. Li, W.B.; Wang, J.X.; Gong, H. Catalytic combustion of VOCs on non-noble metal catalysts. *Catal. Today* **2009**, *148*, 81–87. [CrossRef]
2. Muhammad, S.K.; Shaikh, A.R.; Mohammad, M.H. Catalytic oxidation of volatile organic compounds (VOCs): A review. *Atmos. Environ.* **2016**, *140*, 117–134.
3. Cheng, G. Study on VOCs Emissions Research Present Situation and the Emission Reduction Potential of Key Industries in Hebei Province. Master's Thesis, Hebei University of Science and Technology, Shijiangzhuang, China, 2016.
4. Li, Y. Study on Emission Standard of Air Pollutants for Pharmaceutical Industry Chemical Synthesis Products Category. Master's Thesis, Zhejiang University of Technology, Hangzhou, China, 2016.
5. Atkinson, R. Gas-phase tropospheric chemistry of volatile organic compounds: 1. Alkanes and alkenes. *J. Phys. Chem. Ref. Data* **1997**, *26*, 215–229. [CrossRef]
6. Fontane, H.; Veillerot, M.; Gallo, J.C.; Guillermo, R. *8th International Symposium on Transport and Air Pollution*; Verlag der Technischen Universität Graz: Graz, Austria, 1999.
7. Rivière, E. *CITEPA Report*; Fieldwork Inhabitants: Paris, France, 1998.
8. Duncan, B.N.; Yoshida, Y.; Olson, J.R.; Sillman, S.; Martin, R.V.; Lamsal, L.; Hu, Y.; Pickering, K.E.; Retscher, C.; Allen, D.J.; et al. Application of OMI observations to a space-based indicator of NOx and VOC controls on surface ozone formation. *Atmos. Environ.* **2010**, *44*, 2213–2223. [CrossRef]
9. Atkinson, R. Atmospheric chemistry of VOCs and NO_x. *Atmos. Environ.* **2000**, *34*, 2063–2101.
10. National Bureau of Statistics of the People's Republic of China. Available online: http://www.stats.gov.cn/tjsj/ (accessed on 26 May 2020).
11. Chen, Y. Study on Current and Future Industrial Emission of Volatile Organic Compounds in China. Master's Thesis, South China University of Technology, Guangzhou, China, 2011.
12. Huang, W. Characteristics of Industrial VOCs Emissions and Evaluation of Control Technology in China. Master's Thesis, Zhejiang University, Hangzhou, China, 2016.
13. Ministry of Ecological Environment of the People's Republic of China. *Standards for Emissions of Atmospheric Pollutants from Pharmaceutical Industry (Draft for Comments)*; Ministry of Ecological Environment of the People's Republic of China: Beijing, China, 2017.
14. Lu, Y. Establishment and Application on Assessment System of VOCs's Control Technology of Pharmaceutical Industry. Master's Thesis, Hebei University of Science and Technology, Shijiazhuang, China, 2016.
15. Hu, G. Research of Emission and Control Technology of VOCs and Odorous Gas from Pharmacy Industry in China. Master's Thesis, Nankai University, Tianjing, China, 2013.
16. Li, X. The System of Discharge Standards of Pollutants for Pharmaceutical Industry and a case Study. Master's Thesis, The Chinese Research Academy of Environmental Science, Beijing, China, 2006.
17. He, J. *Fermentology*; China Medical Science Press: Beijing, China, 2009.
18. Wang, Y. *Chemical Pharmaceutical Technology*; Chemical Industry Press: Beijing, China, 2008.
19. Zhou, T.; Ma, Q.; Chen, H. Treatment of chemical synthesized pharmaceutical wastewater by hybrid biological reactor. *Ind. Water Wastewater* **2010**, *41*, 42–45.
20. Chen, P. *Pharmaceutical Technology*; Hubei Science and Technology Press: Wuhan, China, 2008.
21. Qi, X. *Modern Biopharmaceutical Technology*; Chemical Industry Press: Beijing, China, 2009.
22. Cao, G. *Pharmaceutical Engineering of Traditional Chinese Medicine*; Chemical Industry Press: Beijing, China, 2004.
23. Wu, C. The thinking and Countermeasures for the Structuring of Learning Seals Team in Chengdu Zhonghui Pharmaceuticals Company. Master's Thesis, Southwestern University of Finance and Economics, Chengdu, China, 2006.
24. Chen, Y.; Ye, D.; Liu, X.; Wu, J.; Huang, B.; Zheng, Y. Source tracing and characteristics of industrial VOCs emissions in China. *Chin. Environ. Sci.* **2012**, *32*, 48–55.
25. Song, J. Studies on the Adsorption of VOCs by Activated Carbons and the Structure-Function Relationship. Ph.D. Thesis, Central South University, Changsha, China, 2014.
26. Khan, F.I.; Ghoshal, A.K. Removal of Volatile Organic Compounds from polluted air. *J. Loss Prevent. Proc. Ind.* **2000**, *13*, 527–545. [CrossRef]

27. Parmar, G.R.; Rao, N.N. Emerging Control Technologies for Volatile Organic Compounds. *Crit. Rev. Environ. Sci. Technol.* **2009**, *39*, 41–78. [CrossRef]
28. Shah, R.; Thonon, B.; Benforado, D. Opportunities for heat exchanger applications in environmental systems. *Appl. Therm. Eng.* **2000**, *20*, 631–650. [CrossRef]
29. Dunn, R.F.; El-Halwagi, M.M. Selection of Optimal VOC-condensation Systems. *Waste Manag.* **1994**, *14*, 103–113. [CrossRef]
30. Huang, W.; Shi, L.; Hu, Z.; Zheng, Z. Integrated technology of condensation and adsorption for volatile organic compounds recovery. *Chem. Eng.* **2012**, *6*, 13–17.
31. Zhang, X.; Gao, B.; Creamer, A.E.; Cao, C.; Li, Y. Adsorption of VOCs onto engineered carbon materials: A review. *J. Hazard. Mater.* **2017**, *338*, 102–123. [CrossRef]
32. Wang, S.; Zhang, L.; Long, C.; Li, A. Enhanced adsorption and desorption of vocs vapor on novel micro-mesoporous polymeric adsorbents. *J. Coll. Interface Sci.* **2014**, *428*, 185–190. [CrossRef] [PubMed]
33. Serna-Guerrero, R.; Sayari, A. Applications of pore-expanded mesoporoussilica 7, Adsorption of volatile organic compounds. *Environ. Sci. Technol.* **2007**, *41*, 4761–4766. [CrossRef] [PubMed]
34. U.S. Environmental Protection Agency Clean Air Technology Center. Choosing an adsorption system for VOC: Carbon, zeolite, or polymers? In Proceedings of the SPIE—The International Society for Optical Engineering, Research Triangle Park, NC, USA, 1 May 1999.
35. Kujawa, J.; Cerneaux, S.; Kujawski, W. Removal of hazardous volatile organic compounds from water by vacuum pervaporation with hydrophobic ceramic membranes. *J. Membr. Sci.* **2015**, *474*, 11–19. [CrossRef]
36. Li, Y.X.; Chen, J.Y.; Sun, Y.H. Adsorption of multicomponent volatile organic compounds on semi-coke. *Carbon* **2008**, *46*, 858–863.
37. Komori. Preparation of N-acetylmorpholine. *J. Chem. Soc. Jpn. Ind. Chem.* **1959**, *62*, 220–225.
38. Chang, F.T.; Lin, Y.C.; Bai, H.; Pei, B.S. Adsorption and desorption characteristics of semiconductor volatile organic compounds on the thermal swing honeycomb zeolite concentrator. *J. Air Waste Manage. Assoc.* **2003**, *53*, 1384–1390. [CrossRef]
39. Liu, Y. Study on Ceramic Honeycomb Monolithic Adsorbents for VOCs Adsorption. Master's Thesis, South China University of Technology, Guangzhou, China, 2015.
40. Moretti, E.C. Reduce VOC and HAP emissions. *Chem. Eng. Prog.* **2002**, *98*, 30–40.
41. Everaert, K.; Baeyens, J. Catalytic combustion of volatile organic compounds. *J. Hazard. Mater.* **2004**, *B109*, 113–139. [CrossRef]
42. Liotta, L.F. Catalytic oxidation of volatile organic compounds on supported noble metals. *Appl. Catal. B Environ.* **2010**, *100*, 403–412. [CrossRef]
43. Spivey, J.J. Complete Catalytic Oxidation of Volatile Organics. *Ind. Eng. Chem. Res.* **1987**, *26*, 2165–2180. [CrossRef]
44. Zhang, Z.; Jiang, Z.; Shangguan, W. Low-temperature catalysis for VOCs removal in technology andapplication: A state-of-the-art review. *Catal. Today* **2016**, *264*, 270–278. [CrossRef]
45. Barbusinski, K.; Kalemba, K.; Kasperczyk, D.; Urbaniec, K.; Kozik, V. Biological methods for odor treatment: A review. *J. Clean. Prod.* **2017**, *152*, 223–241. [CrossRef]
46. Mudliar, S.; Giri, B.; Padoley, K.; Satpute, D.; Dixit, R.; Bhatt, P.; Pandey, R.; Juwarkar, A.; Vaidya, A. Bioreactors for treatment of VOCs and odours—A review. *J. Environ. Manage.* **2010**, *91*, 1039–1054. [CrossRef]
47. Veerapandian, S.K.P.; Leys, C.; Geyter N., D.; Morent, R. Abatement of VOCs using packed bed non-thermal plasma reactors: A review. *Catalysts* **2017**, *7*, 113. [CrossRef]
48. Son, Y.S. Decomposition of VOCs and odorous compounds by radiolysis: A critical review. *Chem. Eng. J.* **2017**, *316*, 609–622. [CrossRef]
49. Wang, H. Study on Removal of Miced VOCs in Air by Dielectric Barrier Discharge. Ph.D. Thesis, Dalian University of Technology, Dalian, China, 2009.
50. Hao, J. Control Study of VOCs Regioned Joint Prevention and Control of Atmospheric Pollution. Master's Thesis, Hebei University of Technology, Shijiazhuang, China, 2012.
51. Xi, J.; Wu, J.; Hu, H.; Wang, C. Application status of industrial VOCs gas treatment techniques. *China Environ. Sci.* **2012**, *32*, 1955–1960.
52. Hao, Y. Study on The Pharmaceutical and Chemical Industry VOCs and Odor Pollution Characteristics. Master's Thesis, Hebei University of Science and Technology, Shijiazhung, China, 2014.

53. Department of Environmental Protection of Zhejiang Province. *Standards for Emissions of Atmospheric Pollutants from Pharmaceutical Industry in Zhejiang Province*; Department of Environmental Protection of Zhejiang Province: Hangzhou, Chain, 2015–2016.
54. Salar-García, M.J.; Ortiz-Martínez, V.M.; Hernández-Fernández, F.J.; de los Ríos, A.P.; Quesada-Medina, J. Ionic liquid technology to recover volatile organic compounds (VOCs). *J. Hazard. Mater.* **2017**, *321*, 484–499. [CrossRef]
55. Minella, M.; Baudino, M.; Minero, C. A revised photocatalytic transformation mechanism for chlorinated VOCs: Experimental evidence from C_2Cl_4 in the gas phase. *Catal. Today* **2018**, *313*, 114–121. [CrossRef]
56. Papaefthimiou, P.; Ioannides, T.; Verykios, X.E. Combustion of non-halogenated volatile organic compounds over group VIII metal catalysts. *Appl. Catal. B Environ.* **1997**, *13*, 175–184. [CrossRef]
57. Tabakova, T.; Ilieva, L.; Petrova, P. Complete benzene oxidation over mono and bimetallic Au–Pd catalysts supported on Fe-modified ceria. *Chem. Eng. J.* **2015**, *260*, 133–141. [CrossRef]
58. Wang, Y.; Yang, D.; Li, S.; Zhang, L.; Zheng, G.; Guo, L. Layered copper manganese oxide for the efficient catalytic CO and VOCs oxidation. *Chem. Eng. J.* **2018**, *357*, 258–268. [CrossRef]
59. Alifanti, M.; Florea, M.; Pârvulescu, V.I. Ceria-based oxides as supports for $LaCoO_3$ perovskite catalysts for total oxidation of VOC. *Appl. Catal. B Environ.* **2007**, *70*, 400–405. [CrossRef]
60. Spinicci, R.; Faticanti, M.; Marini, P.; De Rossi, S.; Porta, P. Catalytic activity of $LaMnO_3$ and $LaCoO_3$ perovskites towards VOCs combustion. *J. Mol. Catal. A Chem.* **2003**, *197*, 147–155. [CrossRef]
61. Liu, L.; Li, J.; Zhang, H.; Li, L.; Zhou, P.; Meng, X.; Guo, M.; Jia, J.; Sun, T. In situ fabrication of highly active $\gamma\text{-}MnO_2/SmMnO_3$ catalyst for deep catalytic oxidation of gaseous benzene, ethylbenzene, toluene, and o-xylene. *J. Hazard. Mater.* **2019**, *362*, 178–186. [CrossRef]
62. Shu, Y.; Ji, J.; Xu, Y.; Deng, J.; Huang, H.; He, M.; Leung, D.Y.C.; Wu, M.; Liu, S.; Liu, S.; et al. Promotional role of Mn doping on catalytic oxidation of VOCs over mesoporous TiO_2 under vacuum ultraviolet (VUV) irradiation. *Appl. Catal. B Environ.* **2018**, *220*, 78–87. [CrossRef]
63. Ji, J.; Xu, Y.; Huang, H.; He, M.; Liu, S.; Liu, G.; Xie, R.; Feng, Q.; Shu, Y.; Zhan, Y.; et al. Mesoporous TiO_2 under VUV irradiation: Enhanced photocatalytic oxidation for VOCs degradation at room temperature. *Chem. Eng. J.* **2017**, *327*, 490–499. [CrossRef]
64. Yadav, H.M.; Jung, S.C.; Kim, J.S. Visible light photocatalyticperformance of in situ synthesized graphite-SiO_2–TiO_2 composite towards degradation of benzene gas. *J. Nanosci. Nanotechnol.* **2018**, *18*, 2032–2036. [CrossRef]
65. Li, J.; Han, S.; Bai, S.; Han, S.; Song, H.; Pu, Y.; Zhu, X.; Chen, W. Effect of Pt/gamma-Al_2O_3 catalyst on nonthermal plasma decomposition of benzene and byproducts. *Environ. Eng. Sci.* **2011**, *28*, 395–403. [CrossRef]
66. Pei, W.; Liu, Y.; Deng, J.; Zhang, K.; Hou, Z.; Zhao, X.; Dai, H. Partially embedding Pt nanoparticles in the skeleton of 3DOM Mn_2O_3: An effective strategy for enhancing catalytic stability in toluene combustion. *Appl. Catal. B Environ.* **2019**, *256*, 117814–117824. [CrossRef]
67. Hosseini, M.; Barakat, T.; Cousin, R. Catalytic performance of core–shell and alloy Pd–Au nanoparticles for total oxidation of VOC: The effect of metal deposition. *Appl. Catal. B Environ.* **2012**, *111*, 218–224. [CrossRef]
68. Lee, D.S.; Chen, Y.W. The mutual promotional effect of Au–Pd/CeO_2 bimetallic catalysts on destruction of toluene. *J. Taiwan Ins. Chem. Eng.* **2013**, *44*, 40–44. [CrossRef]
69. Yang, X.; Yu, X.; Lin, M.; Ma, X.; Ge, M. Enhancement effect of acid treatment on Mn_2O_3 catalyst for toluene oxidation. *Catal. Today* **2019**, *327*, 254–261. [CrossRef]
70. Huang, H.; Liu, Y.; Tang, W.; Chen, Y. Catalytic activity of nanometer $La_{1-x}Sr_xCoO_3$ ($x = 0$, 0.2) perovskites towards VOCs combustion. *Catal. Commun.* **2008**, *9*, 55–59. [CrossRef]
71. Suárez-Vázquez, S.I.; Gil, S.; García-Vargas, J.M.; Cruz-López, A.; Giroir-Fendler, A. Catalytic oxidation of toluene by $SrTi_{1-X}BXO_3$ (B = Cu and Mn) with dendritic morphology synthesized by one pot hydrothermal route. *Appl. Catal. B Environ.* **2018**, *223*, 201–208. [CrossRef]
72. Guo, M.; Li, K.; Liu, L.; Zhang, H.; Hu, X.; Min, X.; Jia, J.; Sun, T. Resource utilization of spent ternary lithium-ions batteries: Synthesis of highly active manganese-based perovskite catalyst for toluene oxidation. *J. Taiwan Inst. Chem. Eng.* **2019**, *102*, 268–275. [CrossRef]
73. Yang, J.; Li, L.; Yang, X.; Song, S.; Li, J.; Jing, F.; Chu, W. Enhanced catalytic performances of in situ-assembled $LaMnO/\delta\text{-}MnO_2$ hetero-structures for toluene combustion. *Catal. Today* **2019**, *327*, 19–27. [CrossRef]

74. Sihaib, Z.; Puleo, F.; Pantaleo, G.; Parola, V.L.; Valverde, J.L.; Gil, S.; Liotta, L.F.; Giroir-Fendler, A. The effect of citric acid concentration on the properties of LaMnO$_3$ as a catalyst for hydrocarbon oxidation. *Catalysts* **2019**, *9*, 226. [CrossRef]
75. Peng, Y.; Zhang, L.; Chen, L.; Yuan, D.; Wang, G.; Meng, X.; Xiao, F.S. Catalytic performance for toluene abatement over Al-rich Beta zeolite supported manganese oxides. *Catal. Today* **2017**, *297*, 182–187. [CrossRef]
76. Qin, Y.; Qu, Z.; Dong, C.; Wang, Y.; Huang, N. Highly catalytic activity of Mn/SBA-15 catalysts for toluene combustion improved by adjusting the morphology of supports. *J. Environ. Sci.* **2019**, *76*, 208–216. [CrossRef]
77. Lee, Y.E.; Chung, W.C.; Chang, M.B. Photocatalytic oxidation of toluene and isopropanol by LaFeO3/black-TiO$_2$. *Environ. Sci. Pollut. Res.* **2019**, *26*, 20908–20919. [CrossRef]
78. Bettini, S.; Pagano, R.; Semeraro, P.; Ottolini, M.; Salvatore, L.; Marzo, F.; Lovergine, N.; Giancane, G.; Valli, L. SiO$_2$-Coated ZnO Nanoflakes Decorated with Ag Nanoparticles for Photocatalytic Water Oxidation. *Chem. Eur. J.* **2019**, *25*, 14123–14132. [CrossRef] [PubMed]
79. Li, Z.; Pan, X.; Yi, Z. Photocatalytic oxidation of methane over CuO-decorated ZnO nanocatalysts. *J. Mater. Chem. A* **2019**, *7*, 469–475. [CrossRef]
80. Lu, M.; Huang, R.; Wu, J.; Fu, M.; Chen, L.; Ye, D. On the performance and mechanisms of toluene removal by FeO$_x$/SBA-15-assisted non-thermal plasma at atmospheric pressure and room temperature. *Catal. Today* **2015**, *242*, 274–286. [CrossRef]
81. Nevanperä, T.K.; Ojala, S.; Laitinen, T.; Pitkäaho, S.; Saukko, S.; Keiski, R.L. Catalytic Oxidation of Dimethyl Disulfide over Bimetallic Cu–Au and Pt–Au Catalysts Supported on γ-Al$_2$O$_3$, CeO$_2$, and CeO$_2$–Al$_2$O$_3$. *Catalysts* **2019**, *9*, 603. [CrossRef]
82. Rotter, H.; Landau, M.; Herskowitz, M. Combustion of chlorinated VOC on nanostructured chromia aerogel as catalyst and catalyst support. *Environ. Sci. Technol.* **2005**, *39*, 6845–6850. [CrossRef]
83. Dai, Q.; Wang, X.; Lu, G. Low-temperature catalytic destruction of chlorinated VOCs over cerium oxide. *Catal. Commun.* **2007**, *8*, 1645–1649. [CrossRef]
84. Ikhlaq, A.; Kasprzyk-Hordern, B. Catalytic ozonation of chlorinated VOCs on ZSM-5 zeolites and alumina: Formation of chlorides. *Appl. Catal. B Environ.* **2017**, *200*, 274–282. [CrossRef]
85. Fei, Z.; Cheng, C.; Chen, H.; Li, L.; Yang, Y.; Liu, Q.; Chen, X.; Zhang, Z.; Tang, J.; Cui, M.; et al. Construction of uniform nanodots CeO$_2$ stabilized by porous silica matrix for 1,2-dichloroethane catalytic combustion. *Chem. Eng. J.* **2019**, *370*, 916–924. [CrossRef]
86. Stucchi, M.; Galli, F.; Bianchi, C.L.; Pirola, C.; Boffito, D.C.; Biasioli, F.; Capucci, V. Simultaneous photodegradation of VOC mixture by TiO$_2$ powders. *Chemosphere* **2018**, *193*, 198–206. [CrossRef] [PubMed]
87. Zhu, X.; Tu, X.; Mei, D.; Zheng, C.; Zhou, J.; Gao, X.; Luo, Z.; Ni, M.; Cen, K. Investigation of hybrid plasma-catalytic removal of acetone over CuO/-Al$_2$O$_3$ catalysts using response surface method. *Chemosphere* **2016**, *155*, 9–17. [CrossRef] [PubMed]
88. Li, Z.; Meng, X.; Zhang, Z. Fabrication of surface hydroxyl modified g-C$_3$N$_4$ with enhanced photocatalytic oxidation activity. *Catal. Sci. Technol.* **2019**, *9*, 3979–3993. [CrossRef]
89. Wu, Y.; Shi, S.; Yuan, S.; Yuan, S.; Bai, T.; Xing, S. Insight into the enhanced activity of Ag/NiO$_x$-MnO$_2$ for catalytic oxidation of o-xylene at low temperatures. *Appl. Surf. Sci.* **2019**, *479*, 1262–1269. [CrossRef]
90. Yang, S.; Zhao, H.; Dong, F.; Zha, F.; Tang, Z. Highly efficient catalytic combustion of o-dichlorobenzene over three-dimensional ordered mesoporous cerium manganese bimetallic oxides: A new concept of chlorine removal mechanism. *Mol. Catal.* **2019**, *463*, 119–129. [CrossRef]
91. He, D.; Zhao, Y.; Yang, S.; Mei, Y.; Yu, J.; Liu, J.; Chen, D.; He, S.; Luo, Y. Enhancement of catalytic performance and resistance to carbonaceous deposit of lanthanum (La) doped HZSM-5 catalysts for decomposition of methyl mercaptan. *Chem. Eng. J.* **2018**, *336*, 579–586. [CrossRef]
92. Liu, N.; Shi, D.; Zhang, R.; Li, Y.; Chen, B. Highly selective catalytic combustion of acrylonitrile towards nitrogen over Cu-modified zeolites. *Catal. Today* **2018**, *332*, 201–213. [CrossRef]
93. Fiorenza, R.; Crisafulli, C.; Condorelli, G.G.; Lupo, F.; Scire, S. Au–Ag/CeO$_2$ and Au–Cu/CeO$_2$ Catalysts for Volatile Organic Compounds Oxidation and CO Preferential Oxidation. *Catal. Lett.* **2015**, *145*, 1691–1702. [CrossRef]
94. Yao, J.; Lu, H.; Xiao, Y.; Hou, B.; Li, D.; Jia, L. Sub-molten salt-acid treatment of LaCoO$_3$ for a highly active catalyst towards propane combustion. *Catal. Commun.* **2019**, *128*, 10578–10583. [CrossRef]
95. Zhang, K.; Peng, X.; Yang, H.; Wang, X.; Zhang, Y.; Zheng, Y.; Xiao, Y.; Jiang, L. Effect of MnO$_2$ morphology on its catalytic performance in lean methane combustion. *Mater. Res. Bull.* **2018**, *111*, 338–341. [CrossRef]

96. Wei, Y.; Ni, L.; Li, M.; Zhao, J. A template-free method for preparation of MnO$_2$ catalysts with high surface areas. *Catal. Today* **2017**, *297*, 188–192. [CrossRef]
97. Sinquin, G.; Petit, C.; Hindermann, J.P.; Kiennemann, A. Study of the formation of LaMO$_3$ (M = Co, Mn) perovskites by propionates precursors: Application to the catalytic destruction of chlorinated VOCs. *Catal. Today* **2001**, *70*, 183–196. [CrossRef]
98. Alberici, R.M.; Jardim, W.F. Photocatalytic destruction of VOCs in the gas-phase using titanium dioxide. *Appl. Catal. B Environ.* **1997**, *14*, 55–68. [CrossRef]
99. Fujimoto, T.M.; Ponczek, M.; Rochetto, U.L.; Landers, R.; Tomaz, E. Photocatalytic oxidation of selected gas-phase VOCs using UV light, TiO$_2$, and TiO$_2$/Pd. *Environ. Sci. Pollut. Res.* **2017**, *24*, 6390–6396. [CrossRef]
100. Qian, X.; Ren, M.; Yue, D.; Zhu, Y.; Han, Y.; Bian, Z.; Zhao, Y. Mesoporous TiO$_2$ films coated on carbon foam based on waste polyurethane for enhanced photocatalytic oxidation of VOCs. *Appl. Catal. B Environ.* **2017**, *212*, 1–6. [CrossRef]
101. Zhang, L.; Peng, Y.; Zhang, J.; Chen, L.; Meng, X.; Xiao, F.S. Adsorptive and catalytic properties in the removal of volatile organic compounds over zeolite-based materials. *Chin. J. Catal.* **2016**, *37*, 800–809. [CrossRef]
102. Huang, H.; Huang, H.; Zhan, Y.; Liu, G.; Wang, X.; Lu, H.; Xiao, L.; Feng, Q.; Leung, D.Y.C. Efficient degradation of gaseous benzene by VUV photolysis combined with ozone-assisted catalytic oxidation: Performance and mechanism. *Appl. Catal. B Environ.* **2016**, *186*, 62–68. [CrossRef]
103. Song, H.; Peng, Y.; Liu, S.; Bai, S.; Hong, X.; Li, J. The Roles of Various Plasma Active Species in Toluene Degradation by Non-thermal Plasma and Plasma Catalysis. *Plasma Chem. Plasma Process.* **2019**, *39*, 1469–1482. [CrossRef]
104. Wang, L.; Wang, C.; Xie, H.; Zhan, W.; Guo, Y. Catalytic combustion of vinyl chloride over Sr doped LaMnO$_3$. *Catal. Today* **2018**, *327*, 190–195. [CrossRef]
105. Kołodziej, A.; Łojewska, J. Optimization of structured catalyst carriers for VOC combustion. *Catal. Today* **2005**, *105*, 378–384. [CrossRef]
106. Joung, H.J.; Kim, J.H.; Oh, J.S.; You, D.W.; Park, H.O.; Jung, K.W. Catalytic oxidation of VOCs over CNT-supported platinum nanoparticles. *Appl. Surf. Sci.* **2014**, *290*, 267–273. [CrossRef]
107. Kim, S.C.; Shim, W.G. Properties and performance of Pd based catalysts for catalytic oxidation of volatile organic compounds. *Appl. Catal. B Environ.* **2009**, *92*, 429–436. [CrossRef]
108. Bedi, J.; Rosas, J.M.; Rodríguez-Mirasol, J.; Cordero, T. Pd supported on mesoporous activated carbons with high oxidation resistance as catalysts for toluene oxidation. *Appl. Catal. B Environ.* **2010**, *94*, 8–18. [CrossRef]
109. Usón, L.; Colmenares, M.G.; Hueso, J.L.; Sebastián, V.; Balas, F.; Arruebo, M.; Santamaría, J. VOCs abatement using thick eggshell Pt/SBA-15 pellets with hierarchical porosity. *Catal. Today* **2014**, *227*, 179–186. [CrossRef]
110. Carrillo, A.M.; Carriazo, J.G. Cu and Co oxides supported on halloysite for the total oxidation of toluene. *Appl. Catal. B Environ.* **2015**, *164*, 443–452. [CrossRef]
111. Kucherov, A.V.; Sinev, I.M.; Ojala, S.; Keiski, R.L.; Kustov, M. Adsorptive-catalytic removal of CH$_3$OH, CH$_3$SH, and CH$_3$SSCH$_3$ from air over the bifunctional system noble metals/HZSM-5. *Stud. Surf. Sci. Catal.* **2007**, *170*, 1129–1136.
112. Wang, J. Study on Supported Ruthenium Catalysts for the Catalytic Oxidation of VOCs. Ph.D. Thesis, Institute of Process Engineering, CAS, China, 2016.
113. Abdelouahab-Reddam, Z.; Mail, R.E.; Coloma, F.; Sepúlveda-Escribano, A. Platinum supported on highly-dispersed ceria on activated carbon for the total oxidation of VOCs. *Appl. Catal. A Gen.* **2015**, *494*, 87–94. [CrossRef]
114. Piumetti, M.; Fino, D.; Russo, N. Mesoporous manganese oxides prepared by solution combustion synthesis as catalysts for the total oxidation of VOCs. *Appl. Catal. B Environ.* **2015**, *163*, 277–287. [CrossRef]
115. Grbic, B.; Radic, N.; Terlecki-Baricevic, A. Kinetics of deep oxidation of n-hexane and toluene over Pt/Al$_2$O$_3$ catalysts: Oxidation of mixture. *Appl. Catal. B Environ.* **2004**, *50*, 161–166. [CrossRef]
116. Patterson, M.J.; Angove, D.E.; Cant, N.W. The effect of carbon monoxide on the oxidation of four C6 to C8 hydrocarbons over platinum, palladium and rhodium. *Appl. Catal. B Environ.* **2000**, *26*, 47–57. [CrossRef]
117. McCabe, R.W.; Mitchell, P.J. Exhaust-catalyst development for methanol-fueled vehicles: 1. A comparative study of methanol oxidation over alumina-supported catalysts containing group 9, 10, and 11 metals. *Appl. Catal.* **1986**, *27*, 83–98. [CrossRef]

118. Liotta, L.F.; Ousmane, M.; Di Carlo, G.; Pantaleo, G.; Deganello, G.; Marcì, G.; Retailleau, L.; Giroir-Fendler, A. Total oxidation of propene at low temperature over Co_3O_4–CeO_2 mixed oxides: Role of surface oxygen vacancies and bulk oxygen mobility in the catalytic activity. *Appl. Catal. A Gen.* **2008**, *347*, 81–88. [CrossRef]
119. Liotta, L.F.; Ousmane, M.; Di Carlo, G.; Pantaleo, G.; Deganello, G.; Boreave, A.; Giroir-Fendler, A. Catalytic removal of toluene over Co_3O_4–CeO_2 mixed oxide catalysts: Comparison with Pt/Al_2O_3. *Catal. Lett.* **2009**, *127*, 270–276. [CrossRef]
120. Mitsui, T.; Tsutsui, K.; Matsui, T.; Kikuchi, R.; Eguchi, K. Catalytic abatement of acetaldehyde over oxide-supported precious metal catalysts. *Appl. Catal. B Environ.* **2008**, *78*, 158–165. [CrossRef]
121. Zhang, C.; He, H. A comparative study of TiO_2 supported noble metal catalysts for the oxidation of formaldehyde at room temperature. *Catal. Today* **2007**, *126*, 345–350. [CrossRef]
122. Zhang, C.B.; He, H.; Tanaka, K. Catalytic performance and mechanism of a Pt/TiO_2 catalyst for the oxidation of formaldehyde at room temperature. *Appl. Catal. B Environ.* **2006**, *65*, 37–43. [CrossRef]
123. Agarwal, S.K.; Spivey, J.J.; Butt, J.B. Catalyst deactivation during deep oxidation of chlorohydrocarbons. *Appl. Catal. A Gen.* **1992**, *82*, 259–275. [CrossRef]
124. Petrosius, S.C.; Drago, R.S.; Young, V.; Grunewald, G.C. Low-temperature decomposition of some halogenated hydrocarbons using metal oxide/porous carbon catalysts. *J. Am. Chem. Soc.* **1993**, *115*, 6131–6137. [CrossRef]
125. Sedjame, H.J.; Fontaine, C.; Lafaye, G.; Barbier, J.J. On the promoting effect of the addition of ceria to platinum based alumina catalysts for VOCs oxidation. *Appl. Catal. B Environ.* **2014**, *144*, 233–242. [CrossRef]
126. Carabineiro, S.; Chen, X.; Konsolakis, M.; Psarras, A.; Tavares, P.; Orf~ao, J.; Pereira, M.; Figueiredo, J. Catalytic oxidation of toluene on Ce-Co and La-Co mixed oxides synthesized by exotemplating and evaporation methods. *Catal. Today* **2015**, *244*, 161–171. [CrossRef]
127. Carabineiro, S.; Chen, X.; Martynyuk, O.; Bogdanchikova, N.; Avalos-Borja, M.; Pestryakov, A.; Tavares, P.; Orf~ao, J.; Pereira, M.; Figueiredo, J. Gold supported on metal oxides for volatile organic compounds total oxidation. *Catal. Today* **2015**, *244*, 103–114. [CrossRef]
128. Castano, M.H.; Molina, R.; Moreno, S. Catalytic oxidation of VOCs on $MnMgAlO_x$ mixed oxides obtained by auto-combustion. *J. Mol. Catal. A Chem.* **2015**, *398*, 358–367. [CrossRef]
129. Chen, H.; Zhang, H.; Yan, Y. Fabrication of porous copper/manganese binary oxides modified ZSM-5 membrane catalyst and potential application in the removal of VOCs. *Chem. Eng. J.* **2014**, *254*, 133–142. [CrossRef]
130. Solsona, B.; Davies, T.E.; Garcia, T.; Vázquez, I.; Dejoz, A.; Taylor, S.H. Total oxidation of propane using nanocrystalline cobalt oxide and supported cobalt oxide catalysts. *Appl. Catal. B Environ.* **2008**, *84*, 176–184. [CrossRef]
131. Liu, Q.; Wang, L.C.; Chen, M.; Cao, Y.; He, H.Y.; Fan, K.N. Dry citrate precursor synthesized nanocrystalline cobalt oxide as highly active catalyst for total oxidation of propane. *J. Catal.* **2009**, *263*, 104–113. [CrossRef]
132. Tseng, T.K.; Wang, L.; Ho, C.T.; Chu, H. The destruction of dichloroethane over a g-alumina supported manganese oxide catalyst. *J. Hazard. Mater.* **2010**, *178*, 1035–1040. [CrossRef]
133. Krishnamoorthy, S.; Rivas, J.A.; Amiridis, M.D. Catalytic oxidation of 1, 2-dichlorobenzene over supported transition metal oxides. *J. Catal.* **2000**, *193*, 264–272. [CrossRef]
134. Lahousse, C.; Bernier, A.; Grange, P.; Delmon, B.; Papaefthimiou, P.; Ioannides, T.; Verykios, X. Evaluation of g-MnO_2 as a VOC removal catalyst: Comparison with a noble metal catalyst. *J. Catal.* **1998**, *178*, 214–225. [CrossRef]
135. Parida, K.; Samal, A. Catalytic combustion of volatile organic compounds on Indian Ocean manganese nodules. *Appl. Catal. A Gen.* **1999**, *182*, 249–256. [CrossRef]
136. Luo, J.; Zhang, Q.; Huang, A.; Suib, S.L. Total oxidation of volatile organic compounds with hydrophobic cryptomelane-type octahedral molecular sieves. *Microporous Mesoporous Mater.* **2000**, *35–36*, 209–217. [CrossRef]
137. Aguero, F.N.; Scian, A.; Barbero, B.P.; Cadús, L.E. Influence of the support treatment on the behavior of MnO_x/Al_2O_3 catalysts used in VOC combustion. *Catal. Lett.* **2009**, *128*, 268–280. [CrossRef]
138. Kang, M.; Park, E.D.; Kim, J.M.; Yie, J.E. Manganese oxide catalysts for NOx reduction with NH3 at low temperatures. *Appl. Catal. A Gen.* **2007**, *327*, 261–269. [CrossRef]
139. Miranda, B.; Díaz, E.; Ordonez, S.; Vega, A.; Díez, F.V. Oxidation of trichloroethene over metal oxide catalysts: Kinetic studies and correlation with adsorption properties. *Chemosphere* **2007**, *66*, 1706–1715. [CrossRef]

140. Sun, H.; Liu, Z.; Chen, S.; Quan, X. The role of lattice oxygen on the activity and selectivity of the OMS-2 catalyst for the total oxidation of toluene. *Chem. Eng. J.* **2015**, *270*, 58–65. [CrossRef]
141. Cordi, E.M.; O'Neill, P.J.; Falconer, J.L. Transient oxidation of volatile organic compounds on a CuO/Al_2O_3 catalyst. *Appl. Catal. B Environ.* **1997**, *14*, 23–36. [CrossRef]
142. Hutchings, G.J.; Taylor, S.H. Designing oxidation catalysts. *Catal. Today* **1999**, *49*, 105–113. [CrossRef]
143. Heynderickx, P.M.; Thybaut, J.W.; Poelman, H.; Poelman, D.; Marin, G.B. The total oxidation of propane over supported Cu and Ce oxides: A comparison of single and binary metal oxides. *J. Catal.* **2010**, *272*, 109–120. [CrossRef]
144. Sinha, A.K.; Suzuki, K. Novel mesoporous chromium oxide for VOCs elimination. *Appl. Catal. B Environ.* **2007**, *70*, 417–422. [CrossRef]
145. Padilla, A.M.; Corella, J.; Toledo, J.M. Total oxidation of some chlorinated hydrocarbons with commercial chromia based catalysts. *Appl. Catal. B Environ.* **1999**, *22*, 107–121. [CrossRef]
146. Yim, S.D.; Chang, K.-H.; Nam, I.S. Deactivation of chromium oxide catalyst for the removal of perchloroethylene (PCE). *Stud. Surf. Sci. Catal.* **2001**, *139*, 173–180.
147. Gorte, R.J. Ceria in catalysis: From automotive applications to the water-gas shift reaction. *AIChE J.* **2010**, *56*, 1126–1135. [CrossRef]
148. Zimmer, P.; Tschope, A.; Birringer, R. Temperature-programmed reaction spectroscopy of ceria-and Cu/ceria-supported oxide catalyst. *J. Catal.* **2002**, *205*, 339–345. [CrossRef]
149. Huang, H.; Xu, Y.; Feng, Q.; Leung, D.Y.C. Low temperature catalytic oxidation of volatile organic compounds: A review. *Catal. Sci. Technol.* **2015**, *5*, 2649–2669. [CrossRef]
150. Li, H.; Lu, G.; Dai, Q.; Wang, Y.; Guo, Y.; Guo, Y. Hierarchical organization and catalytic activity of high-surface-area mesoporous ceria microspheres prepared via hydrothermal routes. *ACS Appl. Mater. Interfaces* **2010**, *2*, 838–846. [CrossRef]
151. Jones, J.; Ross, J.R. The development of supported vanadia catalysts for the combined catalytic removal of the oxides of nitrogen and of chlorinated hydrocarbons from flue gases. *Catal. Today* **1997**, *35*, 97–105. [CrossRef]
152. Delaigle, R.; Debecker, D.P.; Bertinchamps, F.; Gaigneaux, E.M. Revisiting the behaviour of vanadia-based catalysts in the abatement of (chloro)-aromatic pollutants: Towards an integrated understanding. *Top. Catal.* **2009**, *52*, 501–516. [CrossRef]
153. Solsona, B.; Garcia, T.; Aylon, E.; Dejoz, A.M.; Vazquez, I.; Agouram, S.; Davies, T.E.; Taylor, S.H. Promoting the activity and selectivity of high surface area Ni-Ce-O mixed oxides by gold deposition for VOC catalytic combustion. *Chem. Eng. J.* **2011**, *175*, 271–278. [CrossRef]
154. Delimaris, D.; Ioannides, T. VOC oxidation over MnO_x-CeO_2 catalysts prepared by a combustion method. *Appl. Catal. B Environ.* **2008**, *84*, 303–312. [CrossRef]
155. Morales, M.R.; Barbero, B.P.; Cadús, L.E. Total oxidation of ethanol and propane over Mn-Cu mixed oxide catalysts. *Appl. Catal. B Environ.* **2006**, *67*, 229–236. [CrossRef]
156. Vasile, A.; Bratan, V.; Hornoiu, C.; Caldararu, M.; Ionescu, N.I.; Yuzhakova, T.; Redey, A. Electrical and catalytic properties of cerium etin mixed oxides in CO depollution reaction. *Appl. Catal. B Environ.* **2013**, *140*, 25–31. [CrossRef]
157. Tang, W.; Wu, X.; Li, S.; Li, W.; Chen, Y. Porous Mn-Co mixed oxide nanorod as a novel catalyst with enhanced catalytic activity for removal of VOCs. *Catal. Commun.* **2014**, *56*, 134–138. [CrossRef]
158. Larsson, P.O.; Andersson, A. Complete oxidation of CO, ethanol, and ethyl acetate over copper oxide supported on titania and ceria modified titania. *J. Catal.* **1998**, *179*, 72–89. [CrossRef]
159. Hu, C.; Zhu, Q.; Jiang, Z.; Zhang, Y.; Wang, Y. Preparation and formation mechanism of mesoporous CuOeCeO2 mixed oxides with excellent catalytic performance for removal of VOCs. *Microporous Mesoporous Mater.* **2008**, *113*, 427–434. [CrossRef]
160. Delimaris, D.; Ioannides, T. VOC oxidation over CuO-CeO_2 catalysts prepared by a combustion method. *Appl. Catal. B Environ.* **2009**, *89*, 295–302. [CrossRef]
161. Rao, T.; Shen, M.; Jia, L.; Hao, J.; Wang, J. Oxidation of ethanol over Mn-Ce-O and Mn-Ce-Zr-O complex compounds synthesized by solegel method. *Catal. Commun.* **2007**, *8*, 1743–1747. [CrossRef]
162. Tang, X.; Li, Y.; Huang, X.; Xu, Y.; Zhu, H.; Wang, J.; Shen, W. MnO_x-CeO_2 mixed oxide catalysts for complete oxidation of formaldehyde: Effect of preparation method and calcination temperature. *Appl. Catal. B Environ.* **2006**, *62*, 265–273. [CrossRef]

163. Picasso, G.; Gutierrez, M.; Pina, M.; Herguido, J. Preparation and characterization of Ce-Zr and Ce-Mn based oxides for n-hexane combustion: Application to catalytic membrane reactors. *Chem. Eng. J.* **2007**, *126*, 119–130. [CrossRef]
164. Chen, H.; Sayari, A.; Adnot, A.; Larachi, F. Composition activity effects of Mn-Ce-O composites on phenol catalytic wet oxidation. *Appl. Catal. B Environ.* **2001**, *32*, 195–204. [CrossRef]
165. Yang, P.; Yang, S.; Shi, Z.; Meng, Z.; Zhou, R. Deep oxidation of chlorinated VOCs over CeO_2-based transition metal mixed oxide catalysts. *Appl. Catal. B Environ.* **2015**, *162*, 227–235. [CrossRef]
166. Li, J.; Zhao, P.; Liu, S. SnO_x-MnO_x-TiO_2 catalysts with high resistance to chlorine poisoning for low-temperature chlorobenzene oxidation. *Appl. Catal. A Gen.* **2014**, *482*, 363–369. [CrossRef]
167. Rao, G.R.; Sahu, H.R.; Mishra, B.G. Surface and catalytic properties of Cu-Ce-O composite oxides prepared by combustion method. *Coll. Surf. A Physicochem. Eng. Asp.* **2003**, *220*, 261–269. [CrossRef]
168. Jiang, X.; Lu, G.; Zhou, R.; Mao, J.; Chen, Y.; Zheng, X. Studies of pore structure, temperature-programmed reduction performance, and micro-structure of CuO/CeO_2 catalysts. *Appl. Surf. Sci.* **2001**, *173*, 208–220.
169. Zheng, X.C.; Wu, S.H.; Wang, S.P.; Wang, S.R.; Zhang, S.M.; Huang, W.P. The preparation and catalytic behavior of copperecerium oxide catalysts for low-temperature carbon monoxide oxidation. *Appl. Catal. A Gen.* **2005**, *283*, 217–223. [CrossRef]
170. Pecchi, G.; Reyes, P.; Zamora, R.; Cadus, L.E.; Fierro, J.L.G. Surface properties and performance for VOCs combustion of $LaFe_{1-y}Ni_yO_3$ perovskite oxides. *J. Solid State Chem.* **2008**, *181*, 905–912. [CrossRef]
171. Wei, T. Study on Catalytic Combustion of VOCs over Perovskite Catalysts. Master's Thesis, Zhejiang University of Technology, Zhejiang, China, 2005.
172. Beauchet, R.; Magnoux, P.; Mijoin, J. Catalytic oxidation of volatile organic compounds (VOCs) mixture (isopropanol/o-xylene) on zeolite catalysts. *Catal. Today* **2007**, *124*, 118–123. [CrossRef]
173. Rachapudi, R.; Chintawar, P.S.; Greene, H.L. Aging and Structure/Activity Characteristics of CR–ZSM-5 Catalysts during Exposure to Chlorinated VOCs. *J. Catal.* **1999**, *185*, 58–72. [CrossRef]
174. Muniandy, L.; Adam, F.; Mohamed, A.R.; Iqbal, A.; Rahman, N.R.A. Cu^{2+} coordinated graphitic carbon nitride (Cu-g-C_3N_4) nanosheets from melamine for the liquid phase hydroxylation of benzene and VOCs. *Appl. Surf. Sci.* **2017**, *398*, 43–55. [CrossRef]
175. Blanch-Raga, N.; Palomares, A.E.; Martínez-Triguero, J.; Valencia, S. Cu and Co modified beta zeolite catalysts for the trichloroethylene oxidation. *Appl. Catal. B Environ.* **2016**, *187*, 90–97. [CrossRef]
176. Lopez-Fonseca, R.; Aranzabal, A.; Steltenpohl, P.; Gutierrez-Ortiz, J.I.; Gonzalez Velasco, J.R. Performance of zeolites and product selectivity in the gas-phase oxidation of 1,2-dichloroethane. *Catal. Today* **2000**, *62*, 367–377. [CrossRef]
177. Kullavanijayam, E.; Trimm, D.L.; Cant, N.W. Adsocat: Adsorption/catalytic combustion for VOC and odour control. *Stud. Surf. Sci. Catal.* **2000**, *130*, 569–574.
178. Fujishima, A.; Honda, K. Electrochemical photolysis of water at a semiconductorelectrode. *Nature* **1972**, *238*, 37–38. [CrossRef] [PubMed]
179. Wang, D.; Hou, P.; Yang, P.; Cheng, X. $BiOBr@SiO_2$ flower-like nanospheres chemically-bonded on cement-based materials for photocatalysis. *Appl. Surf. Sci.* **2018**, *430*, 539–548. [CrossRef]
180. Li, Y.; Wu, X.; Li, J.; Wang, K.; Zhang, G. Z-scheme g-C_3N_4@Cs_xWO_3 heterostructure as smart window coating for UV isolating, Vis penetrating, NIR shielding and full spectrum photocatalytic decomposing VOCs. *Appl. Catal. B Environ.* **2018**, *229*, 218–226. [CrossRef]
181. Mishra, A.; Mehta, A.; Kainth, S.; Basu, S. Effect of different plasmonic metals on photocatalytic degradation of volatile organic compounds (VOCs) by bentonite/M-TiO_2 nanocomposites, under UV/visible light. *Appl. Clay Sci.* **2018**, *153*, 144–153. [CrossRef]
182. Song, S.; Lu, C.; Wu, X.; Jiang, S.; Sun, C.; Le, Z. Strong base g-C_3N_4 with perfect structure for photocatalytically eliminating formaldehyde under visible-light irradiation. *Appl. Catal. B Environ.* **2018**, *227*, 145–152. [CrossRef]
183. Bhatkhande, D.S.; Pangarkar, V.G.; Beenackers, A.A.C.M. Photocatalytic degradation for environmental applications—A review. *J. Chem. Technol. Biotechnol. Int. Res. Process Environ. Clean Technol.* **2001**, *77*, 107–116. [CrossRef]
184. Choi, H.; Stathatos, E.; Dionysiou, D.D. Sol–gel preparation of mesoporous photocatalytic TiO_2 films and TiO_2/Al_2O_3 composite membranes for environmental applications. *Appl. Catal. B Environ.* **2006**, *63*, 60–67. [CrossRef]

185. Fujishima, A. TiO$_2$ photocatalysis and related surface phenomena. *Surf. Sci. Rep.* **2008**, *63*, 515–582. [CrossRef]
186. Minero, C. Surface modified photocatalysts, in environmental photochemistry part III. In *The Handbook of Environmental Chemistry*; Bahnemann, D.W., Robertson, K.J., Eds.; Springer: Berlin/Heidelberg, Germany, 2015; Volume 35, pp. 23–44.
187. Tejasvi, R.; Sharma, M.; Upadhyay, K. Passive photo-catalytic destruction of air-borne VOCs in high traffic areas using TiO$_2$-coated flexible PVC sheet. *Chem. Eng. J.* **2015**, *262*, 875–881. [CrossRef]
188. Li, F.B.; Li, X.Z.; Ao, C.H.; Lee, S.C.; Hou, M.F. Enhanced photocatalytic degradation of VOCs using Ln^{3+}–TiO$_2$ catalysts for indoor air purification. *Chemosphere* **2005**, *59*, 787–800. [CrossRef]
189. Héqueta, V.; Raillarda, C.; Debonoa, O.; Thévenet, F.; Locoge, N.; Le Coq, L. Photocatalytic oxidation of VOCs at ppb level using a closed-loop reactor: The mixture effect. *Appl. Catal. B Environ.* **2018**, *226*, 473–486. [CrossRef]
190. Huang, L.; Nakajo, K.; Ozawa, S.; Matsuda, H. Decomposition of dichloromethane in a wire-in-tube pulsed corona reactor. *Environ. Sci. Technol.* **2001**, *35*, 1276–1281. [CrossRef]
191. Norberg, A. Modeling current pulse shape and energy in surface discharges. *IEEE Trans. Ind. Appl.* **1992**, *28*, 498–503. [CrossRef]
192. Rousseau, A.; Dantier, A.; Gatilova, L.; Ionikh, Y.; Röpcke, J.; Tolmachev, Y. On NO$_x$ production and volatile organic compound removal in a pulsed microwave discharge in air. *Plasma Sour. Sci. Technol.* **2005**, *14*, 70–75. [CrossRef]
193. Kogelschatz, U. Dielectric-barrier discharges: Their history, discharge physics, and industrial applications. *Plasma Chem. Plasma Process.* **2003**, *23*, 1–46. [CrossRef]
194. Yamamoto, T.; Ramanathan, K.; Lawless, P.A.; Ensor, D.S.; Newsome, J.R.; Plaks, N.; Ramsey, G.H. Control of volatile organic compounds by an ac energized ferroelectric pellet reactor and a pulsed corona reactor. *IEEE Trans. Ind. Appl.* **1992**, *28*, 528–534. [CrossRef]
195. Ding, H.-X.; Zhu, A.-M.; Yang, X.-F.; Li, C.-H.; Xu, Y. Removal of formaldehyde from gas streams via packed-bed dielectric barrier discharge plasmas. *J. Phys. D App. Phys.* **2005**, *38*, 4160–4167. [CrossRef]
196. Kim, H.H.; Ogata, A.; Futamura, S. Oxygen partial pressure-dependent behavior of various catalysts for the total oxidation of VOCs using cycled system of adsorption and oxygen plasma. *Appl. Catal. B Environ.* **2008**, *79*, 356–367. [CrossRef]
197. Futamura, S.; Yamamoto, T.; Lawless, P.A. Towards understanding of VOC decomposition mechanisms using nonthermal plasmas. In Proceedings of the 1995 Thirtieth IAS Annual Meeting IEEE Conference Record of Industry Applications Conference, Orlando, FL, USA, 8–12 October 1995; Volume 2, pp. 1453–1458.
198. Zheng, C.; Zhu, X.; Gao, X.; Liu, L.; Chang, Q.; Luo, Z.; Cen, K. Experimental study of acetone removal by packed-bed dielectric barrier discharge reactor. *J. Ind. Eng. Chem.* **2014**, *20*, 2761–2768. [CrossRef]
199. Lee, B.Y.; Park, S.H.; Lee, S.C.; Kang, M.; Choung, S.J. Decomposition of benzene by using a discharge plasma-photocatalyst hybrid system. *Catal. Today* **2004**, *93–95*, 769–776. [CrossRef]
200. Gandhi, M.S.; Ananth, A.; Mok, Y.S.; Song, J.I.; Park, K.H. Effect of porosity of α-alumina on non-thermal plasma decomposition of ethylene in a dielectric-packed bed reactor. *Res. Chem. Intermed.* **2014**, *40*, 1483–1493. [CrossRef]
201. Hu, J.; Jiang, N.; Li, J.; Shang, K.; Lu, N.; Wu, Y. Degradation of benzene by bipolar pulsed series surface/packed-bed discharge reactor over MnO$_2$-TiO$_2$/zeolite catalyst. *Chem. Eng. J.* **2016**, *293*, 216–224. [CrossRef]
202. Ogata, A.; Yamanouchi, K.; Mizuno, K.; Kushiyama, S.; Yamamoto, T. Oxidation of dilute benzene in an alumina hybrid plasma reactor at atmospheric pressure. *Plasma Chem. Plasma Process.* **1999**, *19*, 383–394. [CrossRef]
203. Zhu, X.; Gao, X.; Qin, R.; Zeng, Y.; Qu, R.; Zheng, C.; Tu, X. Plasma-catalytic removal of formaldehyde over Cu–Ce catalysts in a dielectric barrier discharge reactor. *Appl. Catal. B Environ.* **2015**, *170–171*, 293–300. [CrossRef]
204. Bradford, M.C.J.; Vannice, M.A. Estimation of CO heats of adsorption on metal surfaces from vibrational spectra. *Ind. Eng. Chem. Res.* **1996**, *35*, 3171–3178. [CrossRef]
205. An, H.T.Q.; Huu, T.P.; Le Van, T.; Cormier, J.M.; Khacef, A. Application of atmospheric non thermal plasma-catalysis hybrid system for air pollution control: Toluene removal. *Catal. Today* **2011**, *176*, 474–477.
206. Futamura, S.; Einaga, H.; Kabashima, H.; Hwan, L.Y. Synergistic effect of silent discharge plasma and catalysts on benzene decomposition. *Catal. Today* **2004**, *89*, 89–95. [CrossRef]

207. Wu, J.; Huang, Y.; Xia, Q.; Li, Z. Decomposition of toluene in a plasma catalysis system with NiO, MnO$_2$, CeO$_2$, Fe$_2$O$_3$, and CuO catalysts. *Plasma Chem. Plasma Process.* **2013**, *33*, 1073–1082. [CrossRef]
208. Zhao, D.-Z.; Li, X.-S.; Shi, C.; Fan, H.-Y.; Zhu, A.-M. Low-concentration formaldehyde removal from air using a cycled storage-discharge (CSD) plasma catalytic process. *Chem. Eng. Sci.* **2011**, *66*, 3922–3929. [CrossRef]
209. Ran, L.; Wang, Z.; Wang, X. The effect of Ce on catalytic decomposition of chlorinated methane over RuO$_x$ catalysts. *Appl. Catal. A gen.* **2014**, *470*, 442–450. [CrossRef]
210. Asilevi, P.J.; Yi, C.W.; Li, J.; Nawaz, M.I.; Wang, H.J.; Yin, L.; Junli, Z. Decomposition of formaldehyde in strong ionization non-thermal plasma at atmospheric pressure. *Int. J. Environ. Sci. Technol.* **2020**, *17*, 765–776. [CrossRef]
211. Zhu, T.; Li, J.; Jin, Y.; Liang, Y.; Ma, G. Decomposition of benzene by non-thermal plasma processing: Photocatalyst and ozone effect. *Int. J. Environ. Sci. Technol.* **2008**, *5*, 375–384. [CrossRef]
212. Urashima, K.; Kostov, K.G.; Chang, J.S.; Okayasu, Y.; Iwaizumi, T.; Yoshimura, K.; Kato, T. Removal of C$_2$F$_6$ from a semiconductor process flue gas by a ferroelectric packed-bed barrier discharge reactor with an adsorber. *IEEE Trans. Ind. Appl.* **2001**, *37*, 1456–1463. [CrossRef]
213. Ogata, A.; Ito, D.; Mizuno, K.; Kushiyama, S.; Gal, A.; Yamamoto, T. Effect of coexisting components on aromatic decomposition in a packed-bed plasma reactor. *Appl. Catal. A Gen.* **2002**, *236*, 9–15. [CrossRef]
214. Ogata, A.; Shintani, N.; Yamanouchi, K.; Mizuno, K.; Kushiyama, S.; Yamamoto, T. Effect of water vapor on benzene decomposition using a nonthermal-discharge plasma reactor. *Plasma Chem. Plasma Process.* **2000**, *20*, 453–467. [CrossRef]
215. Mustafa, M.F.; Fu, X.; Liu, Y.; Abbas, Y.; Wang, H.; Lu, W. Volatile organic compounds (VOCs) removal in non-thermal plasma double dielectric barrier discharge reactor. *J. Hazard. Mater.* **2018**, *347*, 317–324. [CrossRef] [PubMed]
216. Hirota, K.; Sakai, H.; Washio, M.; Kojima, T. Application of electron beams for the treatment of VOC streams. *Ind. Eng. Chem. Res.* **2004**, *43*, 1185–1191. [CrossRef]
217. Zimek, Z. High power accelerators and processing systems for environmental application. Radiation treatment of gaseous and liquid effluents for contaminant removal. *IAEA-TECDOC* **2005**, *1473*, 125–137.
218. Son, Y.S.; Kim, J.; Kim, J.C. Decomposition of acetaldehyde using an electron beam. *Plasma Chem. Plasma Process.* **2014**, *34*, 1233–1245. [CrossRef]
219. Son, Y.S.; Kim, K.H.; Kim, K.J.; Kim, J.C. Ammonia Decomposition Using Electron Beam. *Plasma Chem. Plasma Process.* **2013**, *33*, 617–629. [CrossRef]
220. Han, D.H.; Stuchinskaya, T.; Won, Y.S.; Park, W.S.; Lim, J.K. Oxidative decomposition of aromatic hydrocarbons by electron beam irradiation. *Radiat. Phys. Chem.* **2003**, *67*, 51–60. [CrossRef]
221. Hashimoto, S.; Hakoda, T.; Hirata, K.; Arai, H. Low energy electron beam treatment of VOCs. *Radiat. Phys. Chem.* **2000**, *57*, 485–488. [CrossRef]
222. Sun, Y.; Chmielewski, A.G.; Licki, J.; Bułka, S.; Zimek, Z. Decomposition of organic compounds in simulated industrial off-gas by using electron beam irradiation. *Radiat. Phys. Chem.* **2009**, *78*, 721–723. [CrossRef]
223. Son, Y.S.; Kim, P.; Park, J.H.; Kim, J.; Kim, J.C. Decomposition of trimethylamine by an electron beam. *Plasma Chem. Plasma Process.* **2013**, *33*, 1099–1109. [CrossRef]
224. Son, Y.S.; Park, J.H.; Kim, P.; Kim, J.C. Oxidation of gaseous styrene by electron beam irradiation. *Radiat. Phys. Chem.* **2012**, *81*, 686–692. [CrossRef]
225. Auslender, V.L.; Ryazantsev, A.A.; Spiridonov, G.A. The use of electron beam for solution of some ecological problems in pulp and paper industry. *Radiat. Phys. Chem.* **2002**, *63*, 641–645. [CrossRef]
226. Son, Y.S.; Kim, J.C. Decomposition of sulfur compounds by radiolysis: I. Influential factors. *Chem. Eng. J.* **2015**, *262*, 217–223. [CrossRef]
227. Son, Y.S.; Jung, I.H.; Lee, S.J.; Kim, J.C. Decomposition of sulfur compounds by a radiolysis: III. A hybrid system and field application. *Chem. Eng. J.* **2015**, *274*, 9–16. [CrossRef]
228. Son, Y.S.; Jung, I.H.; Lee, S.J.; Koutrakis, P.; Kim, J.C. Decomposition of sulfur compounds by radiolysis: II. By-products and mechanisms. *Chem. Eng. J.* **2015**, *269*, 27–34. [CrossRef]
229. Kim, J.C. Factors affecting aromatic VOC removal by electron beam treatment. *Radiat. Phys. Chem.* **2002**, *65*, 429–435. [CrossRef]

© 2020 by the authors. Licensee MDPI, Basel, Switzerland. This article is an open access article distributed under the terms and conditions of the Creative Commons Attribution (CC BY) license (http://creativecommons.org/licenses/by/4.0/).

Bimetallic Catalysts for Volatile Organic Compound Oxidation

Review

Roberto Fiorenza

Department of Chemical Sciences, University of Catania, Viale A. Doria 6, 95125 Catania, Italy; rfiorenza@unict.it; Tel.: +39-393-6586864

Received: 25 May 2020; Accepted: 9 June 2020; Published: 12 June 2020

Abstract: In recent years, the impending necessity to improve the quality of outdoor and indoor air has produced a constant increase of investigations in the methodologies to remove and/or to decrease the emission of volatile organic compounds (VOCs). Among the various strategies for VOC elimination, catalytic oxidation and recently photocatalytic oxidation are regarded as some of the most promising technologies for VOC total oxidation from urban and industrial waste streams. This work is focused on bimetallic supported catalysts, investigating systematically the progress and developments in the design of these materials. In particular, we highlight their advantages compared to those of their monometallic counterparts in terms of catalytic performance and physicochemical properties (catalytic stability and reusability). The formation of a synergistic effect between the two metals is the key feature of these particular catalysts. This review examines the state-of-the-art of a peculiar sector (the bimetallic systems) belonging to a wide area (i.e., the several catalysts used for VOC removal) with the aim to contribute to further increase the knowledge of the catalytic materials for VOC removal, stressing the promising potential applications of the bimetallic catalysts in the air purification.

Keywords: VOCs; bimetallic catalysts; air purification

---

1. Introduction

Volatile organic compounds (VOCs) are a wide group of organic compounds characterized to boiling points less than 250 °C at room temperature and at atmospheric pressure [1]. Due to their carcinogenic and toxic nature, most VOCs are considered major causes of air pollution. Indeed, their emission in the environment leads to the formation of secondary dangerous compounds, due to the occurrence of chemical reactions with other airborne pollutants such as NO_x and SO_x, which results in the formation of tropospheric ozone and photochemical smog [2,3]. Long exposure to these pollutants leads to serious problems for human health [4,5]. Global economic and industrial development over the years has caused an exponential increase of anthropogenic VOC emission [6]. VOC discharges include outdoor sources, such as transport, industrial and petrochemical processes, etc., and indoor sources, such as household products, solvents, office materials, cleaning products, domestic cooking, etc. [7]. The emitted VOCs encompass alkanes, paraffins, olefins, aromatics, alcohols, ketones, aldehydes, esters, sulfur/nitrogen-containing VOCs, and halogenated VOCs. Among them, the most common and toxic are benzene, phenol, toluene, styrene, formaldehyde, propylene, and acetone [8], whereas Cl-VOCs and in general halogenated VOCs, due to their inherent stability and toxicity, are also very dangerous [9].

Different technologies have been developed for VOC treatment, and they can be divided into nondestructive and destructive VOC removal. The former include adsorption, membrane separation, and condensation [10–12]. Among these, the adsorption process is considered one of the most efficient treatments, owing to a low energy consumption, relatively low operation cost, and simple operations for the adsorption/regeneration of the adsorbent [12,13]. With adsorption, it is possible to remove, without the generation of dangerous byproducts, a low/medium concentration of VOCs (<1000 mg/m^3) [1]. Regarding destructive (i.e., oxidative) processes, the most commonly used ones are thermal (not-catalytic) combustion and catalytic oxidation, both of which can be applied to treat a medium/high concentration (>5000 mg/m^3) of VOCs [14,15]. In particular, catalytic conversion has some advantages compared to thermal incineration; indeed, this process has become more popular than noncatalytic treatments. Catalytic oxidation allows converting VOCs into less toxic substances, such as carbon dioxide and water, in a temperature range much lower than thermo-oxidation [16,17]. Specifically, with catalytic conversion, the operating temperature range is 200–500 °C or even lower, whereas in thermal incineration, temperatures are higher (800–1200 °C). Lower temperatures permit reducing the production of dioxins and NO$_x$. Furthermore, catalytic oxidation is more versatile and cheaper, especially when it comes to processing low concentrations of organic compounds [18]. In recent years, new technologies have been applied for the elimination of VOCs at low concentrations, namely, the advanced oxidation process (AOP), e.g., photocatalytic degradation, ozone treatment, Fenton oxidation methods [19,20], biodegradation [21], and phytoremediation [22].

Due to the economic and technological advantages of catalytic oxidation, widespread efforts have been committed to the selection of high-performing catalysts for this process. However, considering the large number of organic molecules and the problematic nature of VOCs mixtures, the design and optimization of catalytic materials are challenging tasks. Both noble and transition metals have been widely used as catalysts for either nonhalogenated or halogenated VOCs [23–25]. Notwithstanding their high costs, the supported noble-metal catalysts are widely applied due to their intrinsic features, such as resistance to deactivation, ease of regeneration, and highly catalytic performance [26–28]. These features strictly depend on the synthetic procedure adopted for the preparation of the supported metal catalyst, as well as the type of metal salt precursor, the metal loading, the kind of support, and the particle size [29–31]. Furthermore, VOCs and air/oxygen content, total gas stream rate, and employed reactor (membrane reactor, fixed-bed reactor, etc.) are key parameters that can affect overall catalytic activity [32–34].

In the literature, there are many studies that deeply analyze single or various parameters that influence the final results of catalytic oxidation applied to VOC treatment, including the catalysts used [16,35,36], the nature of VOCs [2,9,37], the combination of different technologies [1,38], the type of reactor [39], or the performances in practical applications [40].

One of the less explored strategies to enhance the catalytic activity of supported noble/transition metal catalysts is the addition of a second metal (noble and/or transition) to the first one.

This work analyzes a little aspect of the VOC catalytic treatment topic: the advantages of using supported bimetallic catalysts with respect to monometallic counterparts, focusing on the morphological, chemicophysical, and textural properties of these peculiar materials, and how these features can influence the catalytic activity.

This review aims to enlarge the scientific panorama about VOC removal through catalytic oxidation, focusing on the bimetallic catalysts, an aspect not yet systematically examined in the literature.

2. Bimetallic Catalysts for VOC Oxidation

Bimetallic nanoparticles (NPs) are a kind of materials formed by two different metals and characterized with peculiar features [41,42]. Specifically, they can show new properties resulting from the combination of features arising from the monometallic counterparts. Usually, the obtained physicochemical properties of bimetallic systems give a holistic result, i.e., the final properties are not the simple additive features of the monometallic analogs, but in many cases, it is possible to exploit a great improvement with new properties due to the presence of synergistic effects [43,44]. Since their application in the field of petrochemistry [45], bimetallic systems are being widely applied in heterogeneous catalysis in various reactions, such as hydrogenation [46,47], reforming [48,49], H$_2$ production and purification [50–53], and oxidation [54–56]. Recently, they have also been applied in the biomedical field [57].

On the basis of the morphology of the bimetallic system, it is possible to classify the main arrangements into three typologies: core–shell or multishell structures, heterostructures, and random or homogeneous alloys (Figure 1) [43,44,58].

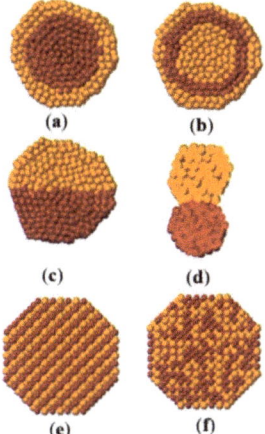

Figure 1. Possible morphologies of bimetallic nanoparticles (one metal in red, the other in dark yellow): (**a**,**b**) core–shell system; (**c**) multishell system; (**d**) subcluster segregated systems; and ordered (**e**) and random (**f**) homogeneous alloys. Reprinted (adapted) from [58], Copyright 2008, American Chemical Society.

Various factors influence the final morphology of bimetallic systems: (1) the intrinsic strength of the bonds between the two metals in comparisons to the strength of the bonds between the two monometallic constituents (in particular, if the resulting alloy bond strength is greater, reciprocal mixing is preferred, and, in the opposite case, the segregation of the two metals is favored); (2) the balance in the surface energy of the metals, where the monometallic element that owns the lower surface energy will move to the surface, establishing a shell structure, while the metal with the smaller atomic sizes will collocate to the core; (3) the mixing is favored when an electron/charge transfer between the metallic counterparts is verified; and (4) magnetic effects can influence the final structure between the two metals [44,58,59].

The morphologies are also strictly affected by the preparation method adopted. In general, the synthesis of the bimetallic systems can be carried out with the solid-state or with the solution methods [60–62]. The latter are preferred because solid-state techniques require high temperatures and long times for annealing procedures, thus decreasing the surface area of the bimetallic catalyst and affecting catalytic performance in a crucial way. By contrast, with solution methods, it is possible to control the nucleation and growth processes by modifying the reaction parameters.

In the main catalytic reactions, the bimetallic system is supported on a specific support [63]. The commonly used techniques for the preparation of supported bimetallic catalysts are: impregnation [64,65], co-precipitation [66,67], deposition–precipitation [50,68], thermal decomposition [69], and chemical reduction [70]. The choice of a proper support is another key feature that strongly influences the catalytic activity of the bimetallic system. In particular, for VOC oxidation, and for oxidation reactions in general, the utilization of an active support (i.e., having redox properties and lattice oxygen mobility), for example, CeO_2 or Fe_2O_3, can contribute to the formation of a synergistic effect involving both the metals and the same support [68,71], whereas other supports, with high surface areas, such as zeolites, silica, alumina, etc., favor the high dispersion of the metal species and, therefore, the interaction between the metals and the same support [16,72].

Since the first works of Haruta and co-workers [73], gold-based catalysts have generated many investigations focused on the catalytic properties of this peculiar metal. With respect to the platinum-group catalysts, gold's properties are in some cases superior; consequently, the supported gold-based materials have found many applications in catalytic reactions dealing with environmental protection and energy production [50,72]. In the specific field of VOC catalytic oxidation, gold-based catalysts play a key role. For these reasons, bimetallic alloys with gold are firstly examined.

2.1. Gold-Based Bimetallic Catalysts

The important role of gold-based catalysts in heterogeneous catalysis was disclosed through the high activity of this metal in CO oxidation and selective oxidations [74,75]. After these studies, many works dealing with this catalyst have emerged, widely expanding the use of gold catalysts [72,76]. With respect to the other commonly used noble metals, namely, Pt and Pd, gold catalysts show peculiar features such as resistance to O_2 poisoning and high selectivity. Another key parameter that strictly affects the catalytic activity of Au catalysts is size-dependence: To be active, gold particles must be, in general, smaller than 5 nm, which balances the proportion of the low coordination surface active sites (edges and corners) [77].

For other applications, as well as for VOC oxidation, the catalytic activity of gold-supported samples is affected by many factors: (a) gold–support interaction; (b) gold loading; (c) the valence state of gold; (d) the adopted synthesis and pre-treatment conditions utilized; and (e) the concentration and nature of the chosen VOC target [78,79]. Among the monometallic gold catalysts employed for VOC oxidation, the Au/CeO_2 sample showed a great performance in the oxidation of oxygenated molecules, i.e., aldehydes, ketones, esters, and alcohols [72]. It was reported that the high activity was related to the enhancement in the surface oxygen reducibility/mobility of the CeO_2 support. The oxidation mechanism followed a Mars–Van Krevelen (MvK) mechanism, where the ceria lattice oxygens were actively involved in the oxidation pathway [68,80] (Figure 2).

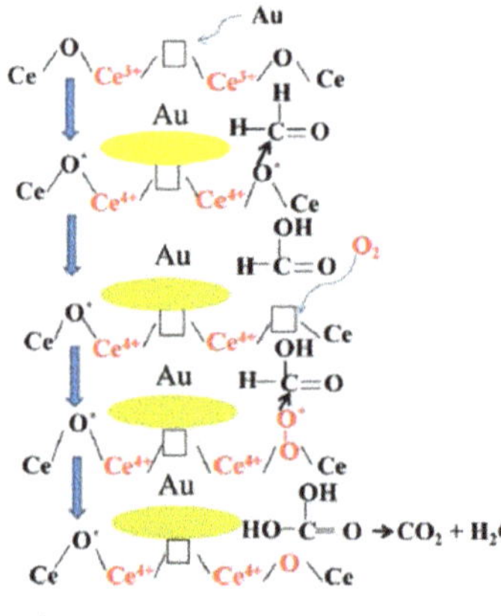

Figure 2. Lattice ceria oxygens of Au/CeO_2 catalyst involved in catalytic oxidation of formaldehyde. Reprinted (adapted) from [80], Copyright 2014, American Chemical Society.

Nevertheless, the application of gold catalysts to industrial level is limited due to certain drawbacks, such as the decrease of activity in the presence of high concentration of moisture and the aggregation of gold nanoparticles at high temperatures [81,82].

Gold-based bimetallic systems are considered a useful solution to overcome the cited limitations, due to the combination of properties from the gold and the second metal. Among the various metals employed to be joined with gold, it has been found that Pd, Ag, and Cu show a high miscibility, and these systems can be prepared with a wide range of methodologies [83,84], whereas Pt, Ru, Co, Fe, and

Ni are only partially or not at all miscible with gold [84]. Consequently, Au-Pd followed by Au-Ag and Au-Cu are the most utilized gold-based bimetallic catalysts for VOC oxidations.

2.1.1. Au-Pd Catalysts

Gold is miscible with palladium in all compositions; consequently, while the formation of gold–palladium alloys is favored, the segregation of single metals was, in fact, avoided [76,83]. In Table 1 are reported some of the experimental results of the application of the Au-Pd systems in catalytic oxidation of various VOCs (T_{90} = temperature at which the 90% of conversion was achieved).

In particular, Hosseini et al. [85] synthetized Au-Pd catalysts supported on mesoporous TiO_2 for the removal of toluene, propene, and a gaseous mixture of both. Interestingly, they found that catalytic activity is influenced by the morphology of the core–shell structure with the best performance shown with the Au-core/Pd-shell. By contrast, with the reverse morphology (Au-shell/Pd-core) the catalytic activity was lower, due to the lower affinity of gold for oxygen adsorption (in this case, the rate determining step of the reaction that followed a Langmuir–Hinshelwood mechanism) caused by the poor ability of gold to polarize oxygen molecules. In the same context, Barakat et al. [86] investigated the catalytic stability of bimetallic Au-Pd/doped TiO_2 samples under severe testing conditions (exposing the catalyst to 110 h of a gaseous toluene/air stream). The bimetallic catalyst maintained a good activity even after a long time. The interaction between the Nb-doped TiO_2 support and the Au-Pd system allowed obtaining a cycle-like activity of the catalyst. This oscillatory behavior was related to the existence of carbonaceous compounds adsorbed on the surface of the spent catalyst that, together with the formed OH radicals, favored the reduction of palladium. The redox process of palladium was linked to the cyclic-like activity of the bimetallic sample (Figure 3).

Table 1. Comparison between different supported Au-Pd bimetallic catalysts in catalytic oxidation of various volatile organic compounds (VOCs).

Catalyst [1]	Preparation Method	Support	VOC	T_{90} (°C)	Ref.
1%Au-0.5%Pd	core–shell	TiO_2	toluene	≈200 °C	[85]
1%Au-1%Pd	chemical reduction	MnO_2	toluene	≈180 °C	[87]
1%Au-1%Pd	chemical reduction	Co_3O_4	toluene	≈160 °C	[88]
1%Au-0.5%Pd	core–shell	TiO_2	propene	≈190 °C	[85]
3%Au-1%Pd	deposition–precipitation	CeO_2-5%Fe_2O_3	benzene	≈95 °C	[89]
1%Au-1%Pd	deposition–precipitation	CeO_2	benzyl alcohol	≈120 °C	[90]
2%Au-2%Pd-0.2%Fe	chemical reduction	Mn_2O_3	o-xylene	≈210 °C	[91]

[1] Nominal concentration, weight percentage (wt%).

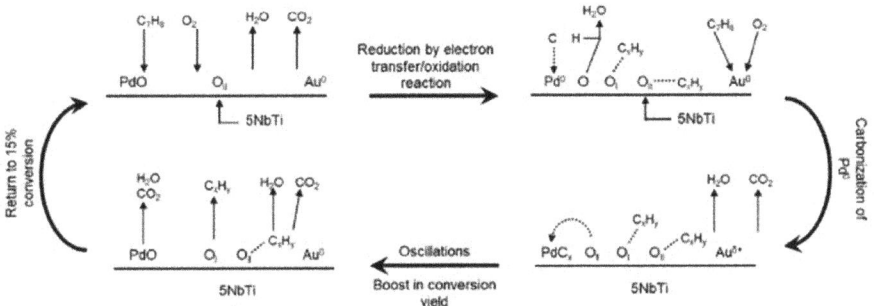

Figure 3. Catalytic oxidation of toluene under ageing condition over Au-Pd/Nb-TiO_2 catalyst. Figure from [86].

Regarding toluene degradation, Xie et al. [88] studied the catalytic performance of the bimetallic gold–palladium system supported on three-dimensionally ordered macroporous (3DOM) Co_3O_4.

The bimetallic sample showed a much higher activity compared to its nonmetallic counterparts, with a T_{90} of about 100 and 30 °C lower compared to monometallic gold and palladium, respectively. The peculiar features of the 3DOM supports (such as higher porosity and ordered pore channels) were also exploit by the same authors [91], who utilized an Au-Pd bimetallic sample prepared via chemical reduction, but employing the Mn_2O_3 as support. The bimetallic catalysts confirmed the higher activity compared to monometallic gold and palladium in the oxidation of different VOCs, such as methane and o-xylene. To further boost catalytic activity, doping with Fe of the Au-Pd/3DOM-Mn_2O_3 catalyst allowed modifying the structural properties of the alloy NPs. With this modification, oxygen activation and the methane adsorption ability were increased, enhancing, as final result, the overall catalytic activity.

Catalytic performance in various mixtures of VOCs (toluene/m-xylene, ethyl acetate/m-xylene acetone/m-xylene, and acetone/ethyl acetate) was examined by Xia et al. [87] on Au-Pd prepared via chemical-reduction-supported αMnO_2 nanotubes. In this case, the authors focused on an MvK-like mechanism with the mutual interaction of the Au-Pd nanoparticles with the α-MnO_2 that improved the mobility/reactivity of the surface lattice oxygen of the support. In catalytic oxidation of VOC mixture, the rate-determining step is the competitive adsorption between the various VOCs on the surface of the catalyst. With the bimetallic catalyst, the authors measured the total oxidation of a single component and of the VOC mixture at T < 300 °C. The same bimetallic sample also showed a high catalytic stability in the long time (50 h) on stream experiments.

Tabakova et al. [89] focused their work on the removal of benzene utilizing an Au-Pd bimetallic system synthetized by deposition–precipitation and supported on Fe-doped CeO_2. The superior performance of the bimetallic catalyst with respect to the monometallic ones was highlighted through comparison of the T_{90}, which was ≈95 °C for the bimetallic sample, ≈180 °C for the Pd/Fe-CeO_2, and ≈190 °C for the Au/Fe-CeO_2. In addition, in this case, the synergistic interaction between the alloy nanoparticles and the support enhanced the mobility of the ceria surface lattice oxygen, further boosted by doping with iron. Moreover, this strong interaction facilitated the nucleation of the noble metal particles on the surface of cerium oxide.

Benzyl alcohol oxidation on Au-Pd bimetallic catalysts was extensively studied by the research group of Prati et al., and an exhaustive comparison of the performance, in this reaction, of the bimetallic Au-Pd systems present in the literature, together with a deep analysis of the various adopted preparation methods are reported in [83]. The same research group [76,83] studied the catalytic behavior of different bimetallic systems composed of Au and various other metals supported on activated carbon prepared using the sol immobilization methodology. Interestingly, they found a structural correlation depending on the second metal utilized. Specifically, an alloy structure was obtained using Pd and Pt, whereas a core–shell morphology was attained with Ru, while with Cu, a phase segregation of this metal, instead of gold, was favored. In the benzyl alcohol tests, the bimetallic synergistic effect was exploited only with copper and palladium.

The strong interaction between gold and palladium, beneficial for catalytic oxidation of benzyl alcohol, was also examined by Li et al. [90] using CeO_2 with different morphology as support. The different morphology of cerium oxide (rod, cube, and polyhedrons), where gold/palladium nanoparticles were deposited–precipitated (Figure 4), affected the catalytic performance. Specifically, the Au-Pd supported on ceria rod showed a higher benzyl alcohol conversion with respect to the samples supported on CeO_2 cubes and CeO_2 polyhedrons and was thus related to the smaller particle size of ceria rod compared to the other CeO_2 supports; moreover, this particular morphology also favored a higher concentration of ceria oxygen defects, enhancing the mobility/reducibility of the ceria surface oxygens. By contrast, the sample supported on the CeO_2 cube exhibited the highest selectivity in benzaldehyde.

Kucherov et al. [92] investigated the performance of mono- and bimetallic gold-based catalysts for the removal of dimethyldisulfide (DMDS), an S-VOC. They corroborate the performance of Au-Pd catalysts also for this type of VOC. The bimetallic sample supported on TiO_2 demonstrated a stable performance and assisted with the removal of DMDS at T < 155 °C with the formation of SO_2 and elemental S.

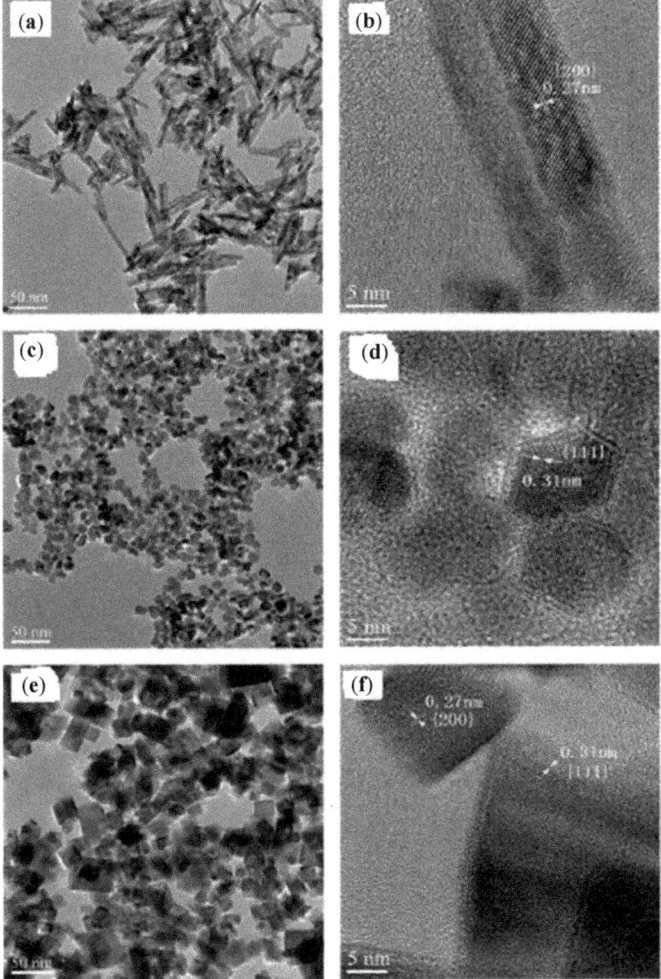

Figure 4. TEM (Transmission electron microscope) and HRTEM (High-resolution transmission electron microscope), images of: Au-Pd/CeO_2-rod (**a,b**); Au-Pd/CeO_2-polyhedron (**c,d**); and Au-Pd/CeO_2-cube (**e,f**) catalysts. Figure from Ref. [90], Copyright 2020, Elsevier.

2.1.2. Au-Ag and Au-Cu Catalysts

The establishment of a strong interaction between gold and silver with the formation of an alloy or of bimetallic clusters was investigated by our research group both in VOC oxidation and in H_2 purification towards the preferential oxidation of CO (PROX reaction) [50,53,68]. In particular, we evaluated the catalytic activity of Au-Ag and Au-Cu bimetallic samples supported on CeO_2 toward the degradation of 2-propanol and ethanol. A higher activity was found of the gold–silver sample with respect to Au-Cu and the monometallic counterparts. The higher activity of the gold–silver system was correlated to a higher mobility/reactivity of ceria surface oxygens, due to a strong synergistic interaction between the gold–silver nanoparticles and the cerium oxide. A linear correlation was stated considering the T_{50} of alcohol oxidation and the TPR (temperature-programmed reduction) initial temperature, i.e., the temperature at which the reduction of ceria surface oxygens started considering the analyzed samples (Figure 5). The Au-Ag/CeO$_2$ catalyst displayed the lowest reduction temperature and T_{50}.

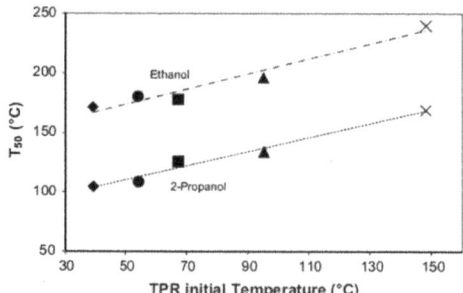

Figure 5. Temperature at which the 50% of conversion of ethanol and 2-propanol was achieved (T_{50}) versus TPR (temperature-programmed reduction) initial temperature: (filled diamond) Au-Ag/CeO$_2$; (filled circle) Au-Cu/CeO$_2$; (filled square) Au/CeO$_2$; (filled triangle) Ag/CeO$_2$; (times) Cu/CeO$_2$. Figure from [68], Copyright 2015, Springer Nature.

Nagy et al. [93] studied the performance of Au-Ag nanoparticles supported on SiO$_2$ synthetized from the adsorption of bimetallic colloids in the oxidation of benzyl alcohol. The authors focused their research on the crucial importance of the molar ratio between the two metals. In particular, a synergistic effect was verified that reflects a higher activity at a low Ag/Au molar ratio (best result Ag/Au = 23/77). For the authors, the synergy is activated by the optimal concentration of the two metals, which increased the activation of both oxygens from gas-phase and from the support. In the same context, a correlation between catalytic activity and the concentration of gold and silver was measured by our research group in the PROX reaction [50] with a higher concentration of gold or silver with respect to the second metal that was detrimental for the overall catalytic performance, whereas the best results were obtained with an approximately equal concentration of gold and silver (1% wt–1% wt).

The crucial importance of the molar ratio between gold and the other metal was also stated in the review of Bracey et al. [94], focused on the Au-Cu system. Specifically, in one of the analyzed works, the following order of reaction in catalytic oxidation of propene is reported: AuCu (1:3 molar ratio)/TiO$_2$ > AuCu (1:1 molar ratio)/TiO$_2$ > AuCu/TiO$_2$ (3:1 molar ratio) > Au/TiO$_2$. The content of copper, in fact, strongly influenced the dispersion of the metal nanoparticles, with a high amount of copper in the alloy that caused a decrease in the size of the metal particles, thus contributing to enhancint the activity and selectivity into propene oxide [95]. In the same review, it was illustrated that, when investigating another reaction, such as selective oxidation of benzyl alcohol to benzaldehyde, the more active bimetallic catalyst was the sample with the higher concentration of gold (the catalyst AuCu/SiO$_2$ with a molar ratio of 4:1). Similarly to the previous examples, the bimetallic catalyst was prepared by impregnation, but in this case, a higher concentration of gold is fundamental to achieve a high selectivity (98%) to benzaldehyde.

The above-discussed literature data on AuCu bimetallic catalysts were mainly focused on the selective oxidation of VOCs, whereas the work of Nevanperä et al. [96] dealt with catalytic oxidation of DMDS with bimetallic gold-based catalysts (Au-Cu and Au-Pt) supported on γ-Al$_2$O$_3$, CeO$_2$, and CeO$_2$-Al$_2$O$_3$ prepared by surface redox reduction. Among the examined supports, the alumina gave the best results, whereas the addition of gold enhanced the catalytic activity of both monometallic copper and platinum samples, Au-Cu catalysts being the most active system. Interestingly, the authors noted that the same Au-Cu catalyst led to the formation of dangerous byproducts, such as carbon monoxide and formaldehyde. This was attributed to the high concentration of reactive surface oxygens favored by the presence of copper oxide and to the dissociation of the oxygen that started at a lower temperature with respect to the monometallic samples, with the consequent modification of the surface acid and basic sites of the bimetallic catalyst. By contrast, selectivity towards CO_2 and H_2O was higher in the Au-Pt sample.

2.1.3. Other Au-Based Bimetallic Catalysts

As discussed in the last examined work, among the other Au-based bimetallic catalysts, the Au-Pt system exhibited promising performance in VOC oxidation [96–98].

Kim et al. [97] investigated catalytic oxidation of toluene employing the Au-Pt/ZnO-Al_2O_3 catalyst prepared by impregnation in air or H_2. They found that the bimetallic sample prepared in air led to an increase of the gold particle size and a decrease of the Pt with respect to the same particles synthetized in H_2 stream, where an inverse correlation was verified (the gold size decreased, and the platinum size increased). Due to the crucial importance of the gold nanoparticles that facilitated the total oxidation of toluene and that increased the reduction of the surface oxygen of the mixed oxide support, the catalytic performance was higher with the bimetallic sample synthetized in H_2 stream and calcined at 400 °C (Pt and Au mean size of about 5 nm). In another study [98], the same authors correlated the catalytic activity of the same bimetallic samples, even in the total oxidation of toluene, to the molar ratio of gold and platinum, finding the following order of activity: Pt75Au25 > Pt67Au33 > Pt100Au0 > Pt50Au50 > Pt33Au67 > Pt25Au75 > Pt0Au100. The small amount of gold promotes the total oxidation of toluene due to the formation of a strong metal–metal interaction.

The good affinity of gold with noble metals was also confirmed via the catalytic performance in VOC removal of the Au-Ru system [99,100]. Sreethawong et al. [99] investigated catalytic oxidation of methanol over gold–ruthenium samples prepared through impregnation and supported on SiO_2. The characterization measurements (TPR, SEM, and XRD) suggested the occurrence of an interaction between the two metals exploited with a particular composition (3.32 wt%Ru–0.61 wt% Au), which led to obtaining a good catalytic activity, notwithstanding the fact that the two metals were not miscible in their bulk phase. Interestingly, if alumina was used as support, the formation of byproducts (methyl formate, formic acid, dimethyl ether, and formaldehyde) other to CO_2 was detected, with an increase of methanol conversion. Catalytic oxidation of methanol was also analyzed by Calzada et al. [100] with Au-Ru/TiO_2 catalysts prepared by deposition–precipitation with urea. The authors highlighted that the synergistic effect between the two metals was activated at a low conversion temperature (from room temperature to T = 50 °C), with a dependence on the bimetallic atomic ratio (the best-performing one was Ru:Au 0.75:1). Interestingly, the DRIFT (Diffuse Reflectance Infrared Fourier Transform Spectroscopy) CO spectra (Figure 6) illustrated as the interaction between the two metals decreased CO adsorption in the Ru surface sites in the Ru-Au 0.75:1 sample, an indication of the modification of the surface gold sites.

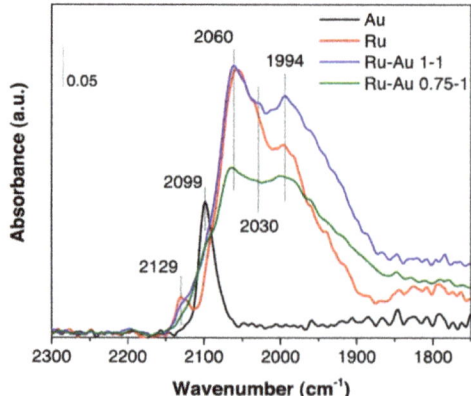

Figure 6. DRIFT (Diffuse Reflectance Infrared Fourier Transform Spectroscopy)-CO spectra of the mono- and bimetallic Ru-Au/TiO$_2$ samples at room temperature. Figure from [100]. Copyright 2017, Elsevier.

The modifications of the bimetallic surface sites due to interaction between the two metals strongly occurring with the Ru:Au 0.75:1 ratio were the reason behind the higher catalytic activity of this bimetallic sample compared to those of the monometallic catalysts and the Ru:Au 1:1 sample. In addition, in this case, the formation of formates as intermediates of the oxidation reaction was verified.

Catalytic activity in the total oxidation of toluene of the Au-Ir bimetallic catalyst supported on TiO$_2$ was studied by Torrente-Murciano et al. [101]. Similarly to the previous case, the synergistic interaction between the two metals allowed them to sensibly decrease the T$_{90}$ that was ≈230 °C for the bimetallic sample, ≈250 °C for the monometallic gold, and ≈270 °C for Ir/TiO$_2$. The key factors that deeply influenced catalytic activity were the strong metals–support interaction exploited with the bimetallic system, which also permitted diminishing the loss of activity due to the metals sintering at high temperatures. Furthermore, the intimate contact between iridium and gold modified the bimetallic surface-active sites enhancing oxygen activation.

In the work cited in Section 2.2, Kucherov et al. [92] demonstrated a good activity of Au-Rh supported on HZSM-5 zeolite into the oxidation of the DMDS in SO$_2$ at 290 °C. In this case, the zeolite support, owing to a high surface area, favored a high dispersion of the metals, a feature that is beneficial for catalytic activity.

Regarding gold–copper bimetallic systems, there are certain studies with other transition metals, utilized together with gold for VOC oxidation. Au-Co and Au-Fe interaction were principally investigated. In the examined works, the gold atoms interacted with the second metal present as a doping agent of the support [102–104].

Solsona et al. [102] synthetized gold nanoparticles anchored on cobalt containing mesoporous silica (UVM-7). The interaction between gold and cobalt permitted increasing catalytic activity in the oxidation of toluene and propane with respect to Au/UVM-7 and Co/UVM-7 catalysts. The presence of gold enhanced the reducibility of cobalt, present as Co$_3$O$_4$ at the Au-Co interface, thus facilitating the redox cycle of cobalt, with an MvK-like mechanism, which boosted catalytic oxidation of VOCs. In the same context, Albonetti et al. [103,104] deeply investigated the catalytic behavior of gold catalysts supported on mesoporous silica (SBA-15) via an iron oxide layer obtaining the Au/FeOx/SBA-15 composite. The good dispersion of nanosized gold favored the incidence of a strong synergism between gold and iron that led to an optimal activity in the combustion of methanol (T$_{90}$ ≈ 140 °C).

As a conclusion of this generic overview of gold-based bimetallic catalysts applied at VOC oxidation, it is possible to recognize some fundamental features of these peculiar catalysts: (a) the essential action of the nanosized gold that is able to establish a metal–metal surface interaction with a wide range of both noble and transition metals; (b) the occurrence of a synergism between the two metals that allows sensibly decreasing the light-off temperatures of VOC oxidation; (c) the synergistic effect, which is not simply the addition of the single characteristics of the corresponding monometallic samples but leads to exploring new physicochemical properties; (d) the mutual interaction between the two metals which also strongly influences metals–support interaction (in particular, if the support is a reducible oxide (CeO_2, MnO_x, CoO_x, etc.), the gold-based bimetallic cluster increases the mobility of the surface oxygen of the support, enhancing, in this way, catalytic oxidation towards an MvK mechanism). If the support is a nonreducible or hardly reducible oxide (TiO_2, SiO_2, and zeolites), the high dispersion of the bimetallic alloy and the modifications of the metals surface active sites allow enhancing oxygen adsorption, improving, as a final result, the overall catalytic performance. The most employed preparation methods of the supported gold-based bimetallic catalysts are impregnation (wet or wetness), deposition–precipitation, and chemical reduction.

2.2. Other Bimetallic Catalysts

Among the other noble metals, the most employed catalysts for catalytic oxidation of VOCs are platinum-based materials [16], and similarly to gold, platinum has shown a good affinity with palladium [105–108]. In general, as can be seen from Table 2, the use of noble-metals-based bimetallic catalysts has allowed obtaining a good performance in the removal of VOCs, whereas the utilization of transition-metals-based materials has led to shift at high temperatures in the total conversion of VOCs. However, especially in recent years, the necessity to reduce the amount of expensive noble metals has led to exploring a new synergism between noble and transition metals.

Table 2. Comparison between different supported bimetallic catalysts in catalytic oxidation of various VOCs.

Catalyst [1]	Preparation Method	Support	VOC	T_{90} (°C)	Ref.
0.2%Pt-0.1%Pd	hydrothermal	Silica MCM-41	toluene	≈170 °C	[105]
0.3%Pt-2%Pd	impregnation	γ-Al_2O_3	benzene	≈225 °C	[106]
2%Ru-5%Co	impregnation	TiO_2	benzene	≈200 °C	[109]
1%Ru-5%Ce	impregnation	TiO_2	chlorobenzene	≈275 °C	[110]
18%Mn-0.1%Pd	impregnation	γ-Al_2O_3	formaldehyde/methanol	≈80 °C	[111]
15%Mn-5%Cu	impregnation	γ-Al_2O_3	toluene	≈350 °C	[112]
1.3%Fe-1.75%Ag	ionic exchange	ZMS-5	ethyl acetate	≈250 °C	[113]

[1] Nominal concentration, weight percentage (wt%).

Fu et al. [105] prepared, for the hydrothermal method, a Pt-Pd bimetallic sample supported on mesoporous silica. Comparably to the catalytic behavior of gold-based samples, the synergism between the two metals allowed obtaining a superior performance in the removal of toluene with respect to the monometallic samples, with an improvement in the reducibility of palladium, involved in the redox cycle PdO→Pd^0, and in oxygen adsorption capability.

The good catalytic activity of the Pt-Pd system was also confirmed by Kim et al. [106] in the degradation of benzene. Catalysts were synthetized via wetness impregnation on γ-Al_2O_3. Metal–metal interaction was favored by the formation of small and uniform particles and, as stated in the previous paragraphs, a specific amount ratio between the two metals (the optimum in the cited work is 0.3 wt% Pt–2%wt Pd). A higher concentration of platinum led to a remarkable decrease in activity due to the blockage of the active sites. The same authors in another paper [107] confirmed with a deep XPS (X-ray photoelectron spectroscopy) analysis the crucial role of the ratio between the metals to avoid the obstruction of the catalyst surface sites. The removal of methanol, acetone, and methylene chloride was instead studied by Sharma et al. [108], utilizing ceramic Raschig rings coated with Pt and Pd on fluorinated carbon. The authors measured a higher activity of the bimetallic catalyst with respect to monometallic ones. Furthermore, the hydrophobic nature of this particular bimetallic catalyst allowed obtaining a 90% of degradation of methanol and acetone at about 150 and 300 °C respectively, whereas

60% of degradation was achieved at 400 °C for the methylene chloride. In this case, a good correlation was established with a semi-empirical Langmuir–Hinshelwood model, which is able to predict the oxidation rate of each VOC in a gas mixture (methanol, acetone, and methylene chloride).

Ethanol adsorption and oxidation were investigated by the research group of Wittayakun et al. [114] with a Pt-M (M = Co, Cu, Mn) sample supported on silica MCM-41. Among the transition metals, cobalt gave the best results, and in particular, the bimetallic 0.5 wt% Pt–15 wt% Co exhibited the best ethanol adsorption and CO_2 desorption. Interestingly, the authors identified two different reaction mechanisms considering the platinum monometallic sample and the bimetallic platinum–cobalt one (Figure 7).

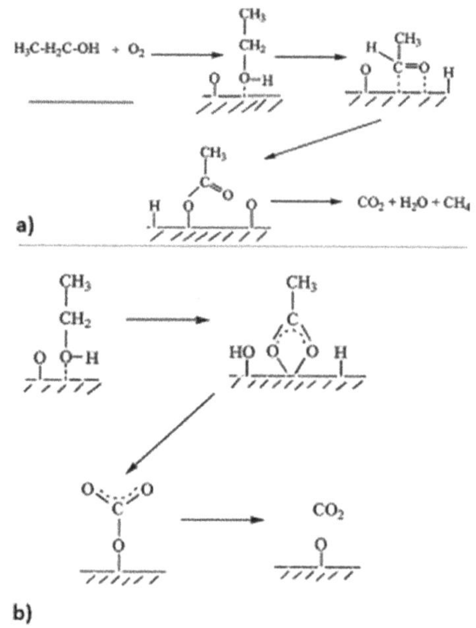

Figure 7. (a) Ethanol oxidation mechanism on 0.5 wt% Pt/MCM-41; and (b) ethanol oxidation mechanism on 0.5 wt% Pt–15 wt% Co/MCM-41. Figure modified from [114]. Copyright 2012, Elsevier.

Specifically, in the monometallic sample after the adsorption of ethanol, a formation was verified of a parallel adsorbed acetaldehyde, further converted into monodentate acetate and at end, dissociated and desorbed as carbon dioxide, methane, and water (Figure 7a). In the bimetallic catalyst, by contrast, the ethoxy species reacted with the adsorbed oxygen to give a bidentate acetate species that was transformed into carbon dioxide (Figure 7b). The modification of the ethanol adsorption led to a higher ethanol conversion with the monometallic platinum sample in comparison with the bimetallic Pt-Co sample that, conversely, showed a higher catalytic stability. Even with the bimetallic platinum-based catalysts, Chantaravitoon et al. [115] examined the performance of a Pt-Sn/γ-Al_2O_3 catalyst prepared with impregnation, for the oxidation of methanol. The authors noted from the temperature-programmed desorption (TPD) measurements of methanol oxidation that on the bimetallic catalyst, methanol decomposed as H_2 and CO and the desorption peaks shifted at higher temperatures, increasing the amount of Sn. In addition, in this case, the monometallic Pt catalyst exhibited a better performance compared to the bimetallic one; however, the addition of a small amount of Sn (<5 wt%) reduced the deactivation of the catalyst in the long-time tests.

In addition, Ru-based bimetallic compounds were discreetly studied for VOC oxidation [109,110,116]. Liu et al. [109] prepared Ru-M (M = Co, Mn, Ce, Fe, Cu) samples supported on TiO_2, evaluating catalytic performance in the degradation of benzene. Among the various metals,

1% wt Ru–5% wt Co showed the best activity; the presence of ruthenium, in fact, increased the reducibility of Co_3O_4. The authors stated also that the presence of water vapor inhibited benzene oxidation at T = 210 °C.

The total oxidation of propene was examined on Ru-Re/γ-Al_2O_3 by Baranowska et al. [116]. As discussed before, this nonconventional combination between these two metals also allowed increasing catalytic stability in the consecutive tests instead of the overall catalytic activity, which remained superior with the monometallic Ru sample. The addition of Re (the best composition being 5% wt Ru–3% wt Re) hampered the formation of RuO_2 agglomerates. In this way, the dispersion of ruthenium is favored, allowing a higher stability compared to monometallic ruthenium catalysts. Ye et al. [110] performed a catalytic test regarding chlorobenzene removal with Ru-Ce/TiO_2 samples prepared via impregnation. Interestingly, on the basis of the crystalline phase of titanium dioxide, the catalytic activity changed. At 280 °C, the bimetallic sample showed a conversion of 91% and 86% if supported on TiO_2 rutile and TiO_2 P25 (80% anatase, 20% rutile), respectively. The mixed crystalline phase of P25 was the best support for the monometallic ruthenium catalyst. By contrast, with respect to the work of Baranowska et al. [116] for this reaction and with the titanium dioxide support, dispersion was not the major parameter that affected catalytic activity; indeed, on 1%wt Ru–5%wt Ce/TiO_2 (rutile), the abundant RuO_2 clusters favored both catalytic activity and stability.

Interaction with Ce/CeO_2 was also investigated by Yue et al. [117] but utilizing palladium. The performance of the bimetallic catalyst Pd-Ce/ZMS-5 synthetized through impregnation was evaluated on the degradation of methyl ethyl ketone (MEK). The presence of cerium oxide considerably increased the acid sites of palladium, enhancing at the same time the re-oxidation of Pd and boosting, in the end, the overall MEK degradation rate through an MvK mechanism.

Another bimetallic catalyst with palladium was prepared by Arias et al. [111]. In this work, the synergistic effect between palladium and manganese was explored utilizing alumina as support. The authors followed the oxidation of a VOC mixture (formaldehyde/methanol), concluding also in this case that an MvK-like mechanism was the reaction pathway, with the interaction between palladium and manganese favoring the oxidation of VOCs due to the activation of the reactive lattice oxygen of PdO and MnO_x.

Various bimetallic samples were tested for catalytic oxidation of VOCs, studying both the physicochemical properties and the catalytic activity of silver-containing samples [113,118–120]. In particular, Jodaei et al. [113,118] tested different Ag-M bimetallic samples supported on ZMS-5 zeolite obtained via ionic exchange. The authors investigated the catalytic combustion of ethyl acetate, finding this order of activity and stability: Fe-Ag/ZSM-5 > Co-Ag/ZSM-5 > Mn-Ag/ZSM-5 > Ag/ZSM-5. The high dispersion of silver was favored by an optimal amount of iron (1.3 wt% Fe–1.75 wt% Ag), thus activating a synergistic effect between the two metals. In the same context, Izadkhah et al. [119] made a theoretical model for the removal of ethyl acetate. In particular, considering the preparation condition, the formulation, and loading of the promoter of silver, with their algorithm, it was possible to identify the optimal catalyst for this reaction. Among the first transition metal series, the bimetallic catalyst that exhibited superior performance compared to the monometallic silver was Fe-Ag/ZSM-5, thus confirming the experimental results of Jodaei et al. [113,118], Ni-Ag/ZSM-5 and V-Ag/ZSM-5.

Complete oxidation of formaldehyde at T < 90 °C was obtained by Qu et al. [120] with Ag-Co/MCM-41 silica. The key feature able to sensibly increase catalytic performance with respect to the monometallic silver was electron transfer between silver and cobalt that enhanced the reducibility of cobalt oxide, increasing, at the same time, the activation of surface oxygen on the bimetallic catalyst. Furthermore, the high metal–metal support interaction (SMMI) at the optimal Ag/Co mass ratio (3:1) favored a faster adsorption–dissociation of formaldehyde on the Ag species with respect to the Co^{3+} sites (Figure 8), thus decreasing the light-off temperature of VOC oxidation.

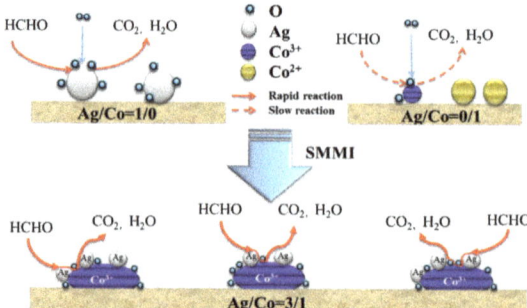

Figure 8. Formaldehyde oxidation on monometallic Ag/MCM-41, Co/MCM-41 and bimetallic Ag-Co/MCM-41. Figure from [120]. Copyright 2014, Elsevier.

On the same support (MCM-41 silica), Pârvulescu et al. [121] synthetized with the hydrothermal method various Co-based bimetallic mesostructures (Co-V, Co-La, Co-Nb) characterized vy a high surface area and narrow pore size distribution. The oxidation of styrene and benzene was deeply influenced by the addition of a second metal component. Indeed, although the addition of La did not result in any synergistic effect, the addition of vanadium favored the oxidation of benzene, whereas the addition of niobium facilitated the removal of styrene, demonstrating that the presence of the second metal changed the surface-active sites of cobalt.

Similarly to cobalt, copper-based bimetallic catalysts also showed a good activity in the removal of VOCs [112,122,123]. Kim et al. [112] found an optimal interaction between Mn and Cu for the total oxidation of toluene. The order of activity considering other transition metals as a second component was: 5% wt Cu–15% wt Mn/γ-Al$_2$O$_3$ > 5% wt Co–15% wt Mn/γ-Al$_2$O$_3$ > 5% wt Ni–15% wt Mn/γ-Al$_2$O$_3$ > 15% wt Mn/γ-Al$_2$O$_3$ > 5% wt Fe–15% wt Mn/γ-Al$_2$O$_3$. The interaction between Mn and Cu favored a high dispersion of manganese, increasing, at the same time, the mobility/reducibility of manganese oxide. The oxidation of toluene was studied recently by Djinović et al. [122], who had examined the performance of monometallic CuO and bimetallic Cu-FeO$_x$ composites supported on KIL-2 silica. The utilization of two reducible oxides allowed increasing the amount and reactivity of oxygen species, which included adsorbed (O$^-$ and O$_2{}^-$) and lattice (O^{2-}) oxygens at the Cu-FeO$_x$ interface, providing a substantial decrease of T$_{90}$ that was ≈350 °C for the bimetallic cluster instead of ≈450 °C of monometallic copper oxide, whereas the FeO$_x$/KIL-2 silica reached only 30% of toluene conversion at 450 °C.

Abdullah et al. [123] investigated the oxidation of a Cl-VOC mixture (dichloromethane (DCM), trichloroethylene (TCE) and trichloromethane (TCM)) with Cu-Cr/ZMS-5. Interestingly, in this case, the presence of water vapor in the gas feed enhanced the total oxidation to CO$_2$. The presence of water vapor favored the formation of reactive carbocations. Furthermore, H$_2$O was beneficial in blocking chlorine-transfer reactions. Indeed, an important deactivation effect was found with the bimetallic catalyst at a higher Cl/H gas feed ratio, and chlorination led to a decrease in metals' reducibility that resulted in a low degradation efficiency. The reaction was driven by an MvK mechanism.

At this point, it is possible to highlight some differences through comparison of the catalytic performance of gold-based bimetallic samples with the others reported above. For the nongold-containing samples category, supports with a high surface area or with a tunable pore size distribution (silica, zeolite, alumina, etc.) were preferred to favor the dispersion of the active metals. Platinum-based samples gave the best results, with a second metal that in many cases enhances catalytic stability rather than overall VOC conversion. Although noble-metals–bimetallic catalysts showed the best performance, in recent years, in order to reduce the high cost of these catalysts, the addition or replacement of at least one of the noble metals with a cheaper transition metal is an interesting approach to reduce the total material cost while maintaining an acceptable catalytic activity.

3. Bimetallic Catalysts for the Photocatalytic Oxidation of VOCs

The urgent request for a "greener" and sustainable industrial chemistry has driven a huge field of research towards alternative ways to treat VOCs instead of catalytic combustion. Among the various AOPs (see the Introduction section), photocatalytic oxidation is the most applied process. With respect to catalytic thermal oxidation, this technique allows working at room temperature, exploiting the chemicophysical processes activated by an appropriate light radiation interacting with the surface of a semiconductor photocatalyst [40,124–126]. Specifically, dangerous organic compounds are oxidized by hydroxyl, and super oxide radicals are generated by the interaction between the photoelectrons and photoholes of the photocatalyst with water and oxygen. These photoelectrons and photoholes are formed when an adequate wavelength ($\lambda \leq$ of the band gap, E_g, of the semiconductor) irradiates the photocatalyst [127,128]. The most used photocatalysts are metal oxides or sulfides such as TiO_2, ZnO, WO_3, ZrO_2, CeO_2, Fe_2O_3, ZnS, and CdS, and among them, TiO_2 and ZnO are the most used [129,130]. Due to its properties, such as its nontoxicity, relatively low cost, and high activity, especially under UV irradiation, titanium dioxide was deeply investigated both in academic and industrial research [124,126]. With a band gap varying from 3.0 to 3.2 eV depending on the crystalline form, TiO_2 is able to exploit only 5% of solar radiation, thus limiting its practical applications. Photothermocatalytic oxidation is a multicatalytic approach that accepts the contemporaneous utilization of a light source to activate the photocatalyst, and thermal heat to boost the conversion of organic molecules and to increase the yield to CO_2. A proper structural and/or chemical modification of titanium dioxide together with this multicatalytic approach can be considered a suitable solution to decrease the total energy consumption, maintaining the high conversion values typical of thermocatalytic oxidation of VOCs [131–133]. Another connected strategy to increase the photocatalytic performance of titanium dioxide under solar/visible light irradiation is doping with metal or nonmetal elements [134–136], and a not yet largely explored strategy is the combination of a bimetallic alloy with TiO_2 [137,138]. The same metals typically employed for catalytic oxidation of VOCs, such as Au, Ag, Pt, Pd, and Cu in the nanoparticle size, if irradiated with (usually) a visible light irradiation, allow exploiting the localized surface plasmon resonance (LSPR) through collective oscillations of the electrons in the surface of the metal nanoparticles. This effect, combined with the photocatalytic properties of TiO_2, is a performance solution to obtain a visible-light-driven photocatalyst [124,139,140]. In addition, for this particular application, bimetallic compounds can help to overcome some of the drawbacks of single metals. For example, the LSPR of some noble metals such as Pd, Pt, and Rh is not efficiently activated by solar irradiation, and a possible combination with the most effective plasmonic metals, such as Au, Ag or Cu, leads to takeing advantages of both the LSPR effect and of the reactive catalytic behavior of the other noble metals [137].

For the above considerations, in VOC photo-oxidation, the most investigated system was the Au-Pd bimetallic compound joined with a semiconductor photocatalyst [141–145]. In these materials, the good affinity of gold and palladium, already discussed in terms of thermocatalytic performance (Section 2.1.1.), is in this case utilized to increase the photocatalytic performances of titanium dioxide or of another semiconductor oxide.

Colmenares et al. [141] synthetized an Au-Pd/TiO_2 photocatalyst with the original technique of sonophotodeposition (Figure 9). The bimetallic sample exhibited high activity (83%) and good selectivity (70%) in the partial oxidation of methanol to methyl formate after 120 min of UV irradiation (125 W mercury lamp λ_{max} = 365 nm). Although the bimetallic catalyst showed a low selectivity to CO_2 ($\approx$30%), demonstrating that this approach is better suitable for selective oxidation than the total oxidation of VOCs, the reported material synthesis and adopted reaction conditions are a fascinating way to obtain results with an energy-efficient procedure and a selective photocatalyst in a short time and under mild conditions.

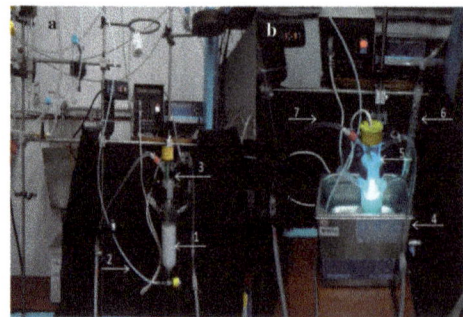

Figure 9. Sonophotodeposition setup before (**a**) and during the deposition procedure (**b**): (1) batch photoreactor; (2) argon line; (3) switched off 6 W UV lamp; (4) ultrasonic bath; (5) switched on 6W UV lamp; (6) reflux condenser; and (7) lamp cooling system (20 °C). Figure from [141], Copyright 2015, John Wiley and Sons.

In a further study, the same research group [142] developed a density functional methodology to analyze the reaction mechanism of the selective photo-oxidation of methanol on the bimetallic Au-Pd/TiO$_2$ sample. The theoretical investigation showed, as with the formation of a synergistic interaction between gold and palladium, a superior photoelectron–hole separation, was verified in comparison with the monometallic samples. Furthermore, it was shown that to favor total photo-oxidation to CO$_2$, the dissociation of molecular oxygen should be driven preferentially on Pd to favor the formation of PdO sites, where complete oxidation (no methyl formate formation) to carbon dioxide occurred.

Cybula et al. [143] investigated the performance of an Au-Pd bimetallic sample supported on rutile TiO$_2$ synthetized with a water in oil microemulsion methodology, in the photocatalytic oxidation of toluene and phenol under visible light irradiation (25 LEDs (λ_{max} = 415 nm)). In particular, the authors focused on the effect of calcination temperature on materials' preparation. The bimetallic sample calcined at 350 °C achieved 65% of toluene degradation and 22% of phenol conversion after 60 min of visible light irradiation. The performances were inferior compared to the photoactivity of monometallic palladium (79% in the toluene degradation and 24% in the phenol removal); however, the synergistic effect combined with a strong metals–support interaction was better exploited in the UV-vis tests, where the intrinsic photoactivity of rutile TiO$_2$ also made a substantial contribution in removal efficiency. In fact, with the bimetallic sample, 100% of phenol degradation was achieved after 60 min of irradiation instead of the 56% of Pd/TiO$_2$.

The interaction of the gold–palladium compound with other semiconductors was examined by the research group of Zhang et al. [144,145]. The photocatalytic oxidation of the trichloroethylene was studied on Au-Pd/BiPO$_4$ nanorods and on Au-Pd/MoO$_3$ nanowires. Interestingly, with the deposition of the Au-Pd alloy on the surface of the BiPO$_4$ nanorod, the photocatalytic degradation rate increased quickly, being about 25 times higher compared to that achieved with bare BiPO$_4$. The authors proposed, on the basis of the characterization measurements, the reaction mechanism illustrated in Figure 10. Under visible light irradiation (solar simulator with a 440 nm cut-off filter), Au-Pd/BiPO$_4$ were excited due to the LSPR of the Au-Pd alloy. An effective charge carrier separation was achieved due to electron transfer from the conduction band (CB) of BiPO$_4$ to the Au-Pd surface interface, whereas the photoholes remained confined in the valence band (VB) of BiPO$_4$. Subsequently, the same photoelectrons present in the surface of the Au-Pd alloy reacted with the oxygens in the gas-phase that were successively reduced into superoxide radicals. These radicals together with the holes in the VB of BiPO$_4$ oxidize the trichloroethylene in water and carbon dioxide.

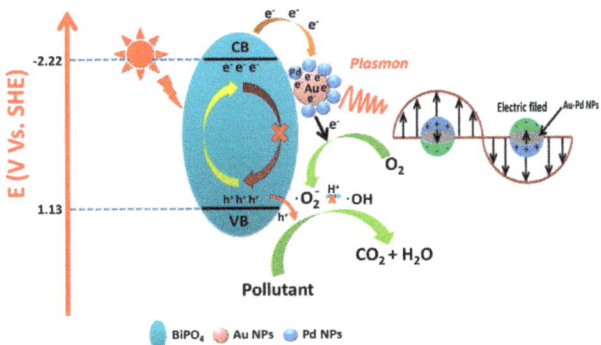

Figure 10. Reaction mechanism of the trichloroethylene oxidation with the Au-Pd/BiPO$_4$ photocatalyst. Figure from [144]. Copyright 2017, Elsevier.

The synergistic effect between gold and palladium nanoparticles that permitted increasing charge carrier separation, enhancing in this way photocatalytic activity, was verified, also employing MoO$_3$ nanowires as a semiconductor photocatalyst [145].

The exploration of the synergistic effect with a bimetallic alloy in the photocatalytic oxidation of VOCs has not yet been largely investigated in the literature, and only fa ew examples are present considering other compounds instead of the Au-Pd system [146–149]. The possibility of favoring a selective oxidation route due to the high selectivity of gold–silver nanoparticles was examined by Han et al. [146] even in partial oxidation of methanol to methyl formate. In this case, the Au-Ag/TiO$_2$ powders prepared via chemical reduction showed good results with a methanol conversion of 90% and a selectivity to methyl formate of about 85% under UV irradiation (500 W high pressure mercury lamp, λ_{max} = 365 nm). Similarly to the Au-Pd systems, the bimetallic alloy enhanced the photoelectron–photohole separation with electron transfer from the conduction band of TiO$_2$ (excited by the UV irradiation) to the gold–silver surface interface.

Another alloy with silver, i.e., Ag-Pt, was studied by Zieli'nska-Jurek et al. [147] in the photo-oxidation of toluene under visible light irradiation (LEDs, λ_{max} = 415 nm). The authors found an interesting correlation regarding the order of deposition of the Ag-Pt/TiO$_2$ photocatalysts prepared by sol–gel. In particular, the best sample (best photoactivity) was the material where the silver precursor was added before the platinum one. It was fundamental, in fact, to obtain particles with a definite size and dispersion (Ag-Pt size between 6–12 nm). In this way, it was possible to increase the toluene degradation rate with respect to the monometallic samples. By contrast, the bimetallic sample prepared with a simultaneous addition of metals precursors on TiO$_2$ gave a lower photoactivity and different metal size and distribution. The authors concluded that platinum size had a greater influence than silver in determining overall photocatalytic activity. Recently, the same research group evaluated photocatalytic performance in both toluene and acetaldehyde degradation and of *Penicillium chrysogenum*, a dangerous fungus present in the indoor environment with Ag-Pt/TiO$_2$ and Cu-Pt/TiO$_2$ samples [148]. Both bimetallic samples showed a higher fungicidal activity under visible light irradiation than bare TiO$_2$, whereas in VOC degradation, the Ag-Pt system was better-performing compared to Cu-Pt. The peculiar activity of both bimetallic samples was ascribed to the interfacial charge transfer process between the two metals and the TiO$_2$ confirmed by the quenching of fluorescence (i.e., intensity diminution of the TiO$_2$ photolumiscence bands) due to the presence of the metal alloy.

Wolski et al. [149] studied the mechanism of methanol photo-oxidation on bimetallic Au-Cu catalyst supported on Nb$_2$O$_5$ with an in operando IR methodology under both UV and visible light irradiation. Interestingly, they found that photocatalytic activity is strictly related to the light sources used and to the number of Brønsted/Lewis acid sites present on the surface of the catalysts. Specifically, under visible light irradiation, the synergism between gold and copper led to an increase in the amount of Brønsted/Lewis acid sites on the niobia, with a consequent higher activity of bimetallic samples compared to that of monometallic and pure Nb_2O_5 samples. Furthermore, the total oxidation to CO_2 was favored. By contrast, with UV light irradiation, the major activation of niobia ($E_g \approx 3.2$ eV) favored selective oxidation into dimethoxymethane, formaldehyde, and methyl formate.

In this short chapter, the state-of-the-art of the application of bimetallic structures as chemical modifiers of conventional and unconventional semiconductor photocatalysts was examined. This approach is relatively new, and the effects of alloy synergism on the photocatalytic process are currently under investigation. The promising results, especially obtained by combining the LSPR effect of both noble and transition metals with semiconductor photoactivity, together with a possible multicatalytic strategy (i.e., a photothermo atalytic approach employing a bimetallic/semiconductor catalyst and a solar/visible light source) could in the future be a fascinating strategy to develop a greener and sustainable technology applied to the removal of volatile organic compounds.

4. Conclusions

In this review, the application of bimetallic catalysts for VOC oxidation was examined in terms of catalytic activity and physicochemical properties. Among the various systems, gold-based bimetallic catalysts exhibited a good performance in the degradation of a wide range of VOCs. The presence of nanosized gold was essential to decreasing the light-off temperature of VOC oxidation, whereas interaction with the second metal allowed increasing the reactivity of the employed support or enhancing oxygen activation. Although platinum-based bimetallic samples usually did not overcome the degradation yields achieved with the monometallic platinum catalysts, they showed a substantial improvement of catalytic stability due to the synergistic effect between platinum and the second noble or transition metal. Finally, the application of the already stated synergisms in catalytic thermo-oxidation, for example, strong Au-Pd interaction, can be successfully transferred to new technologies for VOC abatement, such as photocatalytic oxidation, with the exploitation of new mutual effects such as surface plasmon resonance combined with the high reactivity of noble/transition metals. This can be a promising strategy to achieve significant progress in the technologies applied to the improvement of air quality.

Funding: R.F. thanks the PON project "AIM" founded by the European Social Found (ESF) CUP: E66C18001220005 for the financial support.

Acknowledgments: This review was written during the Italian quarantine due to COVID-19. I want to dedicate this work to all the people who could not stay at home because they fought against the virus. R.F. thanks S. Scirè and C. Crisafulli who have positively transferred the knowledge and the passion for the research in the field of heterogeneous catalysis focused on environmental applications.

Conflicts of Interest: The author declares no conflict of interest.

References

1. Yang, C.; Miao, G.; Pi, Y.; Xia, Q.; Wu, J.; Li, Z.; Xiao, J. Abatement of various types of VOCs by adsorption/catalytic oxidation: A review. *Chem. Eng. J.* **2019**, *370*, 1128–1153. [CrossRef]
2. He, C.; Cheng, J.; Zhang, X.; Douthwaite, M.; Pattisson, S.; Hao, Z. Recent advances in the catalytic oxidation of volatile organic compounds: A review based on pollutant sorts and sources. *Chem. Rev.* **2019**, *119*, 4471–4568. [CrossRef] [PubMed]
3. Amann, M.; Lutz, M. The revision of the air quality legislation in the European Union related to ground-level ozone. *J. Hazard. Mater.* **2000**, *78*, 41–62. [CrossRef]
4. Dumanoglu, Y.; Kara, M.; Altiok, H.; Odabasi, M.; Elbir, T.; Bayram, A. Spatial and seasonal variation and source apportionment of volatile organic compounds (VOCs) in a heavily industrialized region. *Atmos. Environ.* **2014**, *98*, 168–178. [CrossRef]
5. Hu, R.; Liu, G.; Zhang, H.; Xue, H.; Wang, X.; Lam, P.K.S. Odor pollution due to industrial emission of volatile organic compounds: A case study in Hefei, China. *J. Clean. Prod.* **2020**, *246*, 119075. [CrossRef]

6. Liao, H.T.; Chou, C.C.K.; Chow, J.C.; Watson, J.G.; Hopke, P.K.; Wu, C.F. Source and risk apportionment of selected VOCs and PM2.5 species using partially constrained receptor models with multiple time resolution data. *Environ. Pollut.* **2015**, *205*, 121–130. [CrossRef] [PubMed]
7. Dhamodharan, K.; Varma, V.S.; Veluchamy, C.; Pugazhendhi, A.; Rajendran, K. Emission of volatile organic compounds from composting: A review on assessment, treatment and perspectives. *Sci. Total Environ.* **2019**, *695*, 133725. [CrossRef]
8. Tørseth, K.; Aas, W.; Breivik, K.; Fjeraa, A.M.; Fiebig, M.; Hjellbrekke, A.G.; Lund Myhre, C.; Solberg, S.; Yttri, K.E. Introduction to the European Monitoring and Evaluation Programme (EMEP) and observed atmospheric composition change during 1972–2009. *Atmos. Chem. Phys.* **2012**, *12*, 5447–5481. [CrossRef]
9. Huang, B.; Lei, C.; Wei, C.; Zeng, G. Chlorinated volatile organic compounds (Cl-VOCs) in environment—Sources, potential human health impacts, and current remediation technologies. *Environ. Int.* **2014**, *71*, 118–138. [CrossRef]
10. Kumar, V.; Lee, Y.S.; Shin, J.W.; Kim, K.H.; Kukkar, D.; Tsang, Y.F. Potential applications of graphene-based nanomaterials as adsorbent for removal of volatile organic compounds. *Environ. Int.* **2020**, *135*, 105356. [CrossRef]
11. Parmar, G.R.; Rao, N.N. Emerging control technologies for volatile organic compounds. *Crit. Rev. Environ. Sci. Technol.* **2008**, *39*, 41–78. [CrossRef]
12. Zhu, L.; Shen, D.; Luo, K.H. A critical review on VOCs adsorption by different porous materials: Species, mechanisms and modification methods. *J. Hazard. Mater.* **2020**, *389*, 122102. [CrossRef] [PubMed]
13. Zhang, X.; Gao, B.; Creamer, A.E.; Cao, C.; Li, Y. Adsorption of VOCs onto engineered carbon materials: A review. *J. Hazard. Mater.* **2017**, *338*, 102–123. [CrossRef] [PubMed]
14. Kamal, M.S.; Razzak, S.A.; Hossain, M.M. Catalytic oxidation of volatile organic compounds (VOCs)—A review. *Atmos. Environ.* **2016**, *140*, 117–134. [CrossRef]
15. Spivey, J.J. Complete catalytic oxidation of volatile organics. *Ind. Eng. Chem. Res.* **1987**, *26*, 2165–2180. [CrossRef]
16. Liotta, L.F. Catalytic oxidation of volatile organic compounds on supported noble metals. *Appl. Catal. B Environ.* **2010**, *100*, 403–412. [CrossRef]
17. Zhang, Z.; Jiang, Z.; Shangguan, W. Low-temperature catalysis for VOCs removal in technology and application: A state-of-the-art review. *Catal. Today* **2016**, *264*, 270–278. [CrossRef]
18. Huang, H.; Xu, Y.; Feng, Q.; Leung, D.Y.C. Low temperature catalytic oxidation of volatile organic compounds: A review. *Catal. Sci. Technol.* **2015**, *5*, 2649–2669. [CrossRef]
19. Fiorenza, R.; Balsamo, S.A.; D'Urso, L.; Scirè, S.; Brundo, M.V.; Pecoraro, R.; Scalisi, E.M.; Privitera, V.; Impellizzeri, G. CeO_2 for Water Remediation: Comparison of Various Advanced Oxidation Processes. *Catalysts* **2020**, *10*, 446. [CrossRef]
20. Tokumura, M.; Nakajima, R.; Znad, H.T.; Kawase, Y. Chemical absorption process for degradation of VOC gas using heterogeneous gas-liquid photocatalytic oxidation: Toluene degradation by photo-Fenton reaction. *Chemosphere* **2008**, *73*, 768–775. [CrossRef]
21. Malakar, S.; Das, P.S.; Baskaran, D.; Rajamanickam, R. Comparative study of biofiltration process for treatment of VOCs emission from petroleum refinery wastewater—A review. *Environ. Technol. Innov.* **2017**, *8*, 441–461. [CrossRef]
22. Pettit, T.; Irga, P.J.; Torpy, F.R. Towards practical indoor air phytoremediation: A review. *Chemosphere* **2018**, *208*, 960–974. [CrossRef] [PubMed]
23. Jakubek, T.; Hudy, C.; Indyka, P.; Nowicka, E.; Golunski, S.; Kotarba, A. Effect of noble metal addition to alkali-exchanged cryptomelane on the simultaneous soot and VOC combustion activity. *Catal. Commun.* **2019**, *132*, 105807. [CrossRef]
24. Gallastegi-Villa, M.; Aranzabal, A.; Romero-Sáez, M.; González-Marcos, J.A.; González-Velasco, J.R. Catalytic activity of regenerated catalyst after the oxidation of 1,2-dichloroethane and trichloroethylene. *Chem. Eng. J.* **2014**, *241*, 200–206. [CrossRef]
25. De Rivas, B.; Sampedro, C.; García-Real, M.; López-Fonseca, R.; Gutiérrez-Ortiz, J.I. Promoted activity of sulphated Ce/Zr mixed oxides for chlorinated VOC oxidative abatement. *Appl. Catal. B Environ.* **2013**, *129*, 225–235. [CrossRef]
26. Barakat, T.; Rooke, J.C.; Genty, E.; Cousin, R.; Siffert, S.; Su, B.L. Gold catalysts in environmental remediation and water-gas shift technologies. *Energy Environ. Sci.* **2013**, *6*, 371–391. [CrossRef]

27. Matějová, L.; Topka, P.; Kaluža, L.; Pitkäaho, S.; Ojala, S.; Gaálová, J.; Keiski, R.L. Total oxidation of dichloromethane and ethanol over ceria-zirconia mixed oxide supported platinum and gold catalysts. *Appl. Catal. B Environ.* **2013**, *142*, 54–64. [CrossRef]
28. Guo, Y.; Yang, D.P.; Liu, M.; Zhang, X.; Chen, Y.; Huang, J.; Li, Q.; Luque, R. Enhanced catalytic benzene oxidation over a novel waste-derived Ag/eggshell catalyst. *J. Mater. Chem. A* **2019**, *7*, 8832–8844. [CrossRef]
29. Gaálová, J.; Topka, P.; Kaluža, L.; Soukup, K.; Barbier, J. Effect of gold loading on ceria-zirconia support in total oxidation of VOCs. *Catal. Today* **2019**, *333*, 190–195. [CrossRef]
30. Radic, N.; Grbic, B.; Terlecki-Baricevic, A. Kinetics of deep oxidation of n-hexane and toluene over Pt/Al 2 O 3 catalysts: Platinum crystallite size effect. *Appl. Catal. B Environ.* **2004**, *50*, 153–159. [CrossRef]
31. Topka, P.; Dvořáková, M.; Kšírová, P.; Perekrestov, R.; Čada, M.; Balabánová, J.; Koštejn, M.; Jirátová, K.; Kovanda, F. Structured cobalt oxide catalysts for VOC abatement: The effect of preparation method. *Environ. Sci. Pollut. Res.* **2019**, *27*, 7608–7617. [CrossRef]
32. Rodríguez, M.L.; Cadús, L.E.; Borio, D.O. VOCs abatement in adiabatic monolithic reactors: Heat effects, transport limitations and design considerations. *Chem. Eng. J.* **2016**, *306*, 86–98. [CrossRef]
33. Liu, B.; Zhan, Y.; Xie, R.; Huang, H.; Li, K.; Zeng, Y.; Shrestha, R.P.; Kim Oanh, N.T.; Winijkul, E. Efficient photocatalytic oxidation of gaseous toluene in a bubbling reactor of water. *Chemosphere* **2019**, *233*, 754–761. [CrossRef] [PubMed]
34. Perez, V.; Miachon, S.; Dalmon, J.A.; Bredesen, R.; Pettersen, G.; Rader, H.; Simon, C. Preparation and characterisation of a Pt/ceramic catalytic membrane. *Sep. Purif. Technol.* **2001**, *25*, 33–38. [CrossRef]
35. Trovarelli, A.; Llorca, J. Ceria catalysts at nanoscale: How do crystal shapes shape catalysis? *ACS Catal.* **2017**, *7*, 4716–4735. [CrossRef]
36. Li, J.; Liu, H.; Deng, Y.; Liu, G.; Chen, Y.; Yang, J. Emerging nanostructured materials for the catalytic removal of volatile organic compounds. *Nanotechnol. Rev.* **2016**, *5*, 147–181. [CrossRef]
37. Busca, G.; Berardinelli, S.; Resini, C.; Arrighi, L. Technologies for the removal of phenol from fluid streams: A short review of recent developments. *J. Hazard. Mater.* **2008**, *160*, 265–288. [CrossRef]
38. Van Durme, J.; Dewulf, J.; Leys, C.; van Langenhove, H. Combining non-thermal plasma with heterogeneous catalysis in waste gas treatment: A review. *Appl. Catal. B Environ.* **2008**, *78*, 324–333. [CrossRef]
39. Veerapandian, S.K.P.; Leys, C.; De Geyter, N.; Moren, R. Abatement of VOCs using packed bed non-thermal plasma reactors: A review. *Catalysts* **2017**, *7*, 113. [CrossRef]
40. Mamaghani, A.H.; Haghighat, F.; Lee, C.S. Photocatalytic oxidation technology for indoor environment air purification: The state-of-the-art. *Appl. Catal. B Environ.* **2017**, *203*, 247–269. [CrossRef]
41. Stytsenko, V.D. Surface modified bimetallic catalysts: Preparation, characterization, and applications. *Appl. Catal. A Gen.* **1995**, *126*, 1–26. [CrossRef]
42. Rodriguez, J. Physical and chemical properties of bimetallic surfaces. *Surf. Sci. Rep.* **1996**, *24*, 223–287. [CrossRef]
43. Singh, A.K.; Xu, Q. Synergistic catalysis over bimetallic alloy nanoparticles. *ChemCatChem* **2013**, *5*, 652–676. [CrossRef]
44. Sankar, M.; Dimitratos, N.; Miedziak, P.J.; Wells, P.P.; Kiely, C.J.; Hutchings, G.J. Designing bimetallic catalysts for a green and sustainable future. *Chem. Soc. Rev.* **2012**, *41*, 8099–8139. [CrossRef] [PubMed]
45. Sinfelt, J.H. Supported "bimetallic cluster" catalysts. *J. Catal.* **1973**, *29*, 308–315. [CrossRef]
46. Blaser, H.-U.; Malan, C.; Pugin, B.; Spindler, F.; Steiner, H.; Studer, M. Selective hydrogenation for fine chemicals: Recent trends and new developments. *ChemInform* **2003**, *345*, 103–151. [CrossRef]
47. Hu, M.; Jin, L.; Zhu, Y.; Zhang, L.; Lu, X.; Kerns, P.; Su, X.; Cao, S.; Gao, P.; Suib, S.L.; et al. Self-limiting growth of ligand-free ultrasmall bimetallic nanoparticles on carbon through under temperature reduction for highly efficient methanol electrooxidation and selective hydrogenation. *Appl. Catal. B Environ.* **2020**, *264*, 118553. [CrossRef]
48. Zhang, J.; Wang, H.; Dalai, A.K. Development of stable bimetallic catalysts for carbon dioxide reforming of methane. *J. Catal.* **2007**, *249*, 300–310. [CrossRef]
49. Li, L.; Zuo, S.; An, P.; Wu, H.; Hou, F.; Li, G.; Liu, G. Hydrogen production via steam reforming of n-dodecane over NiPt alloy catalysts. *Fuel* **2020**, *262*, 116469. [CrossRef]
50. Fiorenza, R.; Crisafulli, C.; Scirè, S. H2 purification through preferential oxidation of CO over ceria supported bimetallic Au-based catalysts. *Int. J. Hydrogen Energy* **2016**, *41*, 19390–19398. [CrossRef]

51. Fiorenza, R.; Spitaleri, L.; Gulino, A.; Scirè, S. Ru–Pd bimetallic catalysts supported on CeO2-MnOx oxides as efficient systems for H2 purification through CO preferential Oxidation. *Catalysts* **2018**, *8*, 203. [CrossRef]
52. Fiorenza, R.; Scirè, S.; Venezia, A.M. Carbon supported bimetallic Ru-Co catalysts for H2 production through NaBH4 and NH3BH3 hydrolysis. *Int. J. Energy Res.* **2018**, *42*, 1183–1195. [CrossRef]
53. Fiorenza, R.; Spitaleri, L.; Gulino, A.; Sciré, S. High-performing Au-Ag bimetallic catalysts supported on macro-mesoporous CeO2 for preferential oxidation of CO in H2-rich gases. *Catalysts* **2020**, *10*, 49. [CrossRef]
54. Luo, L.; Chen, S.; Xu, Q.; He, Y.; Dong, Z.; Zhang, L.; Zhu, J.; Du, Y.; Yang, B.; Wang, C. Dynamic atom clusters on AuCu nanoparticle surface during CO oxidation. *J. Am. Chem. Soc.* **2020**, *142*, 4022–4027. [CrossRef]
55. Guo, S.; Dong, S.; Wang, E. Three-dimensional Pt-on-Pd bimetallic nanodendrites supported on graphene nanosheet: Facile synthesis and used as an advanced nanoelectrocatalyst for methanol oxidation. *ACS Nano* **2010**, *4*, 547–555. [CrossRef] [PubMed]
56. Zhang, H.; Jin, M.; Xia, Y. Enhancing the catalytic and electrocatalytic properties of Pt-based catalysts by forming bimetallic nanocrystals with Pd. *Chem. Soc. Rev.* **2012**, *41*, 8035–8049. [CrossRef]
57. Nasrabadi, H.T.; Abbasi, E.; Davaran, S.; Kouhi, M.; Akbarzadeh, A. Bimetallic nanoparticles: Preparation, properties, and biomedical applications. *Artif. Cells Nanomed. Biotechnol.* **2016**, *44*, 376–380. [CrossRef]
58. Ferrando, R.; Jellinek, J.; Johnston, R.L. Nanoalloys: From theory to applications of alloy clusters and nanoparticles. *Chem. Rev.* **2008**, *108*, 845–910. [CrossRef] [PubMed]
59. Duan, S.; Wang, R. Bimetallic nanostructures with magnetic and noble metals and their physicochemical applications. *Prog. Nat. Sci. Mater. Int.* **2013**, *23*, 113–126. [CrossRef]
60. Dehghan Banadaki, A.; Kajbafvala, A. Recent advances in facile synthesis of bimetallic nanostructures: An overview. *J. Nanomater.* **2014**, *2014*, 1–28. [CrossRef]
61. Duan, M.; Jiang, L.; Zeng, G.; Wang, D.; Tang, W.; Liang, J.; Wang, H.; He, D.; Liu, Z.; Tang, L. Bimetallic nanoparticles/metal-organic frameworks: Synthesis, applications and challenges. *Appl. Mater. Today* **2020**, *19*, 100564. [CrossRef]
62. Redina, E.A.; Kirichenko, O.A.; Greish, A.A.; Kucherov, A.V.; Tkachenko, O.P.; Kapustin, G.I.; Mishin, I.V.; Kustov, L.M. Preparation of bimetallic gold catalysts by redox reaction on oxide-supported metals for green chemistry applications. *Catal. Today* **2015**, *246*, 216–231. [CrossRef]
63. Alexeev, O.S.; Gates, B.C. Supported bimetallic cluster catalysts. *Ind. Eng. Chem. Res.* **2003**, *42*, 1571–1587. [CrossRef]
64. Bariås, O.A.; Holmen, A.; Blekkan, E.A. Propane dehydrogenation over supported Pt and Pt-Sn catalysts: Catalyst preparation, characterization, and activity measurements. *J. Catal.* **1996**, *158*, 1–12. [CrossRef]
65. Zhou, S.; Kang, L.; Zhou, X.; Xu, Z.; Zhu, M. Pure acetylene semihydrogenation over Ni–Cu bimetallic catalysts: Effect of the Cu/Ni ratio on catalytic performance. *Nanomaterials* **2020**, *10*, 509. [CrossRef]
66. Luisetto, I.; Tuti, S.; Di Bartolomeo, E. Co and Ni supported on CeO2 as selective bimetallic catalyst for dry reforming of methane. *Int. J. Hydrogen Energy* **2012**, *37*, 15992–15999. [CrossRef]
67. Aguirre, A.; Zanella, R.; Barrios, C.; Hernández, S.; Bonivardi, A.; Collins, S.E. Gold stabilized with iridium on ceria–niobia catalyst: Activity and stability for CO oxidation. *Top. Catal.* **2019**, *62*, 977–988. [CrossRef]
68. Fiorenza, R.; Crisafulli, C.; Condorelli, G.G.; Lupo, F.; Scirè, S. Au-Ag/CeO2 and Au-Cu/CeO2 catalysts for volatile organic compounds oxidation and CO preferential oxidation. *Catal. Lett.* **2015**, *145*, 1691–1702. [CrossRef]
69. Xia, S.; Yuan, Z.; Wang, L.; Chen, P.; Hou, Z. Hydrogenolysis of glycerol on bimetallic Pd-Cu/solid-base catalysts prepared via layered double hydroxides precursors. *Appl. Catal. A Gen.* **2011**, *403*, 173–182. [CrossRef]
70. Bönnemann, H.; Braun, G.; Brijoux, W.; Brinkmann, R.; Tilling, A.S.; Seevogel, K.; Siepen, K. Nanoscale colloidal metals and alloys stabilized by solvents and surfactants: Preparation and use as catalyst precursors. *J. Organomet. Chem.* **1996**, *520*, 143–162. [CrossRef]
71. Minicò, S.; Scirè, S.; Crisafulli, C.; Galvagno, S. Influence of catalyst pretreatments on volatile organic compounds oxidation over gold/iron oxide. *Appl. Catal. B Environ.* **2001**, *34*, 277–285. [CrossRef]
72. Scirè, S.; Liotta, L.F. Supported gold catalysts for the total oxidation of volatile organic compounds. *Appl. Catal. B Environ.* **2012**, *125*, 222–246. [CrossRef]
73. Haruta, M. Gold as a novel catalyst in the 21st century: Preparation, working mechanism and applications. *Gold Bull.* **2004**, *37*, 27–36. [CrossRef]

74. Haruta, M.; Yamada, N.; Kobayashi, T.; Iijima, S. Gold catalysts prepared by coprecipitation for low-temperature oxidation of hydrogen and of carbon monoxide. *J. Catal.* **1989**, *115*, 301–309. [CrossRef]
75. Carrettin, S.; McMorn, P.; Johnston, P.; Griffin, K.; Hutchings, G.J. Selective oxidation of glycerol to glyceric acid using a gold catalyst in aqueous sodium hydroxide. *Chem. Commun.* **2002**, *7*, 696–697. [CrossRef]
76. Prati, L.; Villa, A.; Jouve, A.; Beck, A.; Evangelisti, C.; Savara, A. Gold as a modifier of metal nanoparticles: Effect on structure and catalysis. *Faraday Discuss.* **2018**, *208*, 395–407. [CrossRef]
77. Fajín, J.L.C.; Cordeiro, M.N.D.S.; Gomes, J.R.B. Catalytic reactions on model gold surfaces: Effect of surface steps and of surface doping. *Catalysts* **2011**, *1*, 40–51. [CrossRef]
78. Hashmi, A.S.K.; Rudolph, M. Gold catalysis in total synthesis. *Chem. Soc. Rev.* **2008**, *37*, 1766–1775. [CrossRef]
79. Liu, X.Y.; Wang, A.; Zhang, T.; Mou, C.Y. Catalysis by gold: New insights into the support effect. *Nano Today* **2013**, *8*, 403–416. [CrossRef]
80. Xu, Q.; Lei, W.; Li, X.; Qi, X.; Yu, J.; Liu, G.; Wang, J.; Zhang, P. Efficient removal of formaldehyde by nanosized gold on well-defined CeO2 nanorods at room temperature. *Environ. Sci. Technol.* **2014**, *48*, 9702–9708. [CrossRef]
81. Bell, A.T. The impact of nanoscience on heterogeneous catalysis. *Science* **2003**, *299*, 1688–1691. [CrossRef] [PubMed]
82. Daté, M.; Okumura, M.; Tsubota, S.; Haruta, M. Vital role of moisture in the catalytic activity of supported gold nanoparticles. *Angew. Chem. Int. Ed.* **2004**, *43*, 2129–2132. [CrossRef] [PubMed]
83. Villa, A.; Wang, D.; Su, D.S.; Prati, L. New challenges in gold catalysis: Bimetallic systems. *Catal. Sci. Technol.* **2015**, *5*, 55–68. [CrossRef]
84. Louis, C. Chemical preparation of supported bimetallic catalysts. Gold-based bimetallic, a case study. *Catalysts* **2016**, *6*, 110. [CrossRef]
85. Hosseini, M.; Barakat, T.; Cousin, R.; Aboukaïs, A.; Su, B.L.; De Weireld, G.; Siffert, S. Catalytic performance of core-shell and alloy Pd-Au nanoparticles for total oxidation of VOC: The effect of metal deposition. *Appl. Catal. B Environ.* **2012**, *111*, 218–224. [CrossRef]
86. Barakat, T.; Rooke, J.C.; Chlala, D.; Cousin, R.; Lamonier, J.F.; Giraudon, J.M.; Casale, S.; Massiani, P.; Su, B.L.; Siffert, S. Oscillatory behavior of Pd-Au catalysts in toluene total oxidation. *Catalysts* **2018**, *8*, 574. [CrossRef]
87. Xia, Y.; Xia, L.; Liu, Y.; Yang, T.; Deng, J.; Dai, H. Concurrent catalytic removal of typical volatile organic compound mixtures over Au-Pd/α-MnO2 nanotubes. *J. Environ. Sci. (China)* **2018**, *64*, 276–288. [CrossRef]
88. Xie, S.; Deng, J.; Zang, S.; Yang, H.; Guo, G.; Arandiyan, H.; Dai, H. Au-Pd/3DOM Co3O4: Highly active and stable nanocatalysts for toluene oxidation. *J. Catal.* **2015**, *322*, 38–48. [CrossRef]
89. Tabakova, T.; Ilieva, L.; Petrova, P.; Venezia, A.M.; Avdeev, G.; Zanella, R.; Karakirova, Y. Complete benzene oxidation over mono and bimetallic au-pd catalysts supported on fe-modified ceria. *Chem. Eng. J.* **2015**, *260*, 133–141. [CrossRef]
90. Li, X.; Feng, J.; Perdjon, M.; Oh, R.; Zhao, W.; Huang, X.; Liu, S. Investigations of supported Au-Pd nanoparticles on synthesized CeO2 with different morphologies and application in solvent-free benzyl alcohol oxidation. *Appl. Surf. Sci.* **2020**, *505*, 144473. [CrossRef]
91. Xie, S.; Liu, Y.; Deng, J.; Zhao, X.; Yang, J.; Zhang, K.; Han, Z.; Arandiyan, H.; Dai, H. Effect of transition metal doping on the catalytic performance of Au–Pd/3DOM Mn2O3 for the oxidation of methane and o-xylene. *Appl. Catal. B Environ.* **2017**, *206*, 221–232. [CrossRef]
92. Kucherov, A.V.; Tkachenko, O.P.; Kirichenko, O.A.; Kapustin, G.I.; Mishin, I.V.; Klementiev, K.V.; Ojala, S.; Kustov, L.M.; Keiski, R. Nanogold-containing catalysts for low-temperature removal of S-VOC from air. *Top. Catal.* **2009**, *52*, 351–358. [CrossRef]
93. Nagy, G.; Benkó, T.; Borkó, L.; Csay, T.; Horváth, A.; Frey, K.; Beck, A. Bimetallic Au-Ag/SiO2 catalysts: Comparison in glucose, benzyl alcohol and CO oxidation reactions. *React. Kinet. Mech. Catal.* **2015**, *115*, 45–65. [CrossRef]
94. Bracey, C.L.; Ellis, P.R.; Hutchings, G.J. Application of copper-gold alloys in catalysis: Current status and future perspectives. *Chem. Soc. Rev.* **2009**, *38*, 2231–2243. [CrossRef] [PubMed]
95. Chimentão, R.J.; Medina, F.; Fierro, J.L.G.; Llorca, J.; Sueiras, J.E.; Cesteros, Y.; Salagre, P. Propene epoxidation by nitrous oxide over Au–Cu/TiO2 alloy catalysts. *J. Mol. Catal. A Chem.* **2007**, *274*, 159–168. [CrossRef]
96. Nevanperä, T.K.; Ojala, S.; Laitinen, T.; Pitkäaho, S.; Saukko, S.; Keiski, R.L. Catalytic oxidation of dimethyl disulfide over bimetallic Cu–Au and Pt–Au catalysts supported on γ-Al2O3, CeO2, and CeO2–Al2O3. *Catalysts* **2019**, *9*, 603. [CrossRef]

97. Kim, K.J.; Ahn, H.G. Complete oxidation of toluene over bimetallic Pt-Au catalysts supported on ZnO/Al2O3. *Appl. Catal. B Environ.* **2009**, *91*, 308–318. [CrossRef]
98. Kim, K.J.; Boo, S.-I.; Ahn, H.G. Preparation and characterization of the bimetallic Pt-Au/ZnO/Al2O3 catalysts: Influence of Pt-Au molar ratio on the catalytic activity for toluene oxidation. *J. Ind. Eng. Chem.* **2009**, *15*, 92–97. [CrossRef]
99. Sreethawong, T.; Sukjit, D.; Ouraipryvan, P.; Schwank, J.W.; Chavadej, S. Oxidation of oxygenated volatile organic compound over monometallic and bimetallic Ru-Au catalysts. *Catal. Lett.* **2010**, *138*, 160–170. [CrossRef]
100. Calzada, L.A.; Collins, S.E.; Han, C.W.; Ortalan, V.; Zanella, R. Synergetic effect of bimetallic Au-Ru/TiO2 catalysts for complete oxidation of methanol. *Appl. Catal. B Environ.* **2017**, *207*, 79–92. [CrossRef]
101. Torrente-Murciano, L.; Solsona, B.; Agouram, S.; Sanchis, R.; López, J.M.; García, T.; Zanella, R. Low temperature total oxidation of toluene by bimetallic Au-Ir catalysts. *Catal. Sci. Technol.* **2017**, *7*, 2886–2896. [CrossRef]
102. Solsona, B.; Pérez-Cabero, M.; Vázquez, I.; Dejoz, A.; García, T.; Álvarez-Rodríguez, J.; El-Haskouri, J.; Beltrán, D.; Amorós, P. Total oxidation of VOCs on Au nanoparticles anchored on Co doped mesoporous UVM-7 silica. *Chem. Eng. J.* **2012**, *187*, 391–400. [CrossRef]
103. Bonelli, R.; Lucarelli, C.; Pasini, T.; Liotta, L.F.; Zacchini, S.; Albonetti, S. Total oxidation of volatile organic compounds on Au/FeOx catalysts supported on mesoporous SBA-15 silica. *Appl. Catal. A Gen.* **2011**, *400*, 54–60. [CrossRef]
104. Albonetti, S.; Bonelli, R.; Delaigle, R.; Gaigneaux, E.M.; Femoni, C.; Riccobene, P.M.; Scirè, S.; Tiozzo, C.; Zacchini, S.; Trifirò, F. Design of nano-sized FeOx and Au/FeOx catalysts for total oxidation of VOC and preferential oxidation of CO. In *Studies in Surface Science and Catalysis*; Elsevier: Amsterdam, The Netherlands, 2010; Volume 175, pp. 785–788.
105. Fu, X.; Liu, Y.; Yao, W.; Wu, Z. One-step synthesis of bimetallic Pt-Pd/MCM-41 mesoporous materials with superior catalytic performance for toluene oxidation. *Catal. Commun.* **2016**, *83*, 22–26. [CrossRef]
106. Kim, H.S.; Kim, T.W.; Koh, H.L.; Lee, S.H.; Min, B.R. Complete benzene oxidation over Pt-Pd bimental catalyst supported on γ-alumina: Influence of Pt-Pd ratio on the catalytic activity. *Appl. Catal. A Gen.* **2005**, *280*, 125–131. [CrossRef]
107. Kim, H.S.; Min, M.K.; Song, M.W.; Park, J.W.; Min, B.R. XPS analysis of the effect of Pt addition to Pd catalysts for complete benzene oxidation. *React. Kinet. Catal. Lett.* **2004**, *81*, 251–257. [CrossRef]
108. Sharma, R.K.; Zhou, B.; Tong, S.; Chuang, K.T. Catalytic destruction of volatile organic compounds using supported platinum and palladium hydrophobic catalysts. *Ind. Eng. Chem. Res.* **1995**, *34*, 4310–4317. [CrossRef]
109. Liu, X.; Zeng, J.; Shi, W.; Wang, J.; Zhu, T.; Chen, Y. Catalytic oxidation of benzene over ruthenium-cobalt bimetallic catalysts and study of its mechanism. *Catal. Sci. Technol.* **2017**, *7*, 213–221. [CrossRef]
110. Ye, M.; Chen, L.; Liu, X.; Xu, W.; Zhu, T.; Chen, G. Catalytic oxidation of chlorobenzene over ruthenium-ceria bimetallic catalysts. *Catalysts* **2018**, *8*, 116. [CrossRef]
111. De La Peña O'Shea, V.A.; Álvarez-Galván, M.C.; Fierro, J.L.G.; Arias, P.L. Influence of feed composition on the activity of Mn and PdMn/Al 2O3 catalysts for combustion of formaldehyde/methanol. *Appl. Catal. B Environ.* **2005**, *57*, 191–199.
112. Kim, S.C.; Park, Y.K.; Nah, J.W. Property of a highly active bimetallic catalyst based on a supported manganese oxide for the complete oxidation of toluene. *Powder Technol.* **2014**, *266*, 292–298. [CrossRef]
113. Jodaei, A.; Salari, D.; Niaei, A.; Khatamian, M.; Çaylak, N. Preparation of Ag-M (M: Fe, Co and Mn)-ZSM-5 bimetal catalysts with high performance for catalytic oxidation of ethyl acetate. *Environ. Technol.* **2011**, *32*, 395–406. [CrossRef] [PubMed]
114. Rintramee, K.; Föttinger, K.; Rupprechter, G.; Wittayakun, J. Ethanol adsorption and oxidation on bimetallic catalysts containing platinum and base metal oxide supported on MCM-41. *Appl. Catal. B Environ.* **2012**, *115–116*, 225–235. [CrossRef]
115. Chantaravitoon, P.; Chavadej, S.; Schwank, J. Temperature-programmed desorption of methanol and oxidation of methanol on Pt-Sn/Al2O3 catalysts. *Chem. Eng. J.* **2004**, *97*, 161–171. [CrossRef]
116. Baranowska, K.; Okal, J. Performance and stability of the Ru-Re/γ-Al2O3 catalyst in the total oxidation of propane: Influence of the order of impregnation. *Catal. Lett.* **2016**, *146*, 72–81. [CrossRef]

117. Yue, L.; He, C.; Zhang, X.; Li, P.; Wang, Z.; Wang, H.; Hao, Z. Catalytic behavior and reaction routes of MEK oxidation over Pd/ZSM-5 and Pd-Ce/ZSM-5 catalysts. *J. Hazard. Mater.* **2013**, *244*, 613–620. [CrossRef]
118. Jodaei, A.; Niaei, A.; Salari, D. Performance of nanostructure Fe-Ag-ZSM-5 catalysts for the catalytic oxidation of volatile organic compounds: Process optimization using response surface methodology. *Korean J. Chem. Eng.* **2011**, *28*, 1665–1671. [CrossRef]
119. Izadkhah, B.; Nabavi, S.R.; Niaei, A.; Salari, D.; Mahmuodi Badiki, T.; Çaylak, N. Design and optimization of Bi-metallic Ag-ZSM5 catalysts for catalytic oxidation of volatile organic compounds. *J. Ind. Eng. Chem.* **2012**, *18*, 2083–2091. [CrossRef]
120. Qu, Z.; Chen, D.; Sun, Y.; Wang, Y. High catalytic activity for formaldehyde oxidation of AgCo/APTES@MCM-41 prepared by two steps method. *Appl. Catal. A Gen.* **2014**, *487*, 100–109. [CrossRef]
121. Pârvulescu, V.; Tablet, C.; Anastasescu, C.; Su, B.L. Activity and stability of bimetallic Co (V, Nb, La)-modified MCM-41 catalysts. *Catal. Today* **2004**, *93–95*, 307–313. [CrossRef]
122. Djinović, P.; Ristić, A.; Žumbar, T.; Dasireddy, V.D.B.C.; Rangus, M.; Dražić, G.; Popova, M.; Likozar, B.; Zabukovec Logar, N.; Novak Tušar, N. Synergistic effect of CuO nanocrystals and Cu-oxo-Fe clusters on silica support in promotion of total catalytic oxidation of toluene as a model volatile organic air pollutant. *Appl. Catal. B Environ.* **2020**, *268*, 118749. [CrossRef]
123. Abdullah, A.Z.; Bakar, M.Z.A.; Bhatia, S. Combustion of chlorinated volatile organic compounds (VOCs) using bimetallic chromium-copper supported on modified H-ZSM-5 catalyst. *J. Hazard. Mater.* **2006**, *129*, 39–49. [CrossRef] [PubMed]
124. Shayegan, Z.; Lee, C.S.; Haghighat, F. TiO2 photocatalyst for removal of volatile organic compounds in gas phase—A review. *Chem. Eng. J.* **2018**, *334*, 2408–2439. [CrossRef]
125. Mo, J.; Zhang, Y.; Xu, Q.; Lamson, J.J.; Zhao, R. Photocatalytic purification of volatile organic compounds in indoor air: A literature review. *Atmos. Environ.* **2009**, *43*, 2229–2246. [CrossRef]
126. Tsang, C.H.A.; Li, K.; Zeng, Y.; Zhao, W.; Zhang, T.; Zhan, Y.; Xie, R.; Leung, D.Y.C.; Huang, H. Titanium oxide based photocatalytic materials development and their role of in the air pollutants degradation: Overview and forecast. *Environ. Int.* **2019**, *125*, 200–228. [CrossRef] [PubMed]
127. Parrino, F.; Loddo, V.; Augugliaro, V.; Camera-Roda, G.; Palmisano, G.; Palmisano, L.; Yurdakal, S. Heterogeneous photocatalysis: Guidelines on experimental setup, catalyst characterization, interpretation, and assessment of reactivity. *Catal. Rev.* **2019**, *61*, 163–213. [CrossRef]
128. Parrino, F.; Bellardita, M.; García-López, E.I.; Marcì, G.; Loddo, V.; Palmisano, L. Heterogeneous photocatalysis for selective formation of high-value-added molecules: Some chemical and engineering aspects. *ACS Catal.* **2018**, *8*, 11191–11225. [CrossRef]
129. Hoffmann, M.R.; Martin, S.T.; Choi, W.; Bahnemann, D.W. Environmental applications of semiconductor photocatalysis. *Chem. Rev.* **1995**, *95*, 69–96. [CrossRef]
130. Sciré, S.; Palmisano, L. *Cerium Oxide (CeO2): Synthesis, Properties and Applications*; Elsevier: Amsterdam, The Netherlands, 2020.
131. Fiorenza, R.; Bellardita, M.; Palmisano, L.; Sciré, S. A comparison between photocatalytic and catalytic oxidation of 2-Propanol over Au/TiO2-CeO2 catalysts. *J. Mol. Catal. A Chem.* **2016**, *415*, 56–64. [CrossRef]
132. Fiorenza, R.; Condorelli, M.; D'Urso, L.; Compagnini, G.; Bellardita, M.; Palmisano, L.; Sciré, S. Catalytic and photothermo-catalytic applications of TiO2-CoOx composites. *J. Photocatal.* **2020**, *1*, 1. [CrossRef]
133. Bellardita, M.; Fiorenza, R.; Palmisano, L.; Sciré, S. Photocatalytic and photothermocatalytic applications of cerium oxide-based materials. In *Cerium Oxide (CeO2): Synthesis, Properties and Applications*; Elsevier: Amsterdam, The Netherlands, 2020; pp. 109–167.
134. Higashimoto, S.; Tanihata, W.; Nakagawa, Y.; Azuma, M.; Ohue, H.; Sakata, Y. Effective photocatalytic decomposition of VOC under visible-light irradiation on N-doped TiO2 modified by vanadium species. *Appl. Catal. A Gen.* **2008**, *340*, 98–104. [CrossRef]
135. Fiorenza, R.; Bellardita, M.; Sciré, S.; Palmisano, L. Effect of the addition of different doping agents on visible light activity of porous TiO2 photocatalysts. *Mol. Catal.* **2018**, *455*, 108–120. [CrossRef]
136. Fiorenza, R.; Di Mauro, A.; Cantarella, M.; Gulino, A.; Spitaleri, L.; Privitera, V.; Impellizzeri, G. Molecularly imprinted N-doped TiO2 photocatalysts for the selective degradation of o-phenylphenol fungicide from water. *Mater. Sci. Semicond. Process.* **2020**, *112*, 105019. [CrossRef]
137. Sytwu, K.; Vadai, M.; Dionne, J.A. Bimetallic nanostructures: Combining plasmonic and catalytic metals for photocatalysis. *Adv. Phys. X* **2019**, *4*, 1619480. [CrossRef]

138. Arifin, K.; Majlan, E.H.; Wan Daud, W.R.; Kassim, M.B. Bimetallic complexes in artificial photosynthesis for hydrogen production: A review. *Int. J. Hydrogen Energy* **2012**, *37*, 3066–3087. [CrossRef]
139. Verbruggen, S.W. TiO2 photocatalysis for the degradation of pollutants in gas phase: From morphological design to plasmonic enhancement. *J. Photochem. Photobiol. C Photochem. Rev.* **2015**, *24*, 64–82. [CrossRef]
140. Fiorenza, R.; Bellardita, M.; D'Urso, L.; Compagnini, G.; Palmisano, L.; Scirè, S. Au/TiO2-CeO2 catalysts for photocatalytic water splitting and VOCs oxidation reactions. *Catalysts* **2016**, *6*, 121. [CrossRef]
141. Colmenares, J.C.; Lisowski, P.; Łomot, D.; Chernyayeva, O.; Lisovytskiy, D. Sonophotodeposition of bimetallic photocatalysts Pd-Au/TiO2: Application to selective oxidation of methanol to methyl formate. *ChemSusChem* **2015**, *8*, 1676–1685. [CrossRef]
142. Czelej, K.; Cwieka, K.; Colmenares, J.C.; Kurzydlowski, K.J.; Xu, Y.J. Toward a comprehensive understanding of enhanced photocatalytic activity of the bimetallic PdAu/TiO2 catalyst for selective oxidation of methanol to methyl formate. *ACS Appl. Mater. Interfaces* **2017**, *9*, 31825–31833. [CrossRef]
143. Cybula, A.; Nowaczyk, G.; Jarek, M.; Zaleska, A. Preparation and characterization of Au/Pd Modified-TiO2 photocatalysts for phenol and toluene degradation under visible light—The effect of calcination temperature. *J. Nanomater.* **2014**, *2014*, 918607. [CrossRef]
144. Zhang, Y.; Park, S.J. Au–pd bimetallic alloy nanoparticle-decorated BiPO4 nanorods for enhanced photocatalytic oxidation of trichloroethylene. *J. Catal.* **2017**, *355*, 1–10. [CrossRef]
145. Zhang, Y.; Park, S.J. Bimetallic AuPd alloy nanoparticles deposited on MoO3 nanowires for enhanced visible-light driven trichloroethylene degradation. *J. Catal.* **2018**, *361*, 238–247. [CrossRef]
146. Han, C.; Yang, X.; Gao, G.; Wang, J.; Lu, H.; Liu, J.; Tong, M.; Liang, X. Selective oxidation of methanol to methyl formate on catalysts of Au-Ag alloy nanoparticles supported on titania under UV irradiation. *Green Chem.* **2014**, *16*, 3603–3615. [CrossRef]
147. Zielińska-Jurek, A.; Zaleska, A. Ag/Pt-modified TiO2 nanoparticles for toluene photooxidation in the gas phase. *Catal. Today* **2014**, *230*, 104–111. [CrossRef]
148. Wysocka, I.; Markowska-Szczupak, A.; Szweda, P.; Ryl, J.; Endo-Kimura, M.; Kowalska, E.; Nowaczyk, G.; Zielińska-Jurek, A. Gas-phase removal of indoor volatile organic compounds and airborne microorganisms over mono- and bimetal-modified (Pt, Cu, Ag) titanium(IV) oxide nanocomposites. *Indoor Air* **2019**, *29*, 979–992. [CrossRef] [PubMed]
149. Wolski, L.; El-Roz, M.; Daturi, M.; Nowaczyk, G.; Ziolek, M. Insight into methanol photooxidation over mono- (Au, Cu) and bimetallic (AuCu) catalysts supported on niobium pentoxide—An operando-IR study. *Appl. Catal. B Environ.* **2019**, *258*, 117978. [CrossRef]

© 2020 by the author. Licensee MDPI, Basel, Switzerland. This article is an open access article distributed under the terms and conditions of the Creative Commons Attribution (CC BY) license (http://creativecommons.org/licenses/by/4.0/).

Optimized Synthesis Routes of MnO$_x$-ZrO$_2$ Hybrid Catalysts for Improved Toluene Combustion

Xin Huang [1], Luming Li [2,3,*], Rong Liu [4,*], Hongmei Li [2], Li Lan [5] and Weiqi Zhou [1]

1. School of Mechanical Engineering, Chengdu University, Chengdu 610106, China; huangxin@stu.cdu.edu.cn (X.H.); zhouweiqi@stu.cdu.edu.cn (W.Z.)
2. College of Food and Biological Engineering, Chengdu University, Chengdu 610106, China; lihongmei@cdu.edu.cn
3. Institute for Advanced Study, Chengdu University, Chengdu 610106, China
4. School of Preclinical Medicine (Nursing College), Chengdu University, Chengdu 610106, China
5. College of Materials and Mechatronics, Jiangxi Science and Technology Normal University, Nanchang 330013, China; lanlijxstnu@outlook.com
* Correspondence: liluming@cdu.edu.cn (L.L.); liurong02@cdu.edu.cn (R.L.)

Abstract: In this contribution, the three Mn-Zr catalysts with Mn$_x$Zr$_{1-x}$O$_2$ hybrid phase were synthesized by two-step precipitation route (TP), conventional coprecipitation method (CP) and ball milling process (MP). The components, textural and redox properties of the Mn-Zr hybrid catalysts were studied via XRD, BET, XPS, HR-TEM, H$_2$-TPR. Regarding the variation of synthesis routes, the TP and CP routes offer a more obvious advantage in the adjustment of the concentration of Mn$_x$Zr$_{1-x}$O$_2$ solid solution compared to the MP process, which directly commands the content of Mn^{4+} and oxygen vacancy and lattice oxygen, and thereby leads to the enhanced mobility of reactive oxygen species and catalytic activity for toluene combustion. Moreover, the TP-Mn2Zr3 catalyst with the enriched exposure content of 51.4% for the defective (111) lattice plane of Mn$_x$Zr$_{1-x}$O$_2$ exhibited higher catalytic activity and thermal stability for toluene oxidation than that of the CP-Mn2Zr3 sample with a value of 49.3%. This new observation will provide a new perspective on the design of bimetal catalysts with a higher VOCs combustion abatement.

Keywords: Mn-Zr solid solution; toluene; active oxygen; combustion

1. Introduction

Volatile organic compounds (VOCs) have been considered important harmful pollutants and can be transferred into secondary aerogel and ozone via complex photochemical reactions in the atmosphere, which are threatening the ecological environment and human health [1–4]. It is urgent to adopt effective routes to control VOCs emissions. In particular, the toluene with toxicity and carcinogenicity is one of the most common VOCs, be discharged from the petrochemical industry, and be normally selected as the target VOCs to test the activity of catalyst.

Up to now, several techniques, such as adsorption and absorption, thermal incineration, plasma, membrane separation, biological treatment, and catalytic oxidation, have been widely reported for the degradation of VOCs [5]. Among them, catalytic oxidation has been considered one of the most promising technologies because it can completely convert VOCs under relatively low temperatures (<400 °C) into harmless CO$_2$ and H$_2$O with low energy consumption, which has gained a lot of attention either in scientific and industrial fields. The key issue of catalytic oxidation is to design catalysts with high catalytic activity and low cost [6,7].

Given the active components of the VOCs catalysts, catalysts can be divided into noble metal catalysts and non-noble metal catalysts. There is no doubt that the supported noble metal catalysts have higher activity at lower temperatures. However, the disadvantages of noble metal catalysts also need to be taken into consideration, such as high price, scarce resources, poor thermal stability, and sulfur resistance, etc., which fades their industrial application [8–10]. Therefore, it is necessary to develop alternative catalysts with high catalytic activity. The transition metal oxides have many advantages, such as relatively low price, good activity, abundant content in the earth, and environmental friendliness, which have attracted great interest [11]. For example, manganese-based oxides have been confirmed as one of the most potential alternates to noble metal catalysts and exhibited high catalytic activity due to the high mobility of lattice oxygen and the existence of transformation of unstable valence states, such as $Mn^{4+} \leftrightarrow Mn^{3+} \leftrightarrow Mn^{2+}$ [12–15]. In addition, the physical and chemical characteristics of the catalysts, such as morphology [8,16], specific surface area [17], oxygen vacancy [18], reducibility, and reactivity of lattice oxygen [1], oxygen migration rate [14], also determine the efficiency of the catalytic oxidation of VOCs. However, it is difficult to efficiently remove certain VOCs for a single phase of Mn due to the poor thermal stability [12,20]. Some studies have reported that the synthesized Mn-M (M = Co [21], Cu [22], Ce [15,23], Zr [24,25]) composite metal oxides can improve catalytic activity towards VOCs abatement through the strong interaction between the parent material and the modifier. Among them, ZrO_2-based materials show excellent corrosion resistance, high stability, and ionic conductivity, and the ZrO_2-based solid solution keep high catalytic activity [26]. For instance, MnO_x-ZrO_2 bimetal oxides catalysts present a good catalytic activity due to their excellent redox property and thermal stability [27,2].

Choudhary et al. [29] reported that the reactivity of the lattice oxygen of Mn-doped ZrO_2 catalysts depended on the Mn/Zr ratio and calcination temperature; the Mn-doped ZrO_2 (cubic) with the Mn/Zr ratio of 0.25 prepared at a calcination temperature of 600 processed the best methane combustion activity. Gutie'rrez-Ortiz et al. [24] reported that Mn-Zr mixed oxides exhibited better catalytic activity for 1,2-dichloroethane (DCE) and trichloroethylene (TCE) than that of pure zirconia and manganese oxide, which was attributed to the modification effects of surface acid sites coupled with the readily accessible active oxygen. Zeng et al. [30] successfully manufactured an $MnZrO_x$ catalyst with a three-dimensional microporous (3DM) structure, which hold a better reducibility and oxygen mobility, accompanied by higher Mn^{4+}/Mn^{3+} and O_β/O_α, and the special structure significantly promoted the adsorption of reactant molecules and the mobility of lattice oxygen and propane oxidation. Compared with the B-$Mn_{0.6}Zr_{0.4}$ catalyst prepared traditional sol–gel methods, the activation energy of propane combustion on 3DM MnZr decreased from 156.2 kJ·mol^{-1} to 105.0 kJ·mol^{-1}. Moreover, Yang et al. synthesized a series of impurity $Mn_xZr_{1-x}O_2$ catalysts, and it was found that low-valent manganese (Mn^{2+}) can enter zirconium lattice by substituting Zr^{4+} and inducing the formation of oxygen vacancy and its concentration hinges on the doping content, which is conducive to the activation of oxygen on the surface of Mn-doped c-ZrO_2(111) crystal plane, thereby improving the catalytic performance for toluene oxidation [31].

Recently, we found that preparation process would play a critical role in adjusting the structural properties of nanocatalysts in oxidation reaction [32,33]. Particularly, the precipitation sequence determined the structural, redox properties, and textural stability of ZrO_2-based catalysts [34]. Inspired by those points, we constructed Mn-Zr bimetal catalysts based on the controllable catalytic active centers of $Mn_xZr_{1-x}O_2$ solid solution via a modulated precipitation sequence and further evaluated the catalytic capacity the combustion of toluene model molecules, and the synergistic effect between ZrO_2 and manganese oxide was also analyzed.

2. Results and Discussion

2.1. Material Characterization

2.1.1. XRD Analysis

The XRD patterns of the as-prepared Mn-Zr catalysts with different Mn/Zr ratios are shown in Figure 1. Obviously, the intensity and position of the characteristic peaks varies with the Mn/Zr ratio and synthesized routes. The intensity of characteristic peak of $Mn_xZr_{1-x}O_2$ solid solution was increased as the Mn/Zr ratio decreased, in which the diffraction peaks located at 30.42°, 35.28°, 50.75°, and 60.33° correspond to the (111), (200), (220), and (311) crystal planes of $Mn_xZr_{1-x}O_2$ solid solution (JCPDS PDF#77-2157), respectively (Figure 1a). On the contrary, the low Mn/Zr ratio resulted in the mixture phase formation, such as ZrO_2 (JCPDS PDF# 78-0048), or a small amount of solid solution, which is ascribed to the difference in ion radius. The ion radius of Mn^{n+} cations (Mn^{2+} (0.83 Å), Mn^{3+} (0.64 Å) and Mn^{4+} (0.53 Å)) is smaller than Zr^{4+} (0.84 Å). Therefore, it can be speculated that Mn cations readily partially replace Zr cations in the host lattice, whereas it is very difficult [24]. Additionally, the incorporation of Mn dramatically changes the crystal phase of zirconia (Figure 1b). Moreover, the preparation route played an important role in regulating the intensity of characteristic peaks of the $Mn_xZr_{1-x}O_2$ solid solution. The two-step precipitation strategy is easier to obtain a high content of active phase solid solution as well as trace amounts of manganese oxide. Besides, it can be found that TP-Mn2Zr3 has more obvious (111) crystal plane characteristic peaks compared to CP-Mn2Zr3.; the content of (111) crystal plane of TP-Mn2Zr3 sample accounted for 51.4%, which is considered the active center [31], while the CP-Mn2Zr3 is ca. 49.3%. This is the reason why the TP-Mn2Zr3 sample has a better catalytic performance. The sample (MP-Mn2Zr3) prepared by mechanical ball mills was as a reference. The result showed that the separated mixture of ZrO_2 and Mn_2O_3 oxides caused no synergistic effect between MnO_x and ZrO_2, and a poor catalytic activity for toluene oxidation under the ball milling conditions.

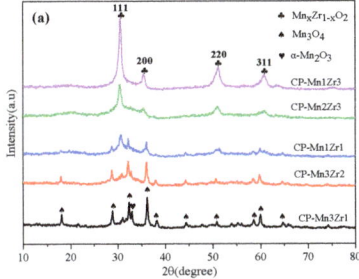

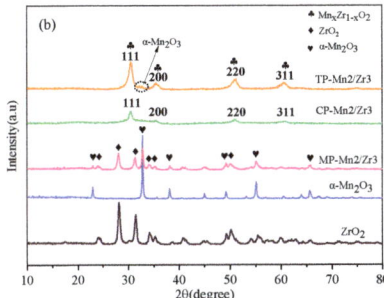

Figure 1. XRD patterns of catalysts with different molar ratios of Mn/Zr (**a**) and different preparation routes (**b**) (two-step precipitation (TP); conventional coprecipitation (CP); ball milling process (MP)).

Moreover, the lattice parameters and grain size of CP-Mn2Zr3, TP-Mn2Zr3, MP-Mn2Zr3, ZrO_2, and α-Mn_2O_3 samples were summarized in Table 1. Compared with MP-Mn2Zr3 and CP-Mn2Zr3, the TP-Mn2Zr3 catalyst has the lowest lattice parameters (5.04 Å) and the smallest grain size (8.4 nm). Indeed, Mn cations with different valence exhibit a different ionic radius. The more Mn^{n+} cations (Mn^{2+}, Mn^{3+}, and Mn^{4+}) with smaller radii (0.83, 0.64, and 0.53 Å, respectively) incorporated into c-zirconia to replace the Zr^{4+} (0.84 Å) can reduce the lattice parameters [28,35,36]. Based on this, we can infer that the improved preparation process is more beneficial to the synthesis of the Mn_xZr_{1-x} solid solution and decreases the grain size.

Table 1. Data obtained from XRD analyses of two-step precipitation (TP), conventional coprecipitation (CP), ball milling process (MP)-Mn2Zr3, ZrO_2, and α-Mn_2O_3 samples.

Samples	a (Å)	b (Å)	c (Å)	v (Å³)	Grain Size (nm)
TP-Mn2Zr3	5.04	-	-	128	8.4
CP-Mn2Zr3	5.05	-	-	128.8	11.3
MP-Mn2Zr3	-	-	-	-	21.5
ZrO_2	5.87	4.86	5.20	148.3	16.6
α-Mn_2O_3	9.42	9.42	9.42	835.9	32.2

[a] The average crystal size was calculated by the Scherrer equation from the XRD data.

2.1.2. BET Analysis

It was accepted that the catalytic activity of nanocatalysts is closely related to the surface texture, such as specific surface area, average pore size, and pore volume. Figure shows the nitrogen adsorption-desorption curves of as-obtained catalysts (TP-Mn2Zr3, CP-Mn2Zr3, and MP-Mn2Zr3), and it can be observed that these catalysts have similar adsorption isotherms, which can be classified as typical type IV adsorption isotherm based on the IUPAC classification [37]. Hysteresis loops demonstrated relative pressure the range of 0.4–1 in all three samples, and there was no adsorption saturation platform in the range of higher relative pressure, suggesting the catalysts have an irregular pore structure that may be caused by the slit-shaped pores. The type IV isotherm is usually basis for judging whether there are mesopores in the catalyst materials [8,38]. Considering the presence of type IV adsorption isotherm and H3 hysteresis loop in the catalyst, it can be determined that irregular mesopores were formed in all the catalysts [39]. Figure shows the pore size distribution of the catalyst, which further confirmed the existence the mesoporous structure in the catalysts. The pore size mainly distributed in the range of 2–40 nm and centered around 10 nm. Specifically, the value of the specific surface area, average pore diameter, and pore volume of the catalyst were listed in Table 2. It was observed that the catalyst prepared by mechanical stirring showed the lowest specific surface area, only 23.7 m²/g, whereas the catalyst doped with Zr obviously demonstrated an improved specific surface area. The specific surface areas of TP-Mn2Zr3 and CP-Mn2Zr3 catalysts were 99.7 and 139.5 m²/g, respectively, which may promote the exposure active sites on the surface of the catalysts [31] and contribute to the improved catalytic performance. Among these oxides, the average pore sizes of TP-Mn2Zr3, CP-Mn2Zr3 and MP-Mn2Zr3 catalysts are 10.8, 5.9, and 13.5 nm, respectively. It was found that the TP-Mn2Zr3 catalyst has the largest pore volume of 0.27 cm³/g, followed by CP-Mn2Zr3 (0.20 cm³/g) and MP-Mn2Zr3 (0.08 cm³/g) catalysts.

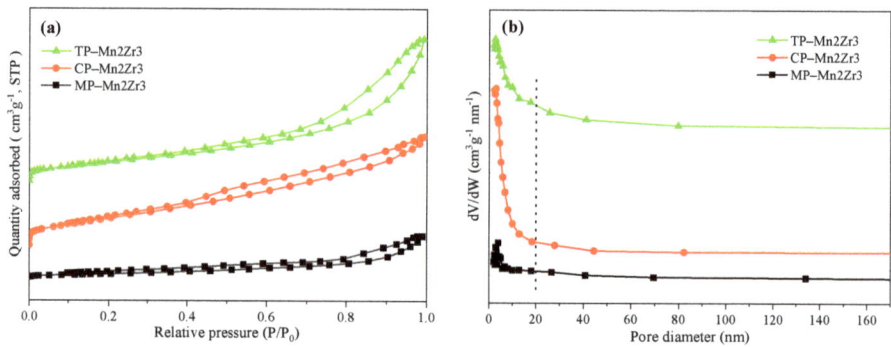

Figure 2. Nitrogen adsorption-desorption isotherms (a) and BJH pore-size distributions (b) of TP-Mn2Zr3, CP-Mn2Zr3, and MP-Mn2Zr3 catalysts.

Table 2. The surface area, average pore diameter, and pore volume of TP-Mn2Zr3, CP-Mn2Zr3, and MP-Mn2Zr3 catalysts.

Samples	S_{BET} (m^2/g)	Average Pore Diameter (nm) [a]	Pore Volume (cm^3/g) [b]
TP-Mn/Zr = 2/3	99.7	10.8	0.27
CP-Mn/Zr = 2/3	139.5	5.9	0.20
MP-Mn/Zr = 2/3	23.7	13.5	0.08

[a] Bases on the total adsorption average pore width (4V/A by BET, A = S_{BET}). [b] Based on the BJH Adsorption cumulative volume.

2.1.3. HRTEM Analysis

In order to further study the morphology of the catalysts, TEM experiments were carried out on the CP-Mn2Zr3 and TP-Mn2Zr3 catalysts, as shown in Figure 3. The results suggested that both CP-Mn2Zr3 and TP-Mn2Zr3 exhibit morphology of nanoparticles (Figure 3a,c) and the lattice spacings of ca. 2.54 and 2.94 Å, well attributed to the (111) crystal plane and (200) crystal plane of the $Mn_xZr_{1-x}O_2$ solid solution, respectively (Figure 3b,d). This is in line with the XRD results. Therefore, it can be inferred that Mn ions are readily embedded in the lattice of ZrO_2 and form Mn-Zr solid solution via co-precipitation or optimized two-step precipitation route and lead to the phase transformation from monoclinic zirconium dioxide into cubic zirconium dioxide [26].

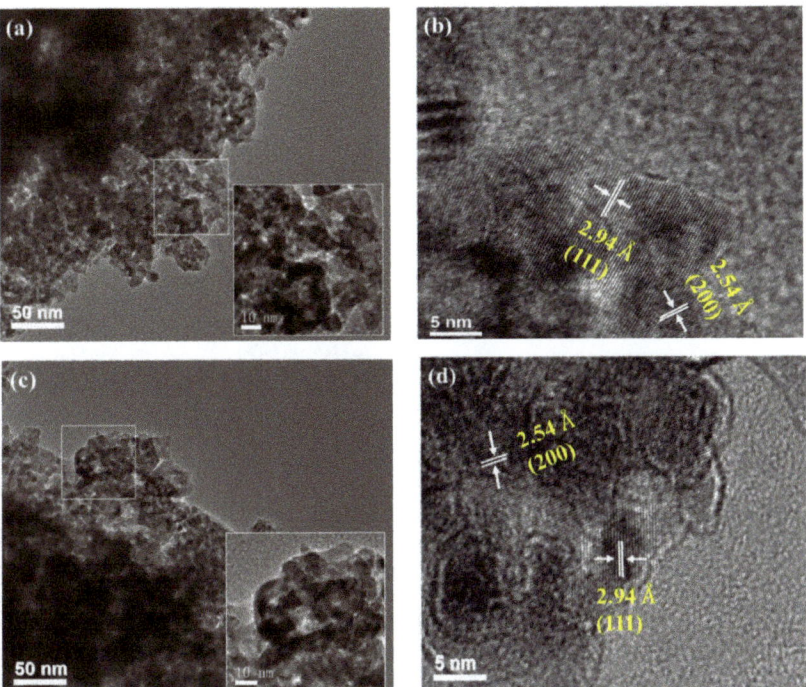

Figure 3. High resolution transmission electron microscope (HRTEM) images of CP-Mn2Zr3 (a,b) and TP-Mn2Zr3 (c,d).

2.1.4. XPS Analysis

XPS experiments were performed to analyze the components and valence states of elements, as shown in Figure 4 and Table 3. It was observed that the surface elements of the catalyst are mainly composed of O (ca. 530 eV), Zr (ca. 183 eV), and Mn (ca. 642 eV) (Figure 4a) [30,40]. Figure 4b shows the XPS spectra of Mn 2p; it was integrated into two characteristic peaks by deconvolution. The characteristic peaks located at 641.7 and 643.4 eV can be attributed to Mn^{3+} and Mn^{4+} species, respectively [1,17,41].

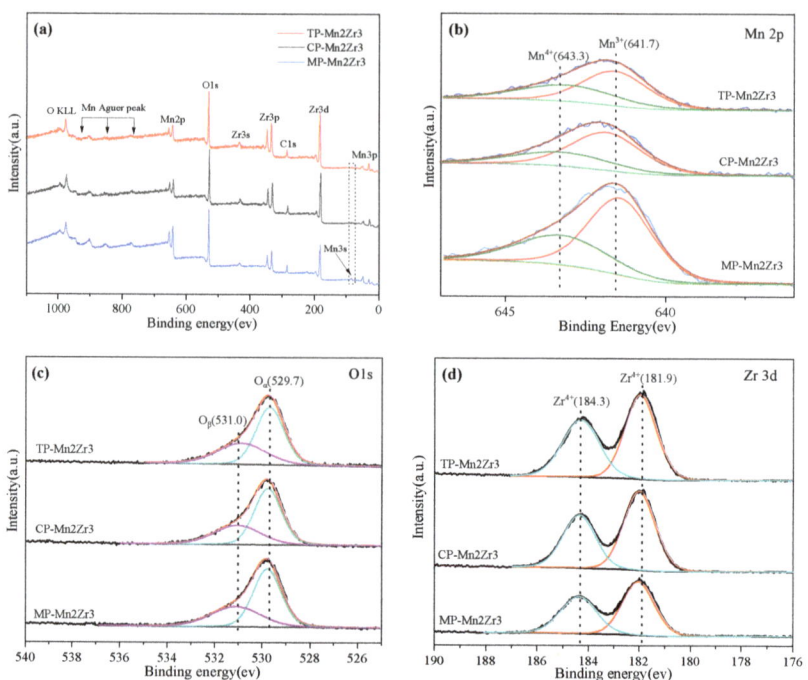

Figure 4. XPS spectra of (a) full spectra, (b) Mn 2p, (c) O 1s, (d) Zr 3d over TP, CP, and MP-Mn2Z

Table 3. Analysis of the surface composition and content of samples based on XPS results.

Samples	Surface Element (at%)			Bind Energy				Surface Element Molar Ratio	
	Mn 2p	Zr 3d	O 1s	Mn^{4+}	Mn^{3+}	O_α	O_β	Mn^{4+}/Mn^{3+}	O_β/O_α
TP-Mn2Zr3	15.92	18.63	65.46	643.3	641.7	529.7	531.0	0.68	0.65
CP-Mn2Zr3	17.37	17.69	64.94	643.5	641.9	529.7	531.0	0.50	0.59
MP-Mn2Zr3	29.70	11.04	59.25	643.4	641.5	529.7	531.1	0.53	0.63

Furthermore, the ratio of Mn^{4+}/Mn^{3+} of the catalyst calculated based on the relati area was closely related to the preparation routes. The TP-Mn2Zr3 catalyst prepared an improved precipitation method had a higher Mn^{4+}/Mn^{3+} value, which reached 0. which followed the MP-Mn2Zr3 (0.53) and CP-Mn2Zr3 catalysts (0.50), respectively. T strong interaction between Mn and Zr in Mn-Zr solid solution can promote the occurrer of the charge transfer process and produce more Mn^{4+} [30], which plays a key role in catalytic degradation of toluene. In other words, the stronger interaction between Mn a Zr in the catalyst of TP-Mn2Zr3 can be conducive to higher catalytic oxidation of tolue Meanwhile, it is widely accepted that the oxygen mobility of manganese-based oxides closely related to the transformation ability of manganese species between different valer states, and the increase of the concentration of high valence metal cations promotes chemical potential and mobility of oxygen [42].

O 1s XPS spectra of the three catalysts were exhibited in Figure 4c, which can be split in two peaks by deconvolution, the bind-energy located at 529.0–530.0 eV and 531.0–532.0 eV c be ascribed to the surface lattice oxygen (O_α) and chemisorbed oxygen and/or defect oxid (O^{2-}, O_2^{2-}, or O^-) (O_β) [28], respectively. It was noted that the content of O_α is more than 6 for the three samples, suggesting the O_α is dominated. Furthermore, the decreasing trend

O_β/O_α value is as follows: TP-Mn2Zr3(0.65) > MP-Mn2Zr3(0.63) > CP-Mn2Zr3 (0.59), which is positively correlated with the value of Mn^{4+}/Mn^{3+} (TP-Mn2Zr3 (0.68) > MP-Mn2Zr3(0.53) > CP-Mn2Zr3 (0.50)). It was reported that both lattice oxygen and chemisorbed oxygen and/or defect oxides are active oxygen species that can participate in the oxidation of toluene, which improves the oxidation performance of the catalyst [19,35]. Herein, the optimized coprecipitation catalyst displays better catalytic activity due to the contribution of lattice oxygen and absorbed oxygen.

The spin-splitting peaks of the Zr 3d orbit can be observed in Figure 4d, centered at 181.9 eV and 184.3 eV and belong to $3d_{5/2}$ and $3d_{3/2}$ orbit, respectively, suggested that Zr cations exist in the catalyst in the form of tetravalent [30]. Meanwhile, it was found that there was hardly a change in the position of the Zr 3D orbital splitting peak in TP-Mn2Zr3, CP-Mn2Zr3, and MP-Mn2Zr3 samples. It was worth noting that the XPS signal peak of Zr in TP-Mn2Zr3 and CP-Mn2Zr3 samples originated from the Mn-Zr solid solution, while the signal peak of Zr in the MP-Mn2Zr3 sample was derived from pure zirconia. This indicates that the Zr^{4+} ion is very stable [43], though the Mn^{n+} ions were entered the framework of ZrO_2. Moreover, the proportion of Mn, Zr, and O in the catalyst is related to the preparation routes (Table 3). The order of the proportion of Mn elements follows: MP-Mn2Zr3 > CP-Mn2Zr3 > TP-Mn2Zr3, while that of Zr and O is the opposite; this suggested that the concentration of Mn-Zr solid solution various with the content of Mn ions entered the framework of ZrO_2.

2.1.5. H_2-TPR Results Analysis

The H_2-TPR experiments were employed to investigate the reducibility of the catalysts. The H_2-TPR cures of TP-Mn2Zr3, CP-Mn2Zr3, and MP-Mn2Zr3 samples in the range of 100–700 °C are shown in Figure 5. There exist three obvious characteristic peaks of hydrogen consumption. It was reported that about 15 wt.% of MnO is soluble in ZrO_2 due to the solidified eutectic temperature [44]. Herein, the three characteristic peaks of catalysts were mainly attributed to the reduction of MnO_x, in which Peak 1 located below 250 °C shows the reduction of amorphous MnO_x, dispersed on the surface of Mn-Zr solid solution, and peak 2 (250–380 °C) should be linked to the reduction of Mn_2O_3 to Mn_3O_4, while peak 3 (380–480 °C) corresponds to the reduction of Mn_3O_4 to MnO, respectively [26,31,45]. Moreover, the TP-Mn2Zr3 and CP-Mn2Zr3 catalysts behave an excellent low-temperature reduction performance compared to that of MP-Mn2Zr3. Especially for the TP-Mn2Zr3 sample, the locations of reduction peaks are separately reduced to 224, 303, and 400 °C, suggesting it possessed a stronger mobility of reactive oxygen species and leads to a better catalytic activity for VOCs abatement [31].

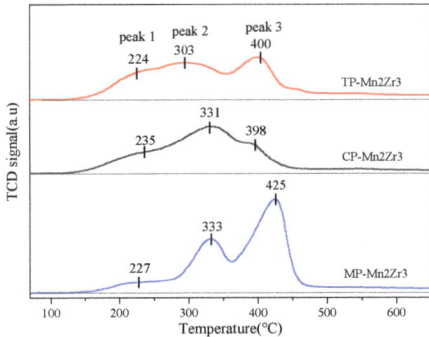

Figure 5. H_2-TPR profiles of TP-Mn2Zr3, CP-Mn2Zr3, and MP-Mn2Zr3 catalysts.

It was found that the reduction temperature of pure zirconia usually appears above 600 °C, and the incorporation of Mn into the framework of zirconia to form a solid solution of $Mn_xZr_{1-x}O_2$ would contribute to the decreasing of reduction temperature [26,29]. In this contribution, the reduction peak of zirconia was not observed on the MP-Mn2Zr catalyst, indicating that $Mn_xZr_{1-x}O_2$ species cannot be formed by the ball milling process. In contrast, the shouldered temperature reduction peak (ca. 400 °C) was displayed in both TP-Mn2Zr3 and CP-Mn2Zr3 catalysts, which may be the result of the synergistic interaction between the surface MnO_x and ZrO_2 alloyed to form Mn-Zr solid solution [31,46], which well corresponds to the XRD results.

2.2. Evaluation of Catalytic Activity

The catalytic oxidation performance of the synthesized catalysts for toluene abatement was assessed. The functional cures between the conversion of toluene on the catalyst and reaction temperature are depicted in Figure 6. All the catalysts can achieve complete catalytic oxidation for toluene below 300 °C. As shown in Figure 6a, the CP-Mn2Zr3 catalyst exhibits better catalytic oxidation activity for toluene. The values of T_{50} (the 50% conversion of 1000 ppm toluene) and T_{90} (the 90% conversion of 1000 ppm toluene) of the CP-Mn2Zr3 catalyst are 270 °C and 278 °C, respectively. In addition, the order of catalytic ability of catalysts is depicted as CP-Mn2Zr3 > CP-Mn3Zr1 > CP-Mn3Zr2 > CP-Mn1Zr1 > CP-Mn1Zr3. It is evident that the catalytic activity was closely linked with the Mn/Zr ratio, which controlled the component of the Mn-Zr solid solution. Thereby, can be inferred that the Mn-Zr solid solution is in the active phase and greatly influences the catalytic activity for toluene.

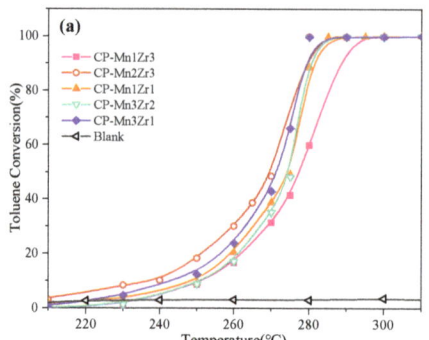 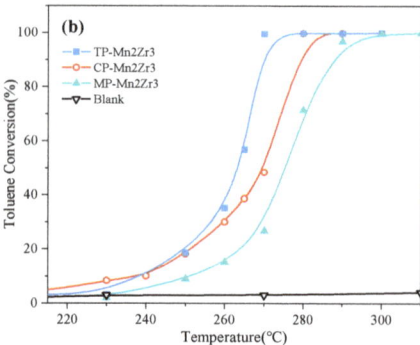

Figure 6. Activity test of catalysts with different molar ratios of Mn/Zr (a) and prepared by different strategies (b).

Moreover, the catalytic activity assessing experiments were also carried out on CP-Mn2Zr3, TP-Mn2Zr3, and MP-Mn2Zr3 catalysts, and the results are depicted in Figure 6b. Obviously, the TP-Mn2Zr3 catalyst exhibits a better catalytic performance with T_{50} and T_{90} values of 263 °C and 269 °C, respectively, followed by CP-Mn2Zr3. The performance of the catalyst with MP-Mn2Zr3 is the lowest, with T_{50} and T_{90} values of 275 °C and 287 °C, which is much lower than the TP-Mn2Zr3 and CP-Mn2Zr3 catalysts. The results clearly show that the TP-Mn2Zr3 and CP-Mn2Zr3 catalysts with the existence of the Mn-Zr solid solution exhibit higher catalytic activity for toluene abatement than that of the MP-Mn2Zr3 catalyst without the formation of the Mn-Zr solution. The outstanding catalytic performance of the TP-Mn2Zr3 catalyst for toluene should be attributed to the more exposed defect (1 1 1) crystal plane of $Mn_xZr_{1-x}O_2$ and the improved capacity of mobility of active oxygen.

The stability of the texture for the Mn-Zr catalysts will determine their potential application to some extent. In this work, five consecutive catalytic light-off cycle tests were performed on the TP-Mn2Zr3 and CP-Mn2Zr3 catalysts, and the results are shown in Figure 7. The value of the T_{90} value of the cycle 2 experiment for the TP-Mn2Zr3 catalyst is 259 °C, which is lower than that of the cycle 1 experiment (269 °C), and runs 4, and 5 on the TP-Mn2Zr3 catalyst maintain a similar value of T_{90} (259 °C ± 1). On the contrary, the CP-Mn2Zr3 catalyst performed inferior cyclic stability, in which the values of T_{90} for cycles 1, 2, 3, 4, and 5 was 278, 269, 264, 264, and 269 °C, respectively. The enhanced oxidation performance for both catalysts after running the first experiment should be attributed to the activation effect at 300 °C [8]. To view the cyclic stability directly, the function curves between cycle times and toluene conversion at 265 °C are depicted as the inset of Figure 7a,b, which imply that the TP-Mn2Zr3 catalyst exhibited better cycle stability. Conversely, the order of cyclic stability of CP-Mn2Zr3 obeys the volcanic model, which implies the inferior stability of CP-Mn2Zr3.

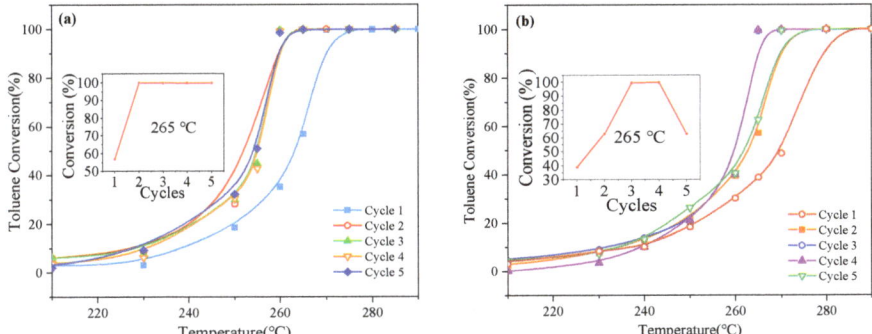

Figure 7. Activity cycle test of the TP-Mn2Zr3 catalyst (**a**) and the CP-Mn2Zr3 catalyst (**b**).

Based on studies reports, the bimetal oxide catalysts involve two reaction mechanisms in the oxidation for toluene, namely the Langmuir–Hinshelwood (L-H) and Mars–van Krevelen (Mv-K) mechanisms. Increasing the temperature, the L-H mechanism is gradually weakened, and the Mv-K mechanism gradually occupies a dominant position. The adsorbed oxygen directly oxidizes the adsorbed organic molecules, which is accorded to the L-H mechanism; on the other hand, lattice oxygen is activated at a higher temperature, and the consumed lattice oxygen can be readily supplemented from gaseous oxygen, which obeys the Mv-K mechanism [42,47].

Figure 8a reveals the effect of different WHSVs on the activity of the TP-Mn2Zr3 catalyst. Obviously, the value of T_{90} is positively correlated with WHSV; in other words, a longer contact time is beneficial to improve the catalytic performance. The longevity experiments for 3000 min were performed on TP-Mn2Zr3 and CP-Mn2Zr3 catalysts, as shown in Figure 8b. It was worth noting that the TP-Mn2Zr3 catalysts maintained a splendid catalytic activity (>90% toluene conversion). While the CP-Mn2Zr3 catalyst declined a lot after 1000 min and the value of toluene degradation was sharply decreased to about 50%, the conversion of toluene was only kept at ca. 33% after the 3000 min combustion reaction.

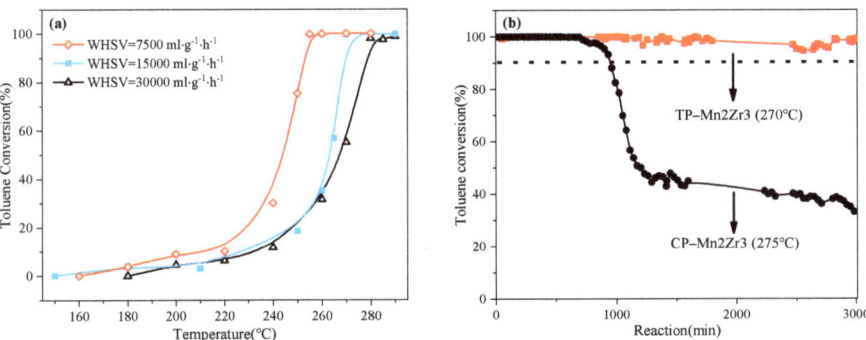

Figure 8. The effect of different WHSVs on the activity of the TP-Mn2Zr3 catalyst (**a**); Longevity test over TP-Mn2Zr3 and CP-Mn2Zr3 in 1000 ppm toluene (**b**).

3. Discussion

It was widely accepted that the physical and chemical properties, such as component of active phase, specific surface area, Mn valence, concentration of adsorbed oxygen/lattice oxygen and reduction ability, can command the activity of the catalyst [48]. For the component factor of the active phase, we found that the ratio of Mn/Zr and their synthesis routes play a vital role in adjusting the content of active centers, and a low molar ratio (<2/3) of Mn/Zr would lead to the formation of a solid solution of $Mn_xZr_{1-x}O_2$. the molar ratio goes up, a second phase, such as Mn_2O_3 and Mn_3O_4, will be formed, depicted in Figure 1a. This is mainly attributed to the solubility limit of manganese in the zirconia lattice [44]. However, a more complex mixed components does not necessary result in a better catalytic activity, as the CP-Mn2Zr3 catalyst presented a better catalytic performance among the samples prepared by the co-precipitation process (Figure 6). Moreover, the improved co-precipitation route can further promote the exposure of the active centers and enhance the catalytic performance. Therefore, it can be inferred that the optimized catalytic activity is linked to components of the active phase and mainly hinges on the ratio of Mn/Zr and their synthesis routes. On the other hand, co-precipitation and improved co-precipitation routes are ready to prepare catalysts with a large specific surface area and enhance their catalytic activities.

In addition, the concentrations of Mn^{4+} and O_α species of oxidizing catalyst were crucial for the oxidative degradation of VOCs [43,49]. The order of the O_β/O_α follow in descending order as Mn^{4+}/Mn^{3+}, which was depicted as TP-Mn2Zr3 > MP-Mn2Zr > CP-Mn2Zr3, indicating the concentration of Mn^{4+} was positively correlated with the content of oxygen vacancy [40]. In combination with the results of catalytic activity and components of active phase, the improved catalytic activity should be attributed to the promoted effects of the solid solution of $Mn_xZr_{1-x}O_2$. Moreover, the shouldered shape of the reduction peak for TP-Mn2Zr3 and CP-Mn2Zr3 catalysts can also account for the formation of $Mn_xZr_{1-x}O_2$ active species, which corresponds well with the results of XPS.

It was reported that the catalytic mechanism of toluene on Mn-based catalysts involves the Mv-K and L-H mechanism. Both lattice oxygen and adsorbed oxygen species can participate in the activation-oxidation process of VOCs [42,47,50]. In this paper, the enhanced catalytic activity of Mn-Zr catalysts may be ascribed to the collective effects of manganese zirconium bimetal because the formation of the solid solution of $Mn_xZr_{1-x}O_2$ would conducive to decrease the formation energy of oxygen vacancy; the oxygen molecule would absorbed and activated to active oxygen to enhance catalytic activity. Furthermore, the oxygen molecule is easily adsorbed on the defective (111) surface of the Mn-Zr catalyst with oxygen vacancy, and the distance of O–O bond was elongated, which probably suggest that the absorbed O_2 is easily activated and broken, which is in favor of the quick replenishment of consumed oxygen molecular during VOCs combustion [31]. This is in good agreement with the H_2-TPR results. Therefore, it is supposed that the enriched exposure of the defective (111) surface of the $Mn_xZr_{1-x}O_2$ active species would lead to better catalytic activity for toluene combustion, and this can account for the outstanding performance of TP-Mn2Zr3.

4. Materials and Methods

4.1. Materials

Zirconium (IV) oxynitrate hydrate ($ZrO(NO_3)_2 \cdot H_2O$) and Manganese nitrate ($Mn(NO_3)$ solution 50 wt.% were purchased from Chengdu Hua Xia Chemical Reagent Co., Ltd (Chengdu, China). Ammonia ($NH_3 \cdot H_2O$) (25~28%) reagent was purchased from Chengdu Ke Long chemicals Co., Ltd (Chengdu, China). All the reagents are an analytical reagent (AR), with no need for further treatment.

4.2. Catalyst Preparation

Firstly, Mn-Zr hybrids catalysts with different ratios were prepared by conventional optimized co-precipitation routes [24,25]. In a typical preparation, manganese nitrate solution (50 wt.%) (3.58, 7.16, 7.16, 10.74, and 10.74 g, respectively) and zirconium (IV) oxynitrate hydrate (7.48, 7.48, 4.99, 4.99, and 2.49 g, respectively) were added into 100 mL deionized water and vigorously stirred for 2 h to form homogeneous solution, respectively. Then ammonia was added dropwise until PH = 9 (Scheme 1a). The obtained precipitates are filtrated and washed until pH = 7. After that, the precipitate was dried in an oven at 90 °C for 5 h, followed by calcination at 550 °C for 5 h under an air atmosphere. Herein the ratios of Mn/Zr were listed as 1/3, 2/3, 1/1, 3/2, and 3/1, marked as CP = Mn1Zr3, CP = Mn2Zr3, CP = Mn1Zr1, CP = Mn3Zr2, and CP = Mn3Zr1, respectively.

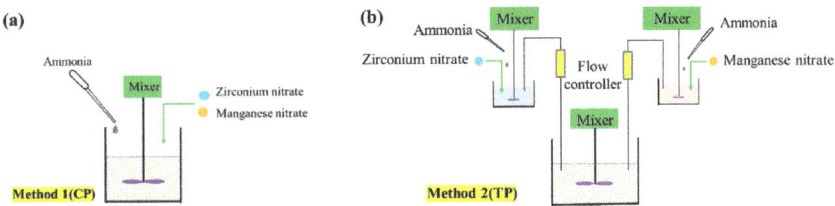

Scheme 1. The synthesis of catalyst by coprecipitation (a), and optimized coprecipitation (b).

The TP-Mn2Zr3 catalysts were prepared by two-step precipitation strategies. Specifically, manganese nitrate solution (50 wt.%) (7.16 g) and zirconium (IV) oxynitrate hydrate (7.48 g) were added into 50 mL deionized water and vigorously stirred for 2 h to form a homogeneous solution, respectively. Then ammonia was added dropwise into the two solutions until pH = 9 (Scheme 1b). We mixed the two solutions under vigorous stirring for 2 h. The washing, filtering, drying and calcining processes were the same procedures for CP-Mn2Zr3. Additionally, the final product is recorded as TP-Mn2Zr3. For comparison, MnO_x and ZrO_2 catalysts were prepared by the above routes. Besides, the MP-Mn2Zr3 sample was prepared by mechanical ball mill of a mixture of MnO_x and ZrO_2. The ball milling experiment was carried out on a planetary ball mill (XQM-0.4). Actually, the stainless-steel balls with different diameters (15, 12, 10, 8, and 5 mm) were mixed and added into a ball milling tank (50 mL) and then stirred for 2 h at 400 r/min. In addition, the total mass of stainless-steel balls was 150 g, and the mass of Mn_2O_3 and ZrO_2 was 1.84 g and 2.16 g, respectively, to keep the molar ratio of Mn/Zr at 2/3.

4.3. Catalyst Activity Evaluation

The catalytic degradation of toluene was assessed in a fixed-bed stainless reactor (id = 7 mm). Prior to the activity test, the catalyst (0.20 g, 40–60 mesh) and quartz sands (0.30 g, 40–60 mesh) were well mixed and then loaded into the center of the reactor and activated at 300 °C for 1 h with a 60 mL/min air flow rate. After that, the temperature was dropped to 150 °C, and 1000 ppm toluene gas was continuously fed into the fixed-bed reactor; the total flow is controlled at 50 mL/min to correspond to a weight hourly space velocity (WHSV) of 15,000 mL g_{cat}^{-1} h^{-1}. In addition, the catalytic activity of TP-Mn2Zr3 was tested under different WHSVs. The exhausted gas from a fixed-bed reactor was analyzed using a gas chromatography device (SP-7890 PLUS, Rui Hong Co, Ltd., Tengzhou, China) with a flame ionization detector (FID) and equipped polyethylene glycol capillary column and a thermal conductivity detector (TCD). The activity of the catalyst was measured by the toluene conversion (X_{con}), which can be calculated as follow:

$$X_{con} = \frac{C_{in} - C_{out}}{C_{in}} \quad (1)$$

where C_{in} stand for the inlet toluene concentration, and C_{out} represent the outlet toluene concentration after 30 min reaction.

4.4. Catalyst Characterization

The investigation of the phase composition of the catalysts was carried out by the Riga DX-2700 (Rigaku, Tokyo, Japan) diffractometer equipment with Cu-Kα (λ = 0.154 nm) as radiation source. The scan started from 10 to 80° with a scanning rate of 0.06°/s. The results of the specific surface area and pore size distribution of the catalysts were obtained by nitrogen adsorption and desorption experiments on V-SorbX800 (Jin Aipu, Beijing China) equipment at −196 °C. The microstructure of the fresh catalysts was investigated by an FEI Tecnai G2 F20 (GCEMarket, Blackwood, NJ, America) Transmission electron microscopy (TEM) equipped with a HAADF detector. The TEM experiment was carried out in an accelerated voltage environment of 2000 volts. The X-ray photoelectron spectroscopy was obtained by the PHI 5000 (Ulvac-PHI, Inc., Kanagawa, Japan) spectrometer with Kα as the radiation. The obtained binding energy was calibrated with C1s (284.8 eV) the internal reference standard. AutoChem2920 (Micromeritics Instrument Corp, Norcross GA, USA) chemisorption analyzer was used to analyze the reduction performance of fresh catalysts. Before the reduction process, 30 mg prepared catalyst was pretreated 300 °C for 1 h in an N_2 atmosphere (30 mL/min) to remove oxygen.

5. Conclusions

In this paper, a series of Mn-Zr catalysts with different Mn/Zr ratios were successfully prepared via co-precipitation and improved co-precipitation routes, and their catalytic performance for toluene combustion was evaluated. It was found that the TP-Mn2Zr catalyst possesses the lowest T_{50} and T_{90} temperature at 263 and 269 °C and exhibits practical cycle stability. Moreover, the relationship between the catalyst performance and texture was deeply investigated via some characterization techniques, including XRD HRTEM, BET, H_2-TPR, and XPS. The results showed that the doping of Mn enables the crystal structure of ZrO_2 to transform from the monoclinic to the cubic phase, exposes stable c-ZrO_2 (111) phase, and increases the specific surface area of Mn-Zr bimetal catalyst via the co-precipitation strategy. In particular, the optimized co-precipitation process enriches more exposure of the defective (111) surface of the $Mn_xZr_{1-x}O_2$ solid solution active oxygen molecular and more oxygen vacancy as well as the higher concentration Mn^{4+}, which is conducive to the mobility of oxygen to improve their catalytic activity toluene combustion. This new observation will provide a promising strategy to design excellent catalysts for VOCs abatement.

Author Contributions: X.H.: completing the experiments and writing—original manuscript. L. (Luming Li): project administration, providing the experiment ideas, and writing—review. R.L., H L.L. (Li Lan) and W.Z.: analysis and agree to take responsibility for the accuracy and authenticity the research work. All authors have read and agreed to the published version of the manuscript

Funding: This research was supported by the Open Fund of the Key Laboratory of Coarse Cereal Processing, Ministry of Agriculture and Rural Affairs, Sichuan Engineering and Technology Research Center of Coarse Cereal Industrialization (2020CC020).

Data Availability Statement: The data of this research is available within the manuscript.

Acknowledgments: We are grateful to the experimental platform provided by the Institute of Environmental Catalysis and Dust Treatment of Chengdu University. Thanks to the TEM characterization provided from Xiamen University and the XPS analyzed provided by Institute for Advanced Study of Chengdu University.

Conflicts of Interest: The authors declare no conflict of interest.

References

1. Liu, Y.; Dai, H.; Du, Y.; Deng, J.; Zhang, L.; Zhao, Z.; Au, C.T. Controlled preparation and high catalytic performance three-dimensionally ordered macroporous LaMnO3 with nanovoid skeletons for the combustion of toluene. *J. Catal.* **2012**, 149–160. [CrossRef]
2. Li, W.B.; Wang, J.X.; Gong, H. Catalytic combustion of VOCs on non-noble metal catalysts. *Catal. Today* **2009**, *148*, 81–87. [CrossRef]

Saqer, S.M.; Kondarides, D.I.; Verykios, X.E. Catalytic Activity of Supported Platinum and Metal Oxide Catalysts for Toluene Oxidation. *Top. Catal.* **2009**, *52*, 517–527. [CrossRef]

Mustafa, M.F.; Fu, X.; Liu, Y.; Abbas, Y.; Wang, H.; Lu, W. Volatile organic compounds (VOCs) removal in non-thermal plasma double dielectric barrier discharge reactor. *J. Hazard. Mater.* **2018**, *347*, 317–324. [CrossRef]

He, C.; Cheng, J.; Zhang, X.; Douthwaite, M.; Pattisson, S.; Hao, Z. Recent Advances in the Catalytic Oxidation of Volatile Organic Compounds: A Review Based on Pollutant Sorts and Sources. *Chem. Rev.* **2019**, *119*, 4471–4568. [CrossRef]

Ma, X.Y.; Yu, X.L.; Yang, X.Q.; Lin, M.Y.; Ge, M.F. Hydrothermal Synthesis of a Novel Double-Sided Nanobrush Co_3O_4 Catalyst and Its Catalytic Performance for Benzene Oxidation. *ChemCatChem* **2019**, *11*, 1214–1221. [CrossRef]

Yang, X.; Yu, X.; Lin, M.; Ge, M.; Zhao, Y.; Wang, F. Interface effect of mixed phase Pt/ZrO_2 catalysts for HCHO oxidation at ambient temperature. *J. Mater. Chem. A* **2017**, *5*, 13799–13806. [CrossRef]

Li, L.; Chu, W.; Liu, Y. Insights into key parameters of MnO_2 catalyst toward high catalytic combustion performance. *J. Mater. Sci.* **2021**, *56*, 6361–6373. [CrossRef]

Xie, S.; Deng, J.; Liu, Y.; Zhang, Z.; Yang, H.; Jiang, Y.; Arandiyan, H.; Dai, H.; Au, C.T. Excellent catalytic performance, thermal stability, and water resistance of 3DOM Mn_2O_3-supported Au–Pd alloy nanoparticles for the complete oxidation of toluene. *Appl. Catal. A Gen.* **2015**, *507*, 82–90. [CrossRef]

Yao, X.; Yu, Q.; Ji, Z.; Lv, Y.; Cao, Y.; Tang, C.; Gao, F.; Dong, L.; Chen, Y. A comparative study of different doped metal cations on the reduction, adsorption and activity of $CuO/Ce_{0.67}M_{0.33}O_2$ (M = Zr^{4+}, Sn^{4+}, Ti^{4+}) catalysts for NO+CO reaction. *Appl. Catal. B Environ.* **2013**, *130–131*, 293–304. [CrossRef]

Chung, W.-C.; Mei, D.-H.; Tu, X.; Chang, M.-B. Removal of VOCs from gas streams via plasma and catalysis. *Catal. Rev.* **2018**, *61*, 270–331. [CrossRef]

Sihaib, Z.; Puleo, F.; Garcia-Vargas, J.M.; Retailleau, L.; Descorme, C.; Liotta, L.F.; Valverde, J.L.; Gil, S.; Giroir-Fendler, A. Manganese oxide-based catalysts for toluene oxidation. *Appl. Catal. B Environ.* **2017**, *209*, 689–700. [CrossRef]

Craciun, R.; Nentwick, B.; Hadjiivanov, K.; Knözinger, H. Structure and redox properties of MnO_x/Yttrium-stabilized zirconia (YSZ) catalyst and its used in CO and CH_4 oxidation. *Appl. Catal. A Gen.* **2003**, *243*, 67–79. [CrossRef]

Kim, S.C.; Shim, W.G. Catalytic combustion of VOCs over a series of manganese oxide catalysts. *Appl. Catal. B Environ.* **2010**, *98*, 180–185. [CrossRef]

Lin, X.; Li, S.; He, H.; Wu, Z.; Wu, J.; Chen, L.; Ye, D.; Fu, M. Evolution of oxygen vacancies in MnO_x-CeO_2 mixed oxides for soot oxidation. *Appl. Catal. B Environ.* **2018**, *223*, 91–102. [CrossRef]

Xie, S.; Liu, Y.; Deng, J.; Zhao, X.; Yang, J.; Zhang, K.; Han, Z.; Dai, H. Three-dimensionally ordered macroporous CeO_2-supported Pd@Co nanoparticles: Highly active catalysts for methane oxidation. *J. Catal.* **2016**, *342*, 17–26. [CrossRef]

Afzal, S.; Quan, X.; Zhang, J. High surface area mesoporous nanocast $LaMO_3$ (M = Mn, Fe) perovskites for efficient catalytic ozonation and an insight into probable catalytic mechanism. *Appl. Catal. B Environ.* **2017**, *206*, 692–703. [CrossRef]

Zhao, Q.; Fu, L.; Jiang, D.; Ouyang, J.; Hu, Y.; Yang, H.; Xi, Y. Nanoclay-modulated oxygen vacancies of metal oxide. *Commun. Chem.* **2019**, *2*, 11. [CrossRef]

Santos, V.P.; Pereira, M.F.R.; Órfão, J.J.M.; Figueiredo, J.L. The role of lattice oxygen on the activity of manganese oxides towards the oxidation of volatile organic compounds. *Appl. Catal. B Environ.* **2010**, *99*, 353–363. [CrossRef]

Tang, X.; Li, Y.; Huang, X.; Xu, Y.; Zhu, H.; Wang, J.; Shen, W. MnO_x–CeO_2 mixed oxide catalysts for complete oxidation of formaldehyde: Effect of preparation method and calcination temperature. *Appl. Catal. B Environ.* **2006**, *62*, 265–273. [CrossRef]

Tang, W.; Wu, X.; Li, S.; Li, W.; Chen, Y. Porous Mn–Co mixed oxide nanorod as a novel catalyst with enhanced catalytic activity for removal of VOCs. *Catal. Commun.* **2014**, *56*, 134–138. [CrossRef]

Morales, M.; Barbero, B.; Cadus, L. Total oxidation of ethanol and propane over Mn-Cu mixed oxide catalysts. *Appl. Catal. B Environ.* **2006**, *67*, 229–236. [CrossRef]

Li, H.; Qi, G.; Tana; Zhang, X.; Huang, X.; Li, W.; Shen, W. Low-temperature oxidation of ethanol over a $Mn_{0.6}Ce_{0.4}O_2$ mixed oxide. *Appl. Catal. B Environ.* **2011**, *103*, 54–61. [CrossRef]

Gutierrez-Ortiz, J.I.; de Rivas, B.; Lopez-Fonseca, R.; Martin, S.; Gonzalez-Velasco, J.R. Structure of Mn-Zr mixed oxides catalysts and their catalytic performance in the gas-phase oxidation of chlorocarbons. *Chemosphere* **2007**, *68*, 1004–1012. [CrossRef] [PubMed]

Fernández López, E.; Sánchez Escribano, V.; Resini, C.; Gallardo-Amores, J.M.; Busca, G. A study of coprecipitated Mn–Zr oxides and their behaviour as oxidation catalysts. *Appl. Catal. B Environ.* **2001**, *29*, 251–261. [CrossRef]

Bulavchenko, O.A.; Vinokurov, Z.S.; Afonasenko, T.N.; Tsyrul'nikov, P.G.; Tsybulya, S.V.; Saraev, A.A.; Kaichev, V.V. Reduction of mixed Mn-Zr oxides: In situ XPS and XRD studies. *Dalton Trans.* **2015**, *44*, 15499–15507. [CrossRef]

Zhao, Q.; Shih, W.Y.; Chang, H.L.; Shih, W.H. Redox Activity and NO Storage Capacity of MnO_x−ZrO_2 with Enhanced Thermal Stability at Elevated Temperatures. *Ind. Eng. Chem. Res.* **2010**, *49*, 1725–1731. [CrossRef]

Zuo, J.; Chen, Z.; Wang, F.; Yu, Y.; Wang, L.; Li, X. Low-Temperature Selective Catalytic Reduction of NO_x with NH_3 over Novel Mn-Zr Mixed Oxide Catalysts. *Ind. Eng. Chem. Res.* **2014**, *53*, 2647–2655. [CrossRef]

Choudhary, V.R.; Uphade, B.S.; Pataskar, S.G. Low temperature complete combustion of dilute methane over Mn-doped ZrO_2 catalysts: Factors influencing the reactivity of lattice oxygen and methane combustion activity of the catalyst. *Appl. Catal. A Gen.* **2002**, *227*, 29–41. [CrossRef]

30. Zeng, K.; Li, X.; Wang, C.; Wang, Z.; Guo, P.; Yu, J.; Zhang, C.; Zhao, X.S. Three-dimensionally macroporous MnZrOx catalysts propane combustion: Synergistic structure and doping effects on physicochemical and catalytic properties. *J. Colloid Interface* **2020**, *572*, 281–296. [CrossRef] [PubMed]
31. Yang, X.; Yu, X.; Jing, M.; Song, W.; Liu, J.; Ge, M. Defective $Mn_xZr_{1-x}O_2$ Solid Solution for the Catalytic Oxidation of Toluene Insights into the Oxygen Vacancy Contribution. *ACS Appl. Mater. Interfaces* **2019**, *11*, 730–739. [CrossRef]
32. Fan, X.; Li, L.; Jing, F.; Li, J.; Chu, W. Effects of preparation methods on $CoAlO_x/CeO_2$ catalysts for methane catalytic combustion. *Fuel* **2018**, *225*, 588–595. [CrossRef]
33. Li, L.; Luo, J.; Liu, Y.; Jing, F.; Su, D.; Chu, W. Self-Propagated Flaming Synthesis of Highly Active Layered Ce-delta-MnO_2 Hybrid Composites for Catalytic Total Oxidation of Toluene Pollutant. *ACS Appl. Mater. Interfaces* **2017**, 21798–21808. [CrossRef] [PubMed]
34. Lan, L.; Li, H.; Chen, S.; Chen, Y. Preparation of CeO_2–ZrO_2–Al_2O_3 composite with layered structure for improved Pd-only three-way catalyst. *J. Mater. Sci.* **2017**, *52*, 9615–9629. [CrossRef]
35. Hou, Z.; Feng, J.; Lin, T.; Zhang, H.; Zhou, X.; Chen, Y. The performance of manganese-based catalysts with $Ce_{0.65}Zr_{0.35}O_2$ support for catalytic oxidation of toluene. *Appl. Surf. Sci.* **2018**, *434*, 82–90. [CrossRef]
36. Azalim, S.; Franco, M.; Brahmi, R.; Giraudon, J.M.; Lamonier, J.F. Removal of oxygenated volatile organic compounds by catalytic oxidation over Zr-Ce-Mn catalysts. *J. Hazard. Mater.* **2011**, *188*, 422–427. [CrossRef] [PubMed]
37. Yang, X.Q.; Ma, X.Y.; Yu, X.L.; Ge, M.F. Exploration of strong metal-support interaction in zirconia supported catalysts for toluene oxidation. *Appl. Catal. B Environ.* **2020**, *263*, 118355. [CrossRef]
38. Han, Z.; Yu, Q.; Teng, Z.; Wu, B.; Xue, Z.; Qin, Q. Effects of manganese content and calcination temperature on Mn/Zr-PILC catalyst for low-temperature selective catalytic reduction of NO_x by NH_3 in metallurgical sintering flue gas. *Environ. Sci. Pollut. Res. Int.* **2019**, *26*, 12920–12927. [CrossRef]
39. Huang, N.; Qu, Z.; Dong, C.; Qin, Y.; Duan, X. Superior performance of $\alpha@\beta$-MnO_2 for the toluene oxidation: Active interface and oxygen vacancy. *Appl. Catal. A Gen.* **2018**, *560*, 195–205. [CrossRef]
40. Mitran, G.; Chen, S.; Seo, D.-K. Role of oxygen vacancies and Mn^{4+}/Mn^{3+} ratio in oxidation and dry reforming over cobalt-manganese spinel oxides. *Mol. Catal.* **2020**, *483*, 110704. [CrossRef]
41. Wang, W.L.; Meng, Q.; Xue, Y.; Weng, X.; Sun, P.; Wu, Z. Lanthanide perovskite catalysts for oxidation of chloroaromatics Secondary pollution and modifications. *J. Catal.* **2018**, *366*, 213–222. [CrossRef]
42. Tang, W.; Li, W.; Li, D.; Liu, G.; Wu, X.; Chen, Y. Synergistic Effects in Porous Mn–Co Mixed Oxide Nanorods Enhance Catalytic Deep Oxidation of Benzene. *Catal. Lett.* **2014**, *144*, 1900–1910. [CrossRef]
43. Zhu, L.; Li, X.; Liu, Z.; Yao, L.; Yu, P.; Wei, P.; Xu, Y.; Jiang, X. High Catalytic Performance of Mn-Doped Ce-Zr Catalysts Chlorobenzene Elimination. *Nanomaterials* **2019**, *9*, 675. [CrossRef]
44. Dravid, V.P.; Ravikumar, V.; Notis, M.R.; Lyman, C.E.; Dhalenne, G.; Revcolevschi, A. Stabilization of Cubic Zirconia with Manganese Oxide. *J. Am. Ceram. Soc.* **1994**, *77*, 2758–2762. [CrossRef]
45. Jiang, H.; Wang, C.; Wang, H.; Zhang, M. Synthesis of highly efficient MnO_x catalyst for low-temperature NH_3-SCR prepared from Mn-MOF-74 template. *Mater. Lett.* **2016**, *168*, 17–19. [CrossRef]
46. Cuervo, M.R.; Diaz, E.; de Rivas, B.; Lopez-Fonseca, R.; Ordonez, S.; Gutierrez-Ortiz, J.I. Inverse gas chromatography a technique for the characterization of the performance of Mn/Zr mixed oxides as combustion catalysts. *J. Chromatogr. A* **2009**, *1216*, 7873–7881. [CrossRef] [PubMed]
47. Qin, Y.; Wang, H.; Dong, C.; Qu, Z. Evolution and enhancement of the oxygen cycle in the catalytic performance of total toluene oxidation over manganese-based catalysts. *J. Catal.* **2019**, *380*, 21–31. [CrossRef]
48. Li, L.; Wahab, M.A.; Li, H.; Zhang, H.; Deng, J.; Zhai, X.; Masud, M.K.; Hossain, M.S.A. Pt-Modulated $CuMnO_x$ Nanosheets Catalysts for Toluene Oxidation. *ACS Appl. Nano Mater.* **2021**, *4*, 6637–6647. [CrossRef]
49. Mo, S.P.; Zhang, Q.; Li, J.Q.; Sun, Y.H.; Ren, Q.M.; Zou, S.B.; Zhang, Q.; Lu, J.H.; Fu, M.L.; Mo, D.Q.; et al. Highly efficient mesoporous MnO_2 catalysts for the total toluene oxidation: Oxygen-Vacancy defect engineering and involved intermediates using in situ DRIFTS. *Appl. Catal. B Environ.* **2020**, *264*, 118464. [CrossRef]
50. Zhou, C.; Zhang, H.; Zhang, Z.; Li, L. Improved reactivity for toluene oxidation on MnO_x/CeO_2-ZrO_2 catalyst by the synthesis cubic-tetragonal interfaces. *Appl. Surf. Sci.* **2021**, *539*, 148188. [CrossRef]

VOCs Photothermo-Catalytic Removal on MnO$_x$-ZrO$_2$ Catalysts

Roberto Fiorenza *, Roberta Agata Farina, Enrica Maria Malannata, Francesca Lo Presti and Stefano Andrea Balsamo

Dipartimento di Scienze Chimiche, Università di Catania, Viale A. Doria 6, 95125 Catania, Italy; roberta.agata.farina@gmail.com (R.A.F.); enrica.malannata@phd.unict.it (E.M.M.); francesca.lopresti@phd.unict.it (F.L.P.); stefano.balsamo@phd.unict.it (S.A.B.)
* Correspondence: rfiorenza@unict.it; Tel.: +39-0957385012

Abstract: Solar photothermo-catalysis is a fascinating multi-catalytic approach for volatile organic compounds (VOCs) removal. In this work, we have explored the performance and the chemico-physical features of non-critical, noble, metal-free MnO$_x$-ZrO$_2$ mixed oxides. The structural, morphological, and optical characterizations of these materials pointed to as a low amount of ZrO$_2$ favoured a good interaction and the ionic exchange between the Mn and the Zr ions. This favoured the redox properties of MnO$_x$ increasing the mobility of its oxygens that can participate in the VOCs oxidation through a Mars-van Krevelen mechanism. The further application of solar irradiation sped up the oxidation reactions promoting the VOCs total oxidation to CO$_2$. The MnO$_x$-5 wt.%ZrO$_2$ sample showed, in the photothermo-catalytic tests, a toluene T$_{90}$ (temperature of 90% of conversion) of 180 °C and an ethanol T$_{90}$ conversion to CO$_2$ of 156 °C, 36 °C, and 205 °C lower compared to the thermocatalytic tests, respectively. Finally, the same sample exhibited 84% toluene conversion and the best selectivity to CO$_2$ in the ethanol removal after 5 h of solar irradiation at room temperature, a photoactivity similar to the most employed TiO$_2$-based materials. The as-synthetized mixed oxide is promising for an improved sustainability in both catalyst design and environmental applications.

Keywords: VOC; photothermo catalysis; toluene; ethanol; manganese oxide; zirconium oxide

1. Introduction

Nowadays, the quality of air, both in indoor and outdoor environments, is an extremely important concern. Furthermore, the COVID-19 emergency has pointed to the necessity of clean air to discourage virus infection. Among the air pollutants, volatile organic compounds (VOCs) include many of the most dangerous substances for both human health and the environment. Different strategies were employed to remove VOCs from the air, and an innovative and sustainable solution is represented by solar photocatalytic or photothermo-catalytic oxidation [1,2]. Compared to the most used catalytic or non-catalytic VOCs combustion, the photocatalytic process allows one to exploit solar irradiation with green advantages to work at milder conditions using renewable energy [2]. However, the performance of the photocatalysts is much lower compared to the catalysts employed for the thermocatalytic removal of VOCs [3], and for these reasons, the multi-catalytic approach of the photothermo-catalysis is a fascinating way to obtain high VOCs removal values of thermocatalysis but at lower temperatures, increasing, at the same time, the energy efficiency of the process. To design performing photothermo catalysts, different properties are required [4,5]. Analogously to photocatalysis, it is necessary to have a semiconductor material that, after solar irradiation, is able to generate photoelectrons and photoholes in its conduction (CB) and valence (VB) bands, respectively. It should have redox properties activated with the temperature; in this way, the superficial/mobile oxygens of the catalyst or of the support can participate in the oxidation of VOCs increasing the overall activity [3,6]. Finally, the photothermo catalysts should be resistant to both long-time irradiation and heating. The preparation of mixed oxides or composites is the best and easiest way to combine all of these features. Indeed, with the formation c suitable heterojunction, it is possible to exploit solar irradiation, decreasing the bandg (E_g) of the main semiconductor oxide, profiting from both the photocatalytic activity the principal oxides and the thermocatalytic activity of the hosted oxide. Moreover, introduction of host ions in the lattice of the main oxide allows one to create defects a oxygen vacancies that favour VOCs oxidation [7,8].

The TiO_2-CeO_2 composites showed promising performance in the photo-therr approach for both VOCs removal and CO_2 reduction [3,9]; however, one of the s effects of the current pandemic situation is the crisis of raw materials exportation, a as a consequence, in 2020, titanium featured in the EU critical raw materials list [Considering that up until now, TiO_2 is the most studied and applied semiconductor, both academia and in industrial research focused on photocatalytic applications, the explorat of unconventional non-critical (photo)catalysts is highly required.

In this work, we have investigated the photothermo-catalytic properties of Mn ZrO_2 mixed oxides, with the aim of finding new and sustainable alternatives to the m common TiO_2-based photocatalysts, and without the addition of noble metal co-cataly usually used in the catalytic and photocatalytic removal of VOCs [11], to obtain even m environmentally friendly catalysts, in the end.

Manganese oxide exists in four stable forms (MnO, MnO_2, Mn_3O_4 and Mn_2O_3), a all of them own a semiconductor electronic structure characterized by the partially fille orbitals which permit the electronic d-d transitions under UV or visible light irradiation [Based on the preparation method, it is common to obtain a non-stoichiometric oxide c mixture of different MnO_x oxides with the +II, +III and/or +IV oxidation states. The h mobility/reducibility of manganese oxide lattice oxygens is particularly useful for VC removal [13,14], whereas the redox properties of MnO_x can be particularly advantaged for the photothermo-catalytic oxidation of VOCs, as well as its low bandgap (in the rar 2.0–3.5 eV depending of the crystalline structure [12,15]) that can allow a more efficient of solar radiation.

Zirconium oxide (zirconia) was largely used as a support of several noble me based catalysts used for the thermocatalytic oxidation of VOCs, due to its high stabil thermal resistance, and ionic conductivity [16,17]. Furthermore, it is a large bandg semiconductor (E_g of about 5.0 eV [18] or lower depending to the zirconia synthes Therefore, its coupling with a lower bandgap semiconductor (as MnO_x) can be a perform and fascinating strategy to reduce the odds of charge recombination (a common reason photocatalysts deactivation) and to synergistically exploit both the thermal stability a the redox properties of MnO_x and ZrO_2 [19,20] together with their photocatalytic featu

We have also determined the chemico-physical and the photocatalytic, thermocataly and photothermo-catalytic activities of MnO_x-ZrO_2 oxides in the oxidation of toluene a ethanol, chosen as VOCs models, due to the high toxicity nature of toluene and to wide use of ethanol as a solvent in many industrial processes and as an octane boos in combustion engines, whose incomplete oxidation can give the emission of dangerc compounds, as acetaldehyde, in the environment [21].

2. Results

2.1. Structural, Morphological, Textural and Optical Properties of the Samples

The XRD patterns of the analysed samples are shown in Figure 1. The precipitation manganese chloride (II) with NaOH and the employed calcination temperature (600 °C 2 h) allowed to obtain the Mn_3O_4. The signals at 2θ = 18.1°, 28.9°, 31.0°, 32.4°, 36.0°, 38 44.3° and 50.8° are, indeed, in accordance with the PDF card. No.: 00-080-0382 of pu Mn_3O_4 (Hausmannite). Bare ZrO_2 was obtained with the ammonia-driven precipitati of zirconyl nitrate. The signals fitted with the PDF card No. 00-079-1771 of zirconiu oxide, with the typical diffraction peaks at 2θ = 30.2°, 35.2° and 50.3°. Interestingly, co-precipitation with NaOH of both metals salt precursors created substantial changes the crystalline structure of manganese oxide. The addition of 5 wt.% of zirconium oxi led to a mixed Mn_2O_3/Mn_3O_4 phase being present the diffraction peak at $2\theta = 32.9°$; that is, the typical fingerprint of Mn_2O_3 (PDF card No. 00-071-0636, [12,22]), together with the signals at $2\theta = 38.2°$ (overlapped with the signal of Mn_3O_4) and 55.1° that are also ascribed to manganese (III) oxide [12,22]. The increase of ZrO_2 content (MnO_x-10%ZrO_2 sample) restored the main presence of Mn_3O_4 with a trace of Mn_2O_3. In both the mixed oxides, the signals related to ZrO_2 are absent, probably due to the low amount of hosted oxide and/or to the good dispersion of zirconium oxide on manganese oxide.

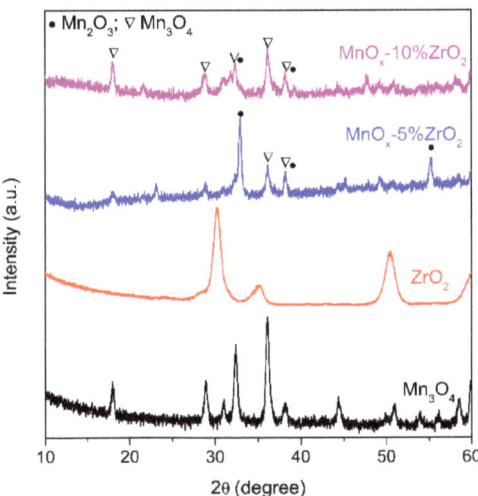

Figure 1. X-ray Diffraction (XRD) patterns of the examined samples.

The ion radius of Zr^{4+} (0.84 Å) is similar of Mn^{2+} (0.83 Å), and this can favour the ionic exchange between these cations [20,23]. On the contrary, the smaller radius of Mn^{3+} (0.64 Å) makes the Zr^{4+}/Mn^{3+} exchange more difficult. Probably, when the amount of ZrO_2 is low, the Zr ions partially replace the Mn^{2+} promoting, in this way, the main presence of Mn^{3+}, whereas a higher amount of ZrO_2 led to a preferential surface covering of the MnO_x instead of a lattice incorporation of the zirconium ions in MnO_x [20,23]. Thus, it can explain the major presence of Mn (III) on MnO_x-5%ZrO_2 sample, and the coexistence of Mn II and III in the MnO_x-10%ZrO_2. The main crystalline size of the samples (Table 1) was determined by applying the Scherrer formula on the principal diffraction peaks of the oxides ($2\theta = 36.0°$ for Mn_3O_4, 30.2° for ZrO_2, 32.9° for MnO_x-5% ZrO_2, whereas for MnO_x-10% ZrO_2, the value was mediated considering both the $2\theta = 32.4°$ and 36.0° signals). The addition of zirconium oxide led to a slight increase of the crystalline size of manganese oxide, whereas the bare ZrO_2 showed the lowest crystalline size (8 nm). However, this latter oxide, in accordance with the surface area values reported in the literature [18], showed the lowest surface area (Table 1, 26.2 m^2/g), whereas the bare Mn_3O_4 exhibited the highest BET surface area (99.6 m^2/g). Compared to the bare Mn_3O_4, the slight increase of the crystallite size of the mixed oxides determined a decrease of their surface area, which were about 85–86 m^2/g for both the MnO_x-ZrO_2 samples (Table 1).

Table 1. Structural, textural and optical properties of the examined samples.

Sample	Crystallite Size (nm) [a]	BET Surface Area (m^2/g)	E$_g$ (eV)
Mn$_3$O$_4$	14.5	99.6	3.29
ZrO$_2$	8.1	26.2	3.02
MnO$_x$-5%ZrO$_2$	17.9	85.4	3.26
MnO$_x$-10%ZrO$_2$	18.2	86.1	3.27

[a] Estimated by XRD.

The SEM-EDX measurements (Figure 2) were performed to evaluate the presence of zirconium oxide on MnO$_x$. The adopted precipitation methods led to, indifferent to the investigated samples, a non-homogenous morphology with spherical particles (Figures 2a and S1). From the EDX elemental analysis (Figure 2b,c, Table 2), it is possible to note that a little surface segregation of zirconium in the MnO$_x$-10%ZrO$_2$ sample was detected, whose zirconium wt.% was 3.7 times higher (instead of twice as expected considering the nominal concentration) compared to the MnO$_x$-5%ZrO$_2$. In accordance with the XRD data, the increase of the amount of ZrO$_2$ led to an enrichment of zirconium oxide on the surface of MnO$_x$, whereas in the MnO$_x$-5%ZrO$_2$ mixed oxide, the ZrO$_2$ was mainly embedded in the lattice of MnO$_x$.

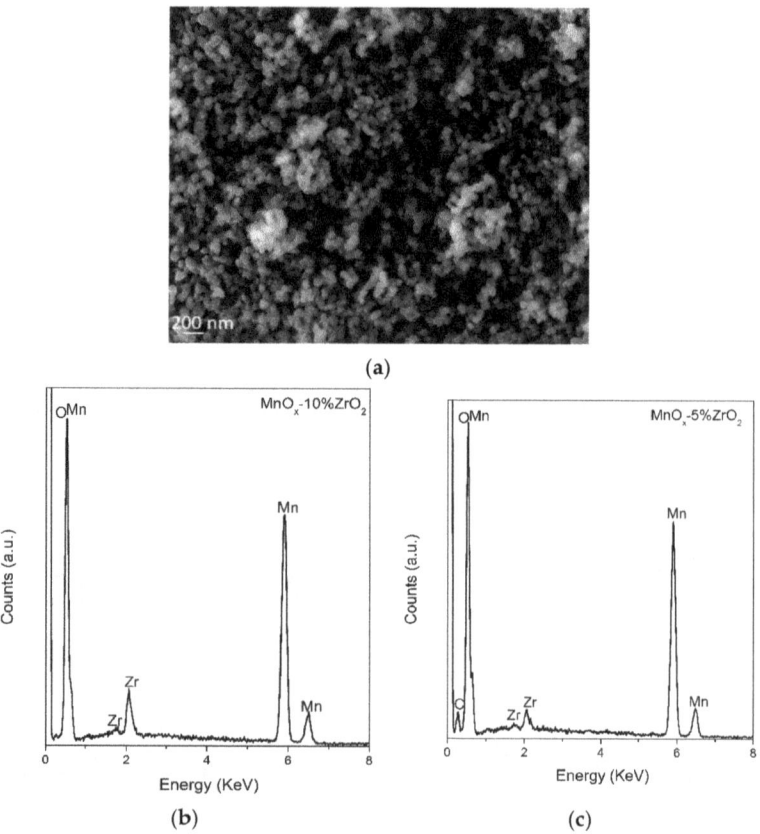

Figure 2. (a) SEM image of the MnO$_x$-5%ZrO$_2$ as representative sample; Energy Dispersive X-ray (EDX) spectra of MnO$_x$-10%ZrO$_2$ (b) and MnO$_x$-5%ZrO$_2$ (c). The EDX spectra were mediated considering four different zones of the samples.

Table 2. EDX elemental analysis of the examined samples. The presence of carbon was due to the carbon tape used to perform the measurements.

Sample	Element	wt.%
MnO$_x$-5%ZrO$_2$	C	1.71
	O	28.32
	Mn	67.51
	Zr	2.46
MnO$_x$-10%ZrO$_2$	C	1.23
	O	22.0
	Mn	67.66
	Zr	9.11

The surface valence state of the components of the catalysts were analysed through X-ray photoelectron spectroscopy (XPS) (Table 3). Interestingly, on the surface of MnO$_x$-based samples, the ratio of the Mn^{3+}/Mn^{2+} ions obtained considering the area of the deconvoluted spectra (see Figure S2, spectra of MnO$_x$-5%ZrO$_2$ as representative sample) was the highest for the MnO$_x$-5%ZrO$_2$ mixed oxide. As pointed to also by the structural properties determined by XRD, in this sample, a strong ionic interaction between the Mn^{2+} and the Zn^{4+} was particularly favoured, leading to an increase in the amount of Mn^{3+} ions. Moreover, in this sample, the ratio between the surface lattice oxygen (O$_\alpha$) located at about 530 eV and the chemisorbed/defective oxygen (O$_\beta$) at 532 eV was also the highest (Table 3, Figure S2), suggesting that the ionic exchange between the zirconium and the manganese ions also promoted a higher concentration of the manganese oxide surface oxygens. These, as reported, can participate in VOCs oxidation, improving the catalytic activity of the catalysts [24]. Finally, the binding energy of the Zr 3d$_{5/2}$ at about 182.0 eV is the typical fingerprint of ZrO$_2$ [25]. The surface atomic percentage of Zr was 3.5 higher (2.77%) on the MnO$_x$-10%ZrO$_2$ compared to MnO$_x$-5%ZrO$_2$ (0.73%) confirming, as too stated by the EDX analysis, the surface covering of zirconia on manganese oxide, verified increasing the amount of ZrO$_2$.

Table 3. XPS analysis and binding energy (BE in eV) of the components of the investigated samples.

Sample	Mn 2p$_{3/2}$ BE	Mn^{3+}/Mn^{2+} Ratio	Zr 3d$_{5/2}$ BE	O 1s BE	O$_\alpha$/O$_\beta$ Ratio
Mn$_3$O$_4$	641.2	0.52		529.8	1.50
ZrO$_2$	/	/	182.0	529.9	1.48
MnO$_x$-5%ZrO$_2$	640.9	0.69	181.9	529.9	1.69
MnO$_x$-10%ZrO$_2$	641.1	0.55	182.0	530.0	1.53

The UV-DRS of the samples were reported in the Figure 3. The MnO$_x$-based samples (Figure 3a) showed a remarkable lower reflectance compared to the bare ZrO$_2$ (Figure 3b), and thus can be highlighted considering also the colours of the as-synthesized powders (dark brown for the MnO$_x$-based materials and white for the zirconium oxide). The optical bandgaps of the semiconductor oxides were estimated plotting the modified Kubelka–Munk function versus hν, as reported in the literature ([26], inset Figure 3b as representative sample). Interestingly, as established by XRD, the good crystallinity of ZrO$_2$ and its nano-size (8 nm, Table 1) allowed us to obtain a ZrO$_2$ with a lower bandgap (3.02 eV) compared to the other E$_g$ reported in the literature for this oxide (about 5.0 eV that, however, can be narrowed down to 2–1.5 eV on the basis of the adopted preparation method [18,27]). No substantial variations were observed comparing the unmodified Mn$_3$O$_4$ and the MnO$_x$-ZrO$_2$ based-oxides, with an E$_g$ of 3.26–3.29 eV (Table 1). Probably, the low amount and the good dispersion of ZrO$_2$ on MnO$_x$ did not alter the bandgap of the manganese oxide. All the manganese oxide-based samples exhibited a similar E$_g$ to TiO$_2$ (3.0–3.2 eV on the basis of the crystalline form [28]); thus, they can exploit the UV-A portion of solar irradiation.

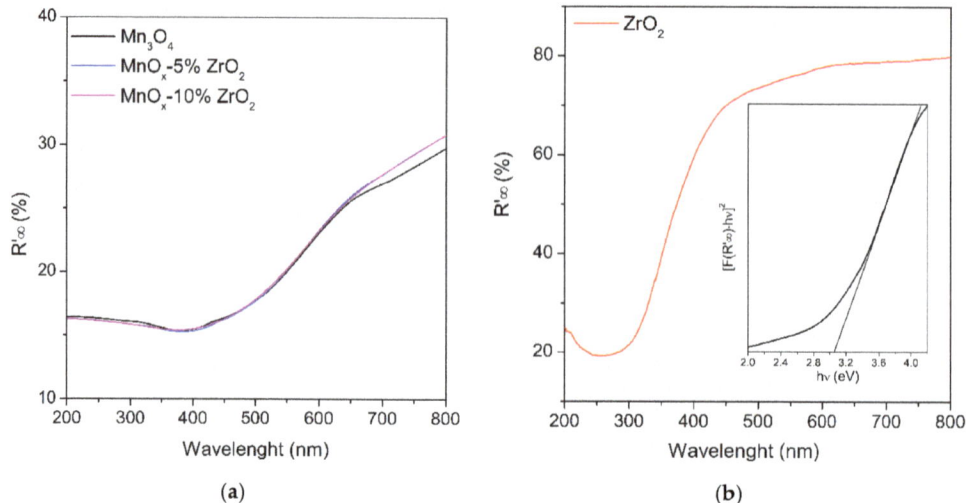

Figure 3. (a) UV-DRS (Diffuse Reflectance Spectroscopy) of the MnOx-based samples; (b) UV-D spectra of bare ZrO_2. In the inset; the estimation of the optical bandgap through the modifi Kubelka–Munk function.

2.2. Photocatalytic, Thermocatalytic and Photothermo-Catalytic Removal of Toluene in Gas Pha

Figure 4a shows the solar photocatalytic activity in the oxidation of toluene at roo temperature after 5 h of irradiation. The highest conversion value was obtained with MnOx-5% ZrO_2 (84%) followed by the MnOx-10% ZrO_2 and the bare Mn_3O_4 (51% a 47%, respectively), whereas pure ZrO_2 exhibited the lowest conversion value (33%). accordance with the literature [3,29], in our experimental condition, the only detec by-products were carbon dioxide and water with traces of benzaldehyde (selectivity in range 1–3%). Although a real comparison with the other reported data for this reaction very difficult, due to the various experimental conditions adopted by the other resear groups (Table 4), the performance of MnOx-5% ZrO_2 mixed oxide is very promising, be similar to (considering the initial concentration of 1000 ppm of toluene) or slightly low than the most used TiO_2-based photocatalysts, or to other unconventional semiconduct (Table 4).

Table 4. Data comparison of the photocatalytic oxidation of toluene.

Catalysts	Experimental Conditions	Toluene Conversion	Ref.
MnOx-5%ZrO2	1000 ppm Toluene, 5 h irradiation solar lamp (300 W, 10.7 mW/cm²), room T, 150 mg catalyst	84%	this work
Brookite TiO2-5% CeO2	1000 ppm Toluene, 2 h irradiation solar lamp (300 W, 10.7 mW/cm²), room T, 150 mg catalyst	25%	[3]
TiO2-C3N4	665 ppm Toluene, 6 h irradiation, solar lamp (300 W, 612 mW/cm²), 100 mg catalyst	93%	[30]
TiO2-MnO2	200 ppmv Toluene, 1 h irradiation, 25 LEDs (λmax = 465 nm)	43%	[31]
0.5% Co/TiO2	150 ppmv Toluene, 140 min irradiation, solar light (1000 mW/cm²), 25 °C	96.5%	[32]
Ag4Bi2O5	220 ppm Toluene, 60 min irradiaton, Xe lamp with a 420 nm cut off filter (300 W, 0.25 mW/cm²), 50 mg catalyst	93.1%	[33]
Fe2O3/In2O3	200 ppm Toluene, 8 h irradiation, Xenon lamp with an optical UV-cutoff filter (500 W, 40 mW/cm²)	88.3%	[34]

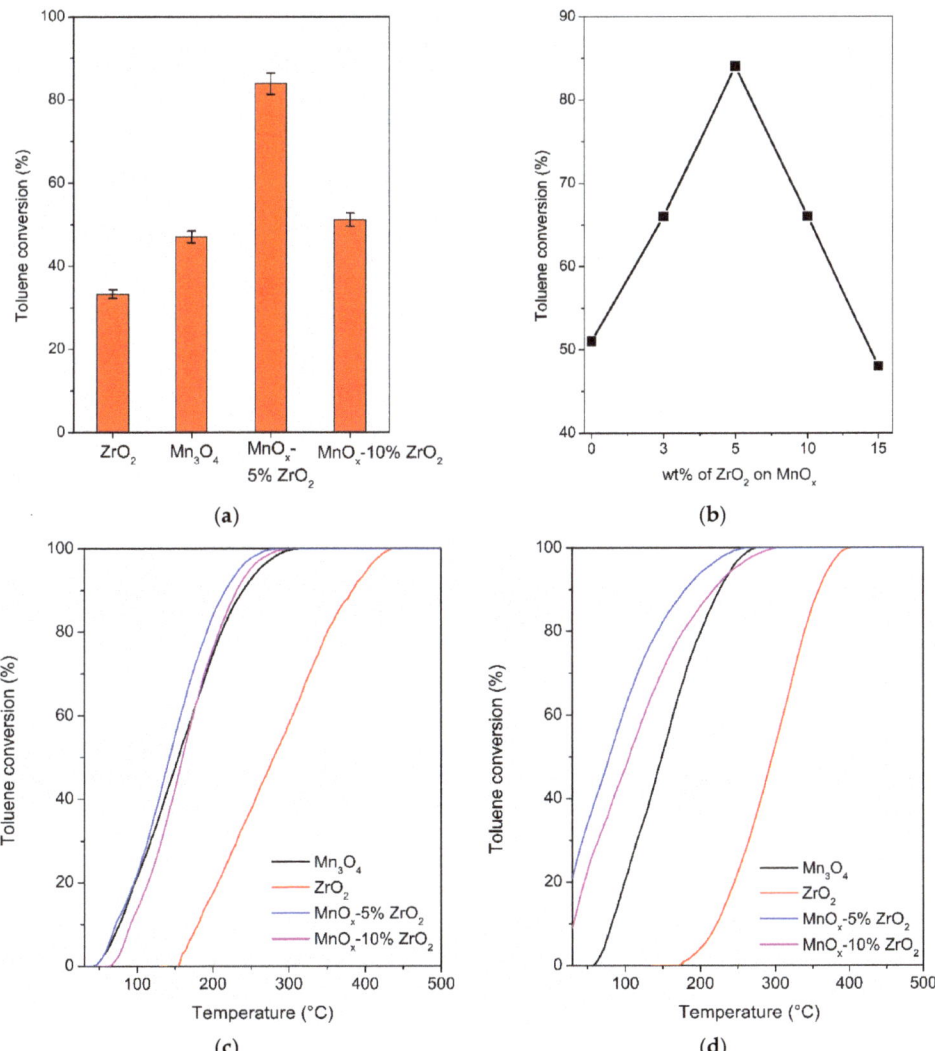

Figure 4. (**a**) Solar photocatalytic oxidation of toluene after 5h of irradiation; (**b**) effect of the wt.% of ZrO_2 on MnO_x in the solar photocatalytic toluene conversion; (**c**) thermocatalytic oxidation of toluene; (**d**) photothermo-catalytic oxidation of toluene on the investigated materials.

It is worth noting that 5 wt.% of zirconia was the best amount to obtain a synergistic positive effect on the MnO_x. Indeed, the samples prepared with the same procedures reported in the par. 4.1 but adding the 3 wt.% and 15 wt.% of zirconium oxide caused a decrease of activity (66% of toluene conversion for MnO_x-3% ZrO_2 and 48 % for MnO_x-15% ZrO_2, i.e., the same conversion value of the bare manganese oxide). The results pointed to a photocatalytic "volcano" trend (Figure 4b). The positive effects of the addition of ZrO_2 on MnO_x reached the maximum with 5% of zirconia, following a progressive decrease at higher amounts. This can be reasonably due, as confirmed by XRD, SEM-EDX and XPS measurements, to a progressive surface coverage of MnO_x, due to the presence of a large amount of ZrO_2. This caused a decrease in the photoactivity considering also the lower photocatalytic performance of bare ZrO_2 compared to manganese oxide (Figure 4a). The detection of a specific amount of the hosted oxide on the main oxide is a typical tre of the mixed oxide-based semiconductors. A large amount of the second compone (hosted oxide) can cover the surface active sites of the main oxide, decreasing the over photocatalytic activity of the photocatalyst [35,36].

Thermal catalytic combustion is the most used process to increase the removal e ciency of toluene. The thermocatalytic activity of the investigated samples is reported the Figure 4c. Moreover, for this catalytic approach, MnO_x-5% ZrO_2 gave the best resu The T_{90} (the temperature at which the 90% of toluene conversion was reached) values we 216 °C, 231 °C, 240 °C and 383 °C for MnO_x-5% ZrO_2, MnO_x-10% ZrO_2, Mn_3O_4, and t bare ZrO_2 respectively, confirming the order of activity measured in the photocatalytic te at room temperature.

To further decrease toluene T_{90}, the solar photothermo-catalytic tests were employ on the same investigated samples (Figure 4d). The solar-assisted thermo catalytic approa allowed one to decrease the T_{90} of 36 °C (180 °C) with respect to thermocatalytic tests on t best mixed oxide, the MnO_x-5% ZrO_2 catalyst, and in general, a decrease of T_{90} was verifi for all the catalysts, with even the same order of activity: MnO_x-5% ZrO_2 > MnO_x-10 ZrO_2 (T_{90} = 217 °C) > Mn_3O_4 (T_{90} = 226 °C) > ZnO_2 (T_{90} = 245 °C). Interestingly, the high T_{90} decrease was verified with the bare zirconium oxide (138 °C lower compared to t thermocatalytic toluene T_{90}) where the activation of the zirconia photocatalytic propert was fundamental to promote the toluene total oxidation.

The positive synergistic effect due to the addition of a small amount of zircor on the MnO_x and the solar multi-catalytic approach led to obtaining a low toluene T considering the absence of noble metals co-catalysts. The obtained value of T_{90} with t MnO_x-5% ZrO_2 sample (180 °C) is comparable or lower with respect to the other MnC based catalysts reported in the literature (in the range 200 °C–270 °C considering an init toluene concentration of 1000 ppm [20,37]).

The influence of the gas hourly space velocity (GSHV) was reported in the Figure S considering the best sample (MnO_x-5% ZrO_2). We have chosen, for all the tests, a GSH of 8×10^4 mL/g_{cat}·h, indeed, as expected, and as reported in the literature [38], with high flow rate; the conversion rate of toluene to CO_2 and water (the only by-produ detected also in all the thermo and photothermo-catalytic tests) was slower, wherea GSHV < 8×10^4 mL/g_{cat}·h did not substantially modify the conversion rate.

2.3. Photocatalytic, Thermocatalytic and Photothermo-Catalytic Removal of Ethanol in Gas Pha

The ethanol being an alcohol was more reactive than the aromatic toluene, but oxidation can give various by-products; the most common in the gas phase oxidation w acetaldehyde [3,31,39], which is also the main by-product detected in all the investigat catalytic approaches here discussed, whereas a very low selectivity (<2%) was detected CO, formic acid, and acetic acid. In the solar photocatalytic tests, MnO_x-5% ZrO_2 confirm its highest activity compared to the other samples (Figure 5) with an ethanol conversi of 98% and the highest selectivity to CO_2 (43%), the most important feature for the VO removal. The mixed oxide with the 10 wt.% of ZrO_2 showed a little decrease of photoactiv (ethanol conversion of 86%) and a higher selectivity to acetaldehyde (60%) with respect CO_2 (36%). These data, in line with the photo-oxidation of toluene (Figure 4a), pointed in our experimental conditions, the 5 wt.% being the optimal amount of zirconia to ha a synergistic effect with MnO_x. Among the bare oxides, the manganese oxide showec higher ethanol conversion, with a higher selectivity to CO_2 compared to ZrO_2. This lat oxide promoted the partial oxidation to acetaldehyde (selectivity of 74%) and consequen exhibited the lowest selectivity to CO_2 (22%).

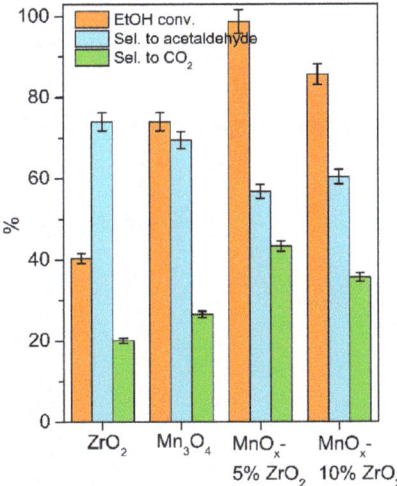

Figure 5. Solar photocatalytic oxidation of ethanol after 5 h of irradiation.

In Table 5 and Figure S4, the data of the thermocatalytic oxidation of ethanol are reported. MnO_x-5%ZrO_2 showed, also for this VOC, the best performance, with the lowest T_{90} (189 °C) and a maximum conversion to acetaldehyde of 66% at 176 °C. Moreover, MnO_x-10%ZrO_2 and Mn_3O_4 showed the same maximum conversion to acetaldehyde, but at a higher temperature (200 °C for MnO_x-10%ZrO_2 and 226 °C for bare manganese oxide). Consequently, MnO_x-5%ZrO_2 also exhibited the lowest T_{90} related to the conversion to CO_2 (361 °C). It is verified also for this approach, a little negative effect of the increased amount of zirconium oxide on the MnO_x, with higher T_{90} of MnO_x-10%ZrO_2 compared to MnO_x-5%ZrO_2. The highest conversion to acetaldehyde was obtained with the bare zirconia (maximum conversion of 98% at 409 °C) confirming, as also detected in the photocatalytic tests at room temperature, the tendency of this catalyst to promote the partial oxidation of ethanol instead of the total combustion.

Table 5. Data of the thermocatalytic oxidation of ethanol on the investigated samples.

Sample	Ethanol Conversion	Conversion to CO_2	Maximum Conversion to Acetaldehyde
Mn_3O_4	T_{10} = 95 °C T_{50} = 167 °C T_{90} = 239 °C	T_{10} = 217 °C T_{50} = 299 °C T_{90} = 411 °C	65% (226 °C)
ZrO_2	T_{10} = 221 °C T_{50} = 292 °C T_{90} = 382 °C	T_{10} = 428 °C T_{50} = 453 °C T_{90} = 474 °C	98% (409 °C)
MnO_x-5%ZrO_2	T_{10} = 45 °C T_{50} = 116 °C T_{90} = 189 °C	T_{10} = 167 °C T_{50} = 249 °C T_{90} = 361 °C	66% (176 °C)
MnO_x-10%ZrO_2	T_{10} = 70 °C T_{50} = 141 °C T_{90} = 214 °C	T_{10} = 182 °C T_{50} = 264 °C T_{90} = 376 °C	66% (200 °C)

Interestingly also for the removal of ethanol, the multi-catalytic reaction (i.e., the photothermo-catalysis) allowed us to improve the performance related to ethanol oxidation (Table 6, Figure S5). With MnO_x-5%ZrO_2, the T_{90} of ethanol conversion was lowered to 34 °C (154 °C), a value that is comparable or lower, considering an initial ethanol concentration of 1000 ppm, with respect to the other MnO_x-based materials reported in literature (in the range 127 °C (initial ethanol concentration of 300 ppm) −200 °C (initial ethanol concentration of 600–1945 ppm) [37,40]). The total oxidation to CO_2 was favoured on this sample, and for this reason, the maximum conversion to acetaldehyde was low (3% at 118 °C), with a decrease of 205 °C of the T_{90} related to the conversion to CO_2 compared to the thermocatalytic tests.

Table 6. Data of the photothermo-catalytic oxidation of ethanol on the investigated samples.

Sample	Ethanol Conversion	Conversion to CO_2	Maximum Conversion to Acetaldehyde
Mn_3O_4	T_{10} = 48 °C T_{50} = 126 °C T_{90} = 214 °C	T_{10} = 150 °C T_{50} = 249 °C T_{90} = 367 °C	58% (187 °C)
ZrO_2	T_{10} = 205 °C T_{50} = 277 °C T_{90} = 369 °C	T_{10} = 397 °C T_{50} = 437 °C T_{90} = 476 °C	91% (378 °C)
MnO_x-5%ZrO_2	T_{10} = 51 °C T_{50} = 123 °C T_{90} = 154 °C	T_{10} = 119 °C T_{50} = 137 °C T_{90} = 156 °C	35% (118 °C)
MnO_x-10%ZrO_2	T_{10} = 74 °C T_{50} = 145 °C T_{90} = 204 °C	T_{10} = 123 °C T_{50} = 165 °C T_{90} = 207 °C	24% (127 °C)

A further decrease of the maximum conversion to acetaldehyde was verified with MnO_x-10%ZrO_2 (24%), but at a higher temperature (127 °C) compared to the MnO_x-5%ZrO_2, confirming that, with these mixed oxides, the combustion of ethanol was favoured with respect to its partial oxidation. For all the tested samples, similar to the photothermal oxidation of toluene, there was a positive effect of the solar light irradiation, with a contextual decrease of conversion temperatures compared to the thermocatalytic tests (comparison between Tables 5 and 6). The unmodified zirconia remained the less active catalyst, however, the solar-assisted reaction decreased T_{90} of ethanol conversion of 13 compared to the tests without irradiation.

3. Discussion

The mixed oxides MnO_x-ZrO_2 here investigated showed promising performance the removal of VOCs in the gas phase, considering the absence of noble metals co-catalyst and an initial VOCs concentration of 1000 ppm. The amount of zirconium oxide added on manganese oxide is a key parameter to improve the catalytic and the photocatalytic performance. The as-synthesized samples showed a comparable optical bandgap, in the range 3.0–3.3 eV (Table 1), similar to the TiO_2, and able to exploit the UV-A portion the solar light. The addition of zirconia on manganese oxide led to a slight decrease the surface area (Table 1) that, however, did not comprise the catalytic activity of the mixed oxides.

The presence of a small amount of zirconium oxide on MnO_x allowed, as stated by XRD and SEM-EDX, an ionic exchange between Zr^{4+} and Mn^{2+}; this favoured the formation of a synergistic effect between the two oxides, with structural changes in the bulk of MnO_x. These modifications led to increasing the (photo)catalytic activity compared the bare oxides. Indeed, when reducible oxides, i.e., which own mobile/reducible oxygen were used for the oxidation of VOCs, the total oxidation to CO_2 is favoured, because the oxygens can participate in the reaction with a Mars–Van Krevelen (MvK) mechanism [41, The oxygen vacancies on the surface of the oxide will be subsequently filled by the present in the gas phase.

This mechanism was boosted up with the photothermo-catalytic approach because the photocatalytic mechanism generated the superoxide ($O_2^{\bullet-}$) and hydroxyl ($OH^{\bullet}$) radicals [43,44], that being more reactive of the molecular O_2, increased the rate of the total oxidation of VOCs (reactions a–i, Figure 6) and the re-filling of the oxygen vacancies, being the mobile oxygens of MnO_x activated by the heating [13,45]. For these reasons, the conversion temperatures of both toluene and ethanol oxidation were sensibly lower compared to the thermocatalysis, especially with the MnO_x-5%ZrO_2 sample. In this way, it was possible to exploit a double positive effect: (i) the solar irradiation effect: that allowed the formation of more reactive species, (ii) the thermal effect: that activated the redox mobility of the manganese oxide oxygens [13,20,45].

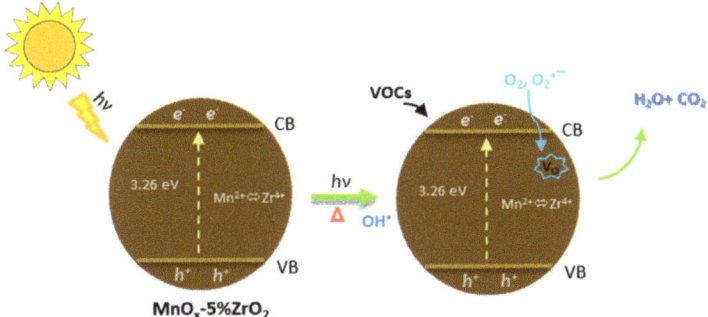

Figure 6. Proposed photothermo-catalytic mechanism. V_o = oxygen vacancy.

Photothermo-catalytic mechanism:

(i) Solar irradiation effect (VB and CB indicate the valence and the conduction bands):
  (a) Charge carriers formation: MnO_x-5%ZrO_2 + hν(solar) → MnO_x-5%ZrO_2 (h_{VB}^+ + e^-_{CB})
  (b) Formation of hydroxyl radical: h^+_{VB} + H_2O (g) → $OH^{\bullet}$ + H^+_{aq}
  (c) Formation of superoxide radical: e^-_{CB} + O_2 → $O_2^{\bullet-}$

(ii) Thermal effect (Vo = oxygen vacancy)
  (d) Oxygen from the mixed oxide: MnO_x-5%ZrO_2 → MnO_x-5%ZrO_2 (V_o) + 1/2 O_2(g)$_{from\ oxide}$
  (e) VOC oxidation: VOC + O_2 (g) + O_2(g)$_{from\ oxide}$ $\xrightarrow{heat}$ CO_2 + H_2O
  (f) Oxygen restoring: MnO_x-5%ZrO_2 (V_o) + 1/2 O_2(g) → MnO_x-5%ZrO_2

(iii) Solar photothermal effect
  (g) MnO_x-5%ZrO_2 + hν $\xrightarrow{heat}$ MnO_x-5%ZrO_2 (V_o) + 1/2 O_2(g)$_{from\ oxide}$ + $OH^{\bullet}$ + $O_2^{\bullet-}$
  (h) Improved VOC oxidation: VOC + O_2 (g) + O_2(g)$_{from\ oxide}$ + $OH^{\bullet}$ + $O_2^{\bullet-}$ → CO_2 + H_2O
  (i) Oxygen speeded up restoring: MnO_x-5%ZrO_2 (V_o) + 1/2 O_2(g) + $O_2^{\bullet}$ → MnO_x-5%ZrO_2

It is worth noting that the reactions (a–c) and (d–f) are also involved in the solar photocatalysis at room temperature and in the bare thermocatalytic tests, respectively. The multi-catalytic effect (reactions g–i) allowed one to increase the performance and to favour the total oxidation of the employed VOCs to CO_2.

Another confirmation of the proposed MvK mechanism was reported in the Figure S3b. In the phothermo-catalytic oxidation of toluene with the MnO_x-5%ZrO_2 sample, the air (more interesting from a practical point of view) was replaced in the gas mixture with the pure oxygen. It is possible to note that the presence of oxygen led to a beneficial effect for the toluene conversion to CO_2, being the T_{90} lower of 25 °C (155 °C) compared to the test with air (180 °C). This can be reasonably ascribed to the easier oxygen restoring on the catalyst surface (reaction i), in an oxygen-rich environment, favouring, in this way, t[?] MvK route.

As stated by the characterization data, the good interaction between the mangane[?] and zirconium oxide (especially at low amount of ZrO_2) improved the photothern[?] catalytic mechanism with the redox process on MnO_x that was favoured by the ior[?] exchange between the zirconium and the manganese ions. On the contrary, an increas[?] amount of zirconium oxide led to a progressive deposition of the hosted oxide on t[?] surface of MnO_x covering, in this way, the surface-active sites of manganese oxide [35,3[?] For these reasons, the optimal performance was obtained with 5 wt.% of ZrO_2. In t[?] contest, the mobility of the surface oxygens of the MnO_x-5%ZrO_2 sample was favour[?] by the MnO_x redox properties, and consequently, it is strictly related to its reducibil[?] Furthermore, the amount of the surface oxygens on MnO_x-5%ZrO_2 was higher compared[?] the other samples, as detected by XPS. To have a further confirmation of the high reducil[?] ity/mobility of the surface oxygens of MnO_x-5%ZrO_2, the H_2-temperature-programm[?] reduction (TPR) measurements were carried out, and the sample profiles were reported[?] Figure 7. In accordance with the literature data [20,46], the TPR profiles of the MnO_x-bas[?] samples were characterized to broad reduction peaks, due to the occurrence of sever[?] reduction processes of the Mn ions. As expected, the MnO_x-5%ZrO_2 sample showed t[?] lowest reduction feature (201 °C) attributed to the reduction of Mn_2O_3 to Mn_3O_4 [4[?] 111 °C and 117 °C lower compared to the same reduction feature of Mn_3O_4 and MnC[?] 10%ZrO_2, respectively. This reduction peak was also more intense for the MnO_x-5%Zr[?] compared to the other MnO_x-based samples confirming, as detected by XRD and X[?] the major presence of Mn^{3+} ions on MnO_x-5%ZrO_2. The higher temperature reducti[?] signals in the range 300–480° were ascribed to the further reduction of Mn_3O_4 to MnO [4[?] Moreover, in this case, the sample with 5 wt.% of ZrO_2 showed the highest reducibility (i[?] the lowest peak temperature). This is connected to the highest mobility/reducibility of t[?] surface oxygens of MnO_x-5%ZrO_2, which favours the MvK mechanism, and therefore[?] better VOCs abatement. The reduction temperature of bare ZrO_2 started at a temperatu[?] above 500 °C [47], and for this reason, in our analysis (in the range 50–550°C), its reducti[?] peak was not complete.

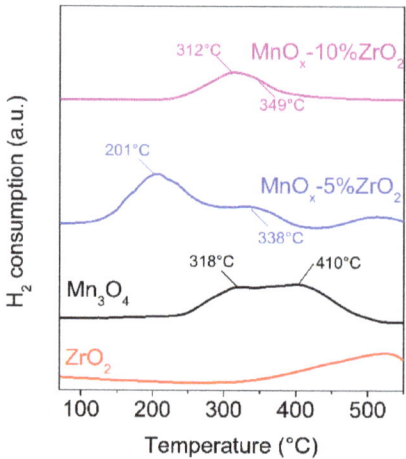

Figure 7. H_2-TPR (Temperature programmed reduction) profiles of the investigated samples.

Between the photocatalytic, the thermocatalytic and the photothermo-catalytic moval of VOCs, although the solar photocatalytic reaction has the advantages of work room temperature and that with the MnO_x-5%ZrO_2, it reached a similar activity of t[?] most used TiO_2-based materials (Table 4); to have a complete VOCs removal, it is necessa[?]

to have contextual heating. For this purpose, the solar photothermo-catalysis can be an optimal solution to obtain the good performance of the thermocatalysis, but with an energy saving, due to the lower temperature required for the VOCs conversion. Indeed, the best sample (MnO_x-5%ZrO_2) tested in our experimental conditions showed a decrease of 36 °C and 34 °C of the toluene and ethanol T_{90} conversion compared to the thermocatalytic tests favouring in both the reactions; the total oxidation to CO_2 (the T_{90} of ethanol conversion to CO_2 was lowered of 205 °C, Tables 5 and 6).

Finally, the stability in the time-on steam toluene removal of MnO_x-5%ZrO_2 was good (Figure 8, toluene solar photothermo-oxidation) and pointed to the MnO_x-ZrO_2 catalyst being a promising versatile material for application in thermocatalysis, photocatalysis, and photothermo-catalysis.

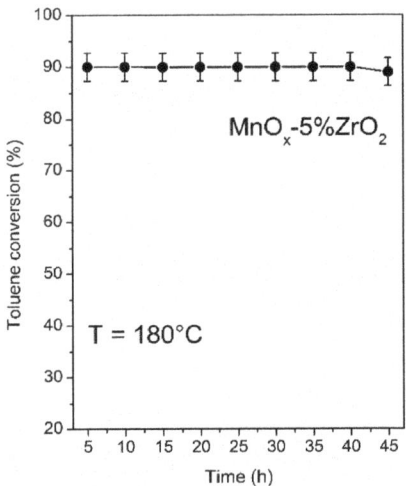

Figure 8. Stability test of MnO_x-5%ZrO_2 catalyst in the toluene solar photothermo-catalytic oxidation.

4. Materials and Methods

4.1. Catalysts Synthesis

Bare manganese oxide was prepared by chemical precipitation with NaOH (1 M) (Panreac Química SLU, Castellar del Vallès (Barcelona), Spain). In particular, a certain amount of manganese (II) chloride tetrahydrate (Sigma-Aldrich, Buchs, Switzerland) was dissolved in demineralized water and heated at 70 °C. After the NaOH was added dropwise until the pH = 10. Successively, the solution was stirred and kept at 70 °C for 2 h. After digestion for 24 h, the slurry was filtered and dried at 120 °C overnight. Finally, the resultant powders were calcined in air at 600 °C for 2 h.

A similar procedure was followed for the bare ZrO_2. In this case, the zirconyl nitrate hydrate (Fluka, Buchs, Switzerland) and ammonia (as precipitant agent, 25–28%, Sigma-Aldrich, Buchs, Switzerland) were used, following the same procedures reported above, and the same thermal treatments (drying at 120 °C, and calcination at 600 °C for 2 h).

For the MnO_x-ZrO_2 mixed oxides, the NaOH-driven precipitation was employed using the required stoichiometric amount of zirconyl nitrate hydrate to obtain the chosen nominal concentration in weight percentage (wt.%) of ZrO_2. Moreover, in this case, the samples were dried at 120 °C and calcined in air at 600 °C for 2 h.

4.2. Catalysts Characterization

The sample structures were determined through the X-ray powder diffraction (XRD) using a Smartlab Rigaku diffractometer (Rigaku Europe SE, Hugenottenallee 167 Neu-Isenburg 63263, Germany) in Bragg–Brentano mode, equipped with a rotating anode of Cu

Kα radiation operating at 45 kV and 200 mA. The surface morphology was examined w[ith] field emission scanning electron microscopy (FE-SEM) using a ZEISS SUPRA 55 VP (C[arl] Zeiss QEC Gmb, Garching b. München, Germany). The composition of the powders w[as] carried out by the energy dispersive X-ray (EDX) analysis using an INCA-Oxford (Oxf[ord] Instruments plc, Tubney Woods, Abingdon, Oxfordshire, United Kingdom) windowl[ess] detector, and a resolution of 127 eV determined using the half-height amplitude (FWH[M) of the Kα of Mn.

The BET surface area values were determined by N_2 adsorption–desorption m[ea]surements with a Sorptomatic 1990 instrument (Thermo Quest, Milano, Italy). Before measurements, the catalysts were outgassed overnight at 200 °C.

The UV-vis Diffuse Reflectance spectra (UV-Vis DRS, Diffuse Reflectance Spectrosco[py]) measurements were performed with a Jasco V- 670 spectrometer (Jasco Europe S.R[L], Cremella, Italy) provided with an integration sphere and using barium sulphate (Flu[ka], Buchs, Switzerland) as standard. The estimation of the optical band gap of the samp[le] was determined using the Kubelka–Munch function [26].

The X-ray photoelectron spectroscopy (XPS) was performed with a K-alpha X-[ray] photoelectron instrument (Thermo Fisher Scientific, Waltham, MA, USA), employing the [C] 1s peak at 284.9 eV (of adventitious carbon) as reference.

The H_2-TPR (Temperature programmed reduction) profiles of the samples were [ob]tained using a home-made flow equipment (gas-mixture 5 vol.% H_2 in Ar) and a TC[D] detector, following the procedures reported in ref. [48].

4.3. Photo, Thermo and Photothermo-Catalytic Oxidation of VOCs

The thermocatalytic removal of VOCs in gas phase and atmospheric pressure was c[ar]ried out in a fixed bed flow reactor packed with the powder catalysts (0.15 g, 80–140 mes[h]) using the same experimental conditions described in the ref. [3]. A heating ramp of 10 [°C] was used in all the tests from room temperature to 500 °C. To assure a steady-state bef[ore] the catalytic measurements, the gas mixture (1000 ppm VOCs; 10 vol.% air, rest He) w[as] flowed on the catalyst for 30 min. No substantial contribution due to the adsorption proc[ess] was detected. The reaction products were analysed by a gas chromatography (Sm[art] IQ+ Thermo Onix, Thermo Fisher Scientific 168 Third Avenue, Waltham, MA, USA 02[)] utilizing a packed column with 10% FFAP on Chromosorb W (from Merck KGaA, Da[rm]stadt, Germany) with a FID (Flame Ionization Detector), coupled with a quadrupole m[ass] spectrometer (VG quadropoles, Fergutec B.V. Dragonder 13 C, 5554 GM Valkenswaa[rd] The Netherlands).

The photocatalytic and the photothermo-catalytic tests were performed with the sa[me] instruments described above. The simultaneous irradiation in the phothermo-catalytic te[st] was made with an artificial solar lamp (OSRAM Vitalux 300 W, 10.7 mW/cm^2, OSRA[M] Opto Semiconductors GmbH, Leibniz, Regensburg Germany). In the photocatalytic test[s a] fan located near the reactor allowed us to maintain a constant temperature, avoiding [the] overheating effects due to lamp emission.

5. Conclusions

The MnO_x-ZrO_2 mixed oxides exhibited promising performance in the removal [of] toluene and ethanol in the gas phase, especially in the multi-catalytic solar photothern[al] approach. The ionic interaction between the manganese and the zirconium ions exploi[ted] with the addition of a low amount of zirconium oxide allowed us to boost up the Mars–v[an] Krevelen mechanism of the VOCs oxidation favouring the total oxidation of VOCs to C[O2]. Furthermore, with the photothermo-catalysis, a decrease of the conversion temperatu[re] compared to the thermocatalysis was verified, and MnO_x-5 wt.%ZrO_2 also showed go[od] stability. Finally, with the same catalyst in the solar photocatalytic tests at room temperat[ure] a similar activity of the most used TiO_2-based materials was obtained, pointing to the f[act] that this sample can be promising for the VOCs remediation technologies, being cheap[,] not critical, and performing considering the absence of noble metal co-catalysts.

Supplementary Materials: The following supporting information can be downloaded at: https://www.mdpi.com/article/10.3390/catal12010085/s1, Figure S1: SEM images of the other investigated samples; Figure S2: XPS characterization of the MnO$_x$-5% ZrO$_2$ sample; Figure S3: Photothermocatalytic oxidation of toluene: influence of different parameters on MnO$_x$-5% ZrO$_2$ sample; Figure S4: thermocatalytic oxidation of ethanol; Figure S5: Photothermo-catalytic oxidation of ethanol.

Author Contributions: Conceptualization, R.F.; investigation, R.A.F., R.F., F.L.P., E.M.M., S.A.B.; writing—original draft preparation, R.F.; writing—review and editing, R.F.; supervision, R.F. All authors have read and agreed to the published version of the manuscript.

Funding: This research received no external funding.

Data Availability Statement: The data presented in this study are available on request from the corresponding author.

Acknowledgments: The authors thank S. Scirè (University of Catania) for the use of the laboratory facilities and the Bio-nanotech Research and Innovation Tower (BRIT) laboratory of the University of Catania for the Smartlab diffractometer facility. R.F. thanks the PON "AIM" of the European Social Found for the support.

Conflicts of Interest: The authors declare no conflict of interest.

References

Li, Y.; Wu, S.; Wu, J.; Hu, Q.; Zhou, C. Photothermocatalysis for efficient abatement of CO and VOCs. *J. Mater. Chem. A* **2020**, *8*, 8171–8194. [CrossRef]

Boyjoo, Y.; Sun, H.; Liu, J.; Pareek, V.K.; Wang, S. A review on photocatalysis for air treatment: From catalyst development to reactor design. *Chem. Eng. J.* **2017**, *310*, 537–559. [CrossRef]

Bellardita, M.; Fiorenza, R.; D'Urso, L.; Spitaleri, L.; Gulino, A.; Compagnini, G.; Scirè, S.; Palmisano, L. Exploring the photothermocatalytic performance of brookite TiO$_2$-CeO$_2$ composites. *Catalysts* **2020**, *10*, 765. [CrossRef]

Ma, R.; Sun, J.; Li, D.H.; Wei, J.J. Review of synergistic photo-thermo-catalysis: Mechanisms, materials and applications. *Int. J. Hydrogen Energy* **2020**, *45*, 30288–30324. [CrossRef]

Keller, N.; Ivanez, J.; Highfield, J.; Ruppert, A.M. Photo-/thermal synergies in heterogeneous catalysis: Towards low-temperature (solar-driven) processing for sustainable energy and chemicals. *Appl. Catal. B Environ.* **2021**, *296*, 120320. [CrossRef]

Abidi, M.; Assadi, A.A.; Bouzaza, A.; Hajjaji, A.; Bessais, B.; Rtimi, S. Photocatalytic indoor/outdoor air treatment and bacterial inactivation on Cu$_x$O/TiO$_2$ prepared by HiPIMS on polyester cloth under low intensity visible light. *Appl. Catal. B Environ.* **2019**, *259*, 118074. [CrossRef]

Li, Q.; Li, F. Recent advances in surface and interface design of photocatalysts for the degradation of volatile organic compounds. *Adv. Colloid Interface Sci.* **2020**, *284*, 102275. [CrossRef]

Zhang, S.; Pu, W.; Chen, A.; Xu, Y.; Wang, Y.; Yang, C.; Gong, J. Oxygen vacancies enhanced photocatalytic activity towards VOCs oxidation over Pt deposited Bi$_2$WO$_6$ under visible light. *J. Hazard. Mater.* **2020**, *384*, 121478. [CrossRef] [PubMed]

Fiorenza, R.; Bellardita, M.; Balsamo, S.A.; Spitaleri, L.; Gulino, A.; Condorelli, M.; D'Urso, L.; Scirè, S.; Palmisano, L. A solar photothermocatalytic approach for the CO$_2$ conversion: Investigation of different synergisms on CoO-CuO/brookite TiO$_2$-CeO$_2$ catalysts. *Chem. Eng. J.* **2022**, *428*, 131249. [CrossRef]

Lewicka, E.; Guzik, K.; Galos, K. On the possibilities of critical raw materials production from the EU's primary sources. *Resources* **2021**, *10*, 50. [CrossRef]

Fiorenza, R. Bimetallic catalysts for volatile organic compound oxidation. *Catalysts* **2020**, *10*, 661. [CrossRef]

Ristig, S.; Cibura, N.; Strunk, J. Manganese oxides in heterogeneous (photo)catalysis: Possibilities and challenges. *Green* **2015**, *5*, 23–41. [CrossRef]

Feng, Y.; Wang, C.; Wang, C.; Huang, H.; Hsi, H.-C.; Duan, E.; Liu, Y.; Guo, G.; Dai, H.; Deng, J. Catalytic stability enhancement for pollutant removal via balancing lattice oxygen mobility and VOCs adsorption. *J. Hazard. Mater.* **2022**, *424*, 127337. [CrossRef] [PubMed]

Cai, T.; Liu, Z.; Yuan, J.; Xu, P.; Zhao, K.; Tong, Q.; Lu, W.; He, D. The structural evolution of MnO$_x$ with calcination temperature and their catalytic performance for propane total oxidation. *Appl. Surf. Sci.* **2021**, *565*, 150596. [CrossRef]

Yang, R.; Fan, Y.; Ye, R.; Tang, Y.; Cao, X.; Yin, Z.; Zeng, Z. MnO$_2$-based materials for environmental applications. *Adv. Mater.* **2021**, *33*, 1–53.

Kondratowicz, T.; Drozdek, M.; Michalik, M.; Gac, W.; Gajewska, M.; Kuśtrowski, P. Catalytic activity of Pt species variously dispersed on hollow ZrO$_2$ spheres in combustion of volatile organic compounds. *Appl. Surf. Sci.* **2020**, *513*, 145788. [CrossRef]

Scirè, S.; Liotta, L.F. Supported gold catalysts for the total oxidation of volatile organic compounds. *Appl. Catal. B Environ.* **2012**, *125*, 222–246. [CrossRef]

Hassan, N.S.; Jalil, A.A. A review on self-modification of zirconium dioxide nanocatalysts with enhanced visible-light-driven photodegradation of organic pollutants. *J. Hazard. Mater.* **2022**, *423*, 126996. [CrossRef] [PubMed]

19. Chen, G.; Wang, Z.; Lin, F.; Zhang, Z.; Yu, H.; Yan, B.; Wang, Z. Comparative investigation on catalytic ozonation of VOCs different types over supported MnO catalysts. *J. Hazard. Mater.* **2020**, *391*, 122218. [CrossRef] [PubMed]
20. Huang, X.; Li, L.; Liu, R.; Li, H.; Lan, L.; Zhou, W. Optimized synthesis routes of MnO_x-ZrO_2 hybrid catalysts for improv toluene combustion. *Catalysts* **2021**, *11*, 1037. [CrossRef]
21. González, U.; Schifter, I.; Díaz, L.; González-Macías, C.; Mejía-Centeno, I.; Sánchez-Reyna, G. Assessment of the use of ethan instead of MTBE as an oxygenated compound in Mexican regular gasoline: Combustion behavior and emissions. *Environ. Mo Assess.* **2018**, *190*, 700. [CrossRef]
22. Niu, X.; Wei, H.; Tang, K.; Liu, W.; Zhao, G.; Yang, Y. Solvothermal synthesis of 1D nanostructured Mn_2O_3: Effect of Ni^{2+} a Co^{2+} substitution on the catalytic activity of nanowires. *RSC Adv.* **2015**, *5*, 66271–66277. [CrossRef]
23. Gutiérrez-Ortiz, J.I.; de Rivas, B.; López-Fonseca, R.; Martín, S.; González-Velasco, J.R. Structure of Mn–Zr mixed oxides cataly and their catalytic performance in the gas-phase oxidation of chlorocarbons. *Chemosphere* **2007**, *68*, 1004–1012. [CrossR [PubMed]
24. Santos, V.P.; Pereira, M.F.R.; Órfão, J.J.M.; Figueiredo, J.L. The role of lattice oxygen on the activity of manganese oxides towa the oxidation of volatile organic compounds. *Appl. Catal. B Environ.* **2010**, *99*, 353–363. [CrossRef]
25. Zeng, K.; Li, X.; Wang, C.; Wang, Z.; Guo, P.; Yu, J.; Zhang, C.; Zhao, X.S. Three-dimensionally macroporous MnZrO catalysts propane combustion: Synergistic structure and doping effects on physicochemical and catalytic properties. *J. Colloid Interface* **2020**, *572*, 281–296. [CrossRef] [PubMed]
26. López, R.; Gómez, R. Band-gap energy estimation from diffuse reflectance measurements on sol–gel and commercial TiO_2 comparative study. *J. Sol-Gel Sci. Technol.* **2012**, *61*, 1–7. [CrossRef]
27. Mishra, S.; Debnath, A.; Muthe, K.; Das, N.; Parhi, P. Rapid synthesis of tetragonal zirconia nanoparticles by microwa solvothermal route and its photocatalytic activity towards organic dyes and hexavalent chromium in single and binary compone systems. *Colloids Surf. A Physicochem. Eng. Asp.* **2021**, *608*, 125551. [CrossRef]
28. Yamakata, A.; Vequizo, J.J.M. Curious behaviors of photogenerated electrons and holes at the defects on anatase, rutile, a brookite TiO_2 powders: A review. *J. Photochem. Photobiol. C Photochem. Rev.* **2019**, *40*, 234–243. [CrossRef]
29. Maira, A.J.; Yeung, K.L.; Soria, J.; Coronado, J.M.; Belver, C.; Lee, C.Y.; Augugliaro, V. Gas-phase photo-oxidation of toluene usi nanometer-size TiO_2 catalysts. *Appl. Catal. B Environ.* **2001**, *29*, 327–336. [CrossRef]
30. Yu, J.; Caravaca, A.; Guillard, C.; Vernoux, P.; Zhou, L.; Wang, L.; Lei, J.; Zhang, J.; Liu, Y. Carbon nitride quantum dots modifi TiO_2 inverse opal photonic crystal for solving indoor vocs pollution. *Catalysts* **2021**, *11*, 464. [CrossRef]
31. Nevárez-Martínez, M.C.; Kobylanski, M.P.; Mazierski, P.; Wółkiewicz, J.; Trykowski, G.; Malankowska, A.; Kozak, M.; Espino Montero, P.J.; Zaleska-Medynska, A. Self-organized TiO_2-MnO_2 nanotube arrays for efficient photocatalytic degradation toluene. *Molecules* **2017**, *22*, 564. [CrossRef] [PubMed]
32. Almomani, F.; Bhosale, R.; Shawaqfah, M. Solar oxidation of toluene over Co doped nano-catalyst. *Chemosphere* **2020**, *255*, 1268 [CrossRef]
33. Chen, A.; Chen, G.; Wang, Y.; Lu, Y.; Chen, J.; Gong, J. Fabrication of novel $Ag_4Bi_2O_{5-x}$ towards excellent photocatalytic oxidati of gaseous toluene under visible light irradiation. *Environ. Res.* **2021**, *197*, 111130. [CrossRef]
34. Zhang, F.; Li, X.; Zhao, Q.; Zhang, Q.; Tadé, M.; Liu, S. Fabrication of α-Fe_2O_3/In_2O_3 composite hollow microspheres: A no hybrid photocatalyst for toluene degradation under visible light. *J. Colloid Interface Sci.* **2015**, *457*, 18–26. [CrossRef] [PubMed]
35. Khanmohammadi, M.; Shahrouzi, J.R.; Rahmani, F. Insights into mesoporous MCM-41-supported titania decorated with Cu nanoparticles for enhanced photodegradation of tetracycline antibiotic. *Environ. Sci. Pollut. Res.* **2021**, *28*, 862–879. [CrossRef]
36. Fiorenza, R.; Bellardita, M.; Scirè, S.; Palmisano, L. Effect of the addition of different doping agents on visible light activity porous TiO_2 photocatalysts. *Mol. Catal.* **2018**, *455*, 108–120. [CrossRef]
37. Wu, P.; Jin, X.; Qiu, Y.; Ye, D. Recent progress of thermocatalytic and photo/thermocatalytic oxidation for VOCs purification ov manganese-based oxide catalysts. *Environ. Sci. Technol.* **2021**, *55*, 4268–4286. [CrossRef]
38. Mulka, R.; Odoom-Wubah, T.; Tan, K.B.; Huang, J.; Li, Q. Biogenic $MnxO_y$ as an efficient catalyst in the catalytic abatement benzene: From kinetic to mathematical modeling. *Mol. Catal.* **2021**, *510*, 111643. [CrossRef]
39. Vorontsov, A. Selectivity of photocatalytic oxidation of gaseous ethanol over pure and modified TiO_2. *J. Catal.* **2004**, *221*, 102–1 [CrossRef]
40. Dai, Y.; Men, Y.; Wang, J.; Liu, S.; Li, S.; Li, Y.; Wang, K.; Li, Z. Tailoring the morphology and crystal facet of Mn_3O_4 for high efficient catalytic combustion of ethanol. *Colloids Surf. A Physicochem. Eng. Asp.* **2021**, *627*, 127216. [CrossRef]
41. Li, J.-J.; Yu, E.-Q.; Cai, S.-C.; Chen, X.; Chen, J.; Jia, H.-P.; Xu, Y.-J. Noble metal free, CeO_2/$LaMnO_3$ hybrid achieving efficie photo-thermal catalytic decomposition of volatile organic compounds under IR light. *Appl. Catal. B Environ.* **2019**, *240*, 141–1 [CrossRef]
42. Morales, M.R.; Yeste, M.P.; Vidal, H.; Gatica, J.M.; Cadus, L.E. Insights on the combustion mechanism of ethanol and n-hexa in honeycomb monolithic type catalysts: Influence of the amount and nature of Mn-Cu mixed oxide. *Fuel* **2017**, *208*, 637–6 [CrossRef]
43. Chen, L.; Chen, P.; Wang, H.; Cui, W.; Sheng, J.; Li, J.; Zhang, Y.; Zhou, Y.; Dong, F. Surface lattice oxygen activation on Sr_2Sb_2 enhances the photocatalytic mineralization of toluene: From reactant activation, intermediate conversion to product desorpti *ACS Appl. Mater. Interfaces* **2021**, *13*, 5153–5164. [CrossRef] [PubMed]

Parrino, F.; Palmisano, L. Reactions in the presence of irradiated semiconductors: Are they simply photocatalytic? *Mini-Rev. Org. Chem.* **2018**, *15*, 157–164. [CrossRef]

Azalim, S.; Franco, M.; Brahmi, R.; Giraudon, J.-M.; Lamonier, J.-F. Removal of oxygenated volatile organic compounds by catalytic oxidation over Zr–Ce–Mn catalysts. *J. Hazard. Mater.* **2011**, *188*, 422–427. [CrossRef]

Stobbe, E.R.; de Boer, B.A.; Geus, J.W. The reduction and oxidation behaviour of manganese oxides. *Catal. Today* **1999**, *47*, 161–167. [CrossRef]

Jabłońska, M. TPR study and catalytic performance of noble metals modified Al_2O_3, TiO_2 and ZrO_2 for low-temperature NH_3-SCO. *Catal. Commun.* **2015**, *70*, 66–71. [CrossRef]

Scirè, S.; Fiorenza, R.; Gulino, A.; Cristaldi, A.; Riccobene, P.M. Selective oxidation of CO in H_2-rich stream over ZSM5 zeolites supported Ru catalysts: An investigation on the role of the support and the Ru particle size. *Appl. Catal. A Gen.* **2016**, *520*, 82–91. [CrossRef]

Photocatalytic Degradation of Fluoroquinolone Antibiotics in Solution by Au@ZnO-rGO-gC₃N₄ Composites

Daniel Machín [1,*], Kenneth Fontánez [2], José Duconge [3], María C. Cotto [3], Florian I. Petrescu [4], Carmen Morant [5] and Francisco Márquez [3,*]

1. Arecibo Observatory, Universidad Ana G. Méndez, Cupey Campus, San Juan, PR 00926, USA
2. Department of Chemistry, University of Puerto Rico, Rio Piedras Campus, San Juan, PR 00925, USA; Kenneth.fontanez@gmail.com
3. Nanomaterials Research Group, Department of Natural Sciences and Technology, Division of Natural Sciences, Technology and Environment, Universidad Ana G. Méndez, Gurabo Campus, Gurabo, PR 00778, USA; jduconge@uagm.edu (J.D.); mcotto48@uagm.edu (M.C.C.)
4. International Federation for the Promotion of Mechanism and Machine Science (IFToMM), Romanian Association for Theory of Machines and Mechanisms (ARoTMM), Bucharest Polytechnic University, 060042 Bucharest, Romania; fitpetrescu@gmail.com
5. Department of Applied Physics, Instituto de Ciencia de Materiales Nicolás Cabrera, Autonomous University of Madrid, 28049 Madrid, Spain; c.morant@uam.es
* Correspondence: machina1@uagm.edu (A.M.); fmarquez@uagm.edu (F.M.)

Abstract: The photocatalytic degradation of two quinolone-type antibiotics (ciprofloxacin and levofloxacin) in aqueous solution was studied, using catalysts based on ZnO nanoparticles, which were synthesized by a thermal procedure. The efficiency of ZnO was subsequently optimized by incorporating different co-catalysts of gC₃N₄, reduced graphene oxide, and nanoparticles of gold. The catalysts were fully characterized by electron microscopy (TEM and SEM), XPS, XRD, Raman, and BET surface area. The most efficient catalyst was 10%Au@ZnONPs-3%rGO-3%gC₃N₄, obtaining degradations of both pollutants above 96%. This catalyst has the largest specific area, and its activity was related to a synergistic effect, involving factors such as the surface of the material and the ability to absorb radiation in the visible region, mainly produced by the incorporation of rGO and gC₃N₄ in the semiconductor. The use of different scavengers during the catalytic process, was used to establish the possible photodegradation mechanism of both antibiotics.

Keywords: ciprofloxacin; levofloxacin; ZnO; gC₃N₄; rGO; Au nanoparticles

1. Introduction

Antibiotics have become emerging pollutants due to their widespread use and persistence in the environment [1–3]. The origin is very varied, although they come mainly from medical treatments, agricultural, livestock, and industrial production [4–7]. The presence of antibiotics in the natural environment presents a serious health risk, since they can lead to the development of antibiotic-resistant bacteria and, in general, promote the destabilization of the natural environment [8,9]. Fluoroquinolone-based antibiotics, particularly ciprofloxacin and levofloxacin, are widely used in human and veterinary medicine [1–3]. They are used to fight bacterial infections like pneumonia, kidney or prostate infections, and even skin infections. In fact, this family of antibiotics is by far the most widely used for medical and veterinary treatments. Due to the chemical structure, after consumption by humans or animals, only between 15 and 20% is metabolized, so the non-metabolized antibiotic is eliminated into the environment through urine or feces [10]. Once released into the natural environment, these antibiotics are highly recalcitrant to degradation, which is why they have been detected in very worrying quantities in wastewater, surface water, ground water, and even in drinking water [11]. In general, some antibiotics can be eliminated from water bodies by different techniques such as adsorption, nanofiltration, coagulation, electrolysis,

or even biodegradation [12,13]. These decontamination techniques have been implement
in wastewater processing plants, especially in developed countries, although these methc
are very expensive, require large spaces and are not very efficient. In fact, already treat
waters continue to show high levels of these pollutants [14]. As an alternative procedu
to those already mentioned above, it is worth highlighting the chemical transformati
through advanced oxidation processes (AOPs), and especially those involving semico
ductor photocatalysts [15]. These processes are based on the catalytic photodegradati
of organic pollutants in aqueous solution, being much more efficient than convention
methods, as well as being friendly with the environment [15]. The AOPs produce high
oxidized species such as radicals and other reactive species, which degrade, through
chain reaction, organic pollutants in solution allowing complete mineralization [16,17].

Among the most widely used semiconductor materials, it is worth mentioning ti
nium oxide (TiO_2) and zinc oxide (ZnO), due to their low cost, non-toxicity, large speci
surface area of some of their forms, and high catalytic activity [18,19]. The main limitati
of both semiconductors is the wide bandgap (ca. 3.2 eV in TiO_2 anatase and 3.0 eV
TiO_2 rutile, and 3.37 eV in ZnO wurtzite), which requires irradiation with UV light for t
photoactivation process [20]. UV radiation from the solar spectrum is limited to about 5%
the total radiation reaching the earth's surface, so these catalysts are clearly inefficient [2
However, alternatives such as doping these semiconductors with metallic nanoparticles a
other additives deposited on the surface have made it possible to reduce bandgap and su
stantially promote electron-hole separation, avoiding rapid recombination processes [2
The incorporation of Au nanoparticles on the surface of semiconductors enables the a
pearance of surface plasmon resonance (SPR), greatly improving the storage capacity a
charge separation which increases the photocatalytic activity of the heterostructure [23,2
Nevertheless, there are many other factors that decisively affect the photocatalytic behavi
of the material, especially morphology, particle size, crystallinity and specific area, co
trolled during the synthesis procedure [25]. Graphene and especially graphene oxide (G
provides scaffolds to anchor other components, due to their two-dimensional structu
and large surface area [26]. There are many examples of nanohybrids prepared with C
with extraordinary optical, electrical, and thermal properties that have a direct effect on t
catalytic activity of the material [27,28]. On the other hand, the incorporation of GO to Zr
has been shown to increase the activity, reducing the photocorrosion of the semiconduct
facilitating the separation of charges and the inhibition of the electron-hole recombinati
processes [29]. Additionally, the two-dimensional structure of GO greatly improves t
interaction with organic pollutants, accelerating the subsequent photocatalytic degrac
tion [30]. Graphitic carbon nitride (gC_3N_4) [31], is an allotropic material of carbon nitric
rich in nitrogen, and with truly extraordinary properties that allow its use in different a
plications such as catalysis, photodegradation of organic pollutants, CO_2 fixation, cataly
and even in energy storage systems [32–34]. Among the most relevant properties, it
worth highlighting its great thermal and chemical stability, its two-dimensional structu
capable of facilitating interaction with other materials, and its simple synthesis method [3
The purpose of doping with noble metal nanoparticles and the manufacture of compl
hybrid composites has been to develop more efficient catalytic heterostructures, capab
of showing reduced bandgap, altered electronic properties that allow efficient generatie
of hole-electron pairs, and inhibition of recombination processes [35]. Over the last fe
years, complex systems have been developed, with improved and increasingly outstandi
properties, which have been applied to many processes, and especially to the degradatie
of highly persistent organic pollutants. In this sense, it is worth highlighting the catalys
based on metallic nanoparticles, dispersed on semiconductors such as TiO_2, ZnO, Fe_2O
ZnS, or CdS [36]. Other catalysts, based on heterostructures formed by rare earth me
oxides, combined with graphitic carbon nitride have also been designed, showing goe
catalytic behavior [37].

Taking into account all the previous research, the objective of this work has be
to obtain catalysts, based on the incorporation of gold nanoparticles on the surface

heterostructures formed by ZnO nanoparticles, to which reduced GO (rGO) and graphitic carbon nitride have been incorporated. These catalysts were used for the photodegradation of ciprofloxacin (CFX) and levofloxacin (LFX) in aqueous solution. The structural properties and morphology of the most active catalyst has been investigated using different spectroscopic and analytical techniques, such as XRD, XPS, Raman, DRS, SEM, TEM, and BET analysis. Finally, based on the results obtained, a possible photodegradation mechanism of the two antibiotics studied is proposed.

2. Results

2.1. Characterization of Catalysts

Three types of catalysts were synthesized, based on Au@ZnONPs, Au@ZnONPs-3%rGO, and Au@ZnONPs-3%rGO-3%gC$_3$N$_4$. In these catalysts, the percentage of rGO and gC$_3$N$_4$ was always maintained at 3%, although percentages of Au nanoparticles of 1%, 5% and 10% were used, thus, a total of nine catalysts were obtained. All these catalysts were used for the photodegradation of CFX and LFX, and the most efficient catalyst (10%Au@ZnONPs-3%rGO-3%gC$_3$N$_4$) was the one that was fully characterized by different techniques.

The BET surface area of the catalysts was analyzed (see Table 1). ZnO nanoparticles showed a relatively large surface area of 24 m^{2}g^{-1}. Incorporation of Au on the surface increased the specific area of the material to 58 m^2/g in the case of 1%Au@ZnONPs, and continued to increase, as a function of Au loading, up to a maximum value of 78 m^2/g in 10%Au@ZnONPs. This behavior, which has been described previously with different metallic nanoparticles, is expected to contribute to the catalytic activity of the material. The incorporation of 3%rGO, which already has a very large surface area, considerably increases the surface area of the material, reaching values as high as 196 m^2/g in the case of 10%Au@ZnONPs-3%rGO. The incorporation of gC$_3$N$_4$ also increased the area of the material, although not as drastically as in the case of rGO, reaching area values of 229 m^2/g in the case of 10%Au@ZnONPs-3%rGO-3%gC$_3$N$_4$. This last catalyst, with the highest specific area, is the most active in the processes studied, as will be seen later.

Table 1. BET surface area of the different as-synthesized materials.

Material	BET Area (m^2/g)
ZnONPs	24
1%Au@ZnONPs	33
5%Au@ZnONPs	49
10%Au@ZnONPs	78
1%Au@ZnONPs-3%rGO	126
5%Au@ZnONPs-3%rGO	161
10%Au@ZnONPs-3%rGO	196
1%Au@ZnONPs-3%rGO-3%gC$_3$N$_4$	143
5%Au@ZnONPs-3%rGO-3%gC$_3$N$_4$	177
10%Au@ZnONPs-3%rGO-3%gC$_3$N$_4$	229

The dispersion of 10%Au on the support of ZnONPs was followed by elemental mapping (see Figure 1). Figure 1b,c show the distribution of Zn and Au, respectively, corresponding to the SEM image of Figure 1a. As can be seen, Au presents some aggregates, although, in general, and considering the high percentage of the metal, a good dispersion is observed. Using this material as the starting catalyst, 3% rGO and 3% gC$_3$N$_4$ were incorporated. Figure 2 shows the TEM and HR-TEM images obtained for the most active catalyst in the photodegradation reactions (10%Au@ZnONPs-3%rGO-3%gC$_3$N$_4$). Figure 2a shows the TEM image of the catalyst, in which the heterostructure can be observed, together with the highly dispersed gold nanoparticles on the surface. The inset image presents the selected area electron diffraction (SAED) pattern of the photocatalyst, demonstrating the crystalline nature of the sample. Figure 2b shows the HRTEM image of a ZnO nanoparticle showing the distinct lattice fringes with interplanar spacing of 0.28 nm, indexed to the (100) crystal plane of ZnO with a hexagonal wurtzite structure [38]. Figure 2c shows a region of the catalyst gC$_3$N$_4$. A crystalline structure is observed, whose lattice distance is approximately 0.33 nm, corresponding to the (002) plane of gC$_3$N$_4$ [39,40]. Figure shows the distribution of gold nanoparticles on the heterostructure, together with the area of the sample that has been identified as rGO. One of these nanoparticles is the one that been magnified in Figure 2e, whose interplanar spacing of ca. 0.23 nm has been indexed to (111) crystal plane of Au [41]. Figure 2f presents the ultra-high resolution detail of gold nanoparticle just over 1 nm in diameter. This particle has been further enlarged appreciate the details of the icosahedral structure. The white lines delineate the boundary between five different crystal domains on the nanoparticle. One of the faces shows interplanar spacing of 0.23 nm, assigned to Au (111).

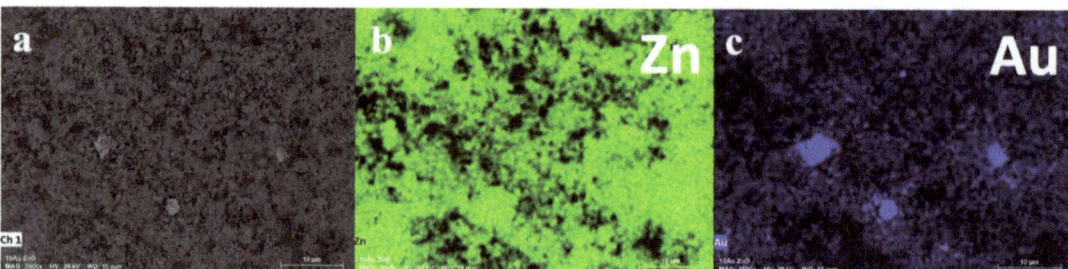

Figure 1. (a) SEM image of 10%Au@ZnONPs, (b) the corresponding elemental mapping of Zn and (c) Au. HV (20 keV), magnification 2500×, and WD (15 mm).

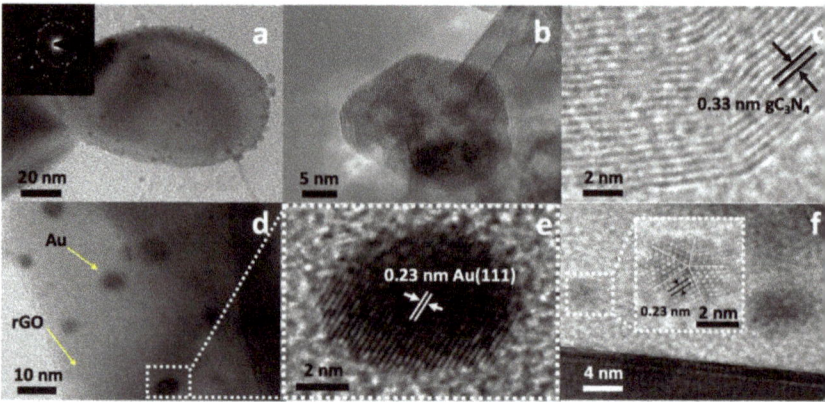

Figure 2. (a) TEM and (b–f) HR-TEM images of 10%Au@ZnONPs-3%rGO-3%gC$_3$N$_4$. (a) Low magnification and SAED pattern of the heterostructure; (b) magnified image of a ZnO nanoparticle (c) crystalline structure of gC$_3$N$_4$; and (d–f) images at different magnification of AuNPs.

10%Au@ZnONPs-3%rGO-3%gC$_3$N$_4$ was also characterized by X-ray diffraction (XRD Figure 3 shows the diffraction pattern of the catalyst, along with that of rGO, gC$_3$N$_4$, and ZnONPs for comparison purposes. There is a broad peak at ca. 23.8° for rGO, assigned the (002) crystal plane, indicating that most of the oxygen functional groups, characteristic of GO, have been removed from the surface [42]. Additionally, rGO shows a second peak about 43° assigned to the (100) plane of the hexagonal carbon structure. The XRD pattern of gC$_3$N$_4$ is shown in Figure 3b. Two broad peaks are observed at ca. 13° and 27.2°, which have been assigned to crystalline planes (100) and (002), respectively [39,40]. The peak shown at 27.2°, and which is also the most intense, corresponds to an interplanar spacing of 0.33 nm, the crystallographic planes which were observed in Figure 1c. The diffraction peaks of ZnONPs (see Figure 3c) can be unambiguously indexed to the ZnO phase hexagonal wurtzite [43], whose peaks are the dominant ones in the 10%Au@ZnONPs-3%rGO-3%gC$_3$N$_4$ catalyst, as observed in Figure 3d. Neither rGO nor gC$_3$N$_4$ are detected in the XRD of the catalyst, possibly because the proportion of these components in the heterostructure is too small. Additionally, a low intensity peak is observed at ca. 38.1°, which has been assigned to Au (111) [41], and whose crystallographic planes were observed in Figure 2e.

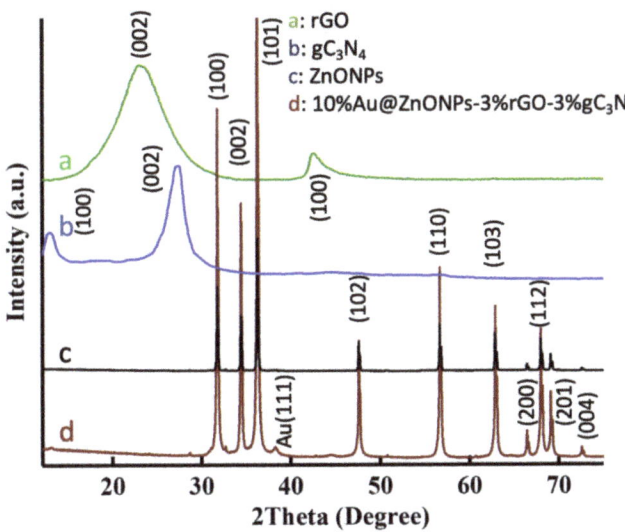

Figure 3. (a) XRD patterns of rGO; (b) gC$_3$N$_4$; (c) ZnONPs; and (d) 10%Au@ZnONPs-3%rGO-3%gC$_3$N$_4$.

Figure 4 shows the Raman spectrum of 10%Au@ZnONPs-3%rGO-3%gC$_3$N$_4$, along with those of rGO, gC$_3$N$_4$, and ZnONPs. As can be seen, rGO is characterized by two broad bands at 1357 and 1600 cm^{-1} (Figure 4a) that have been assigned to bands D (A$_{1g}$ mode), and G (E$_{2g}$ mode of sp^2 carbon atoms), respectively. The intensity ratio of these bands (I$_G$/I$_D$) is ca. 0.95, indicating that the GO reduction is not too high [43]. gC$_3$N$_4$ (Figure 4b) shows two typical bands, similar to those seen in rGO, although slightly displaced. Band D is associated with the possible presence of sp^3 carbon, justified by structural defects and disarrangements, while band G is associated, in the same way as in rGO, with the presence of sp^2 carbon [44]. ZnONPs is characterized by showing two peaks at ca. 566 cm^{-1} and 1143 cm^{-1} (see Figure 4c), which have been assigned to A$_1$-LO and E$_1$-LO vibration modes of ZnO, respectively [45,46]. Both peaks are indicative that ZnONPs has a wurtzite-like crystal structure, as established by XRD. The Raman spectrum of 10%Au@ZnONPs-3%rGO-3%gC$_3$N$_4$ is characterized by showing a main peak at 1042 cm^{-1}, which could come from the peak observed at 1143 cm^{-1} in ZnONPs. Around ca. 630 cm^{-1} an undefined peak is observed, which could also have its origin in the peak observed at 566 cm^{-1} in ZnONPs. The displacements observed in the peaks of the catalyst support the fact that this material is an integrated heterostructure separate from its different components. Additionally, two small peaks are observed at ca. 1301 and 1588 cm^{-1} that must necessarily have their origin in the contribution of rGO and gC$_3$N$_4$.

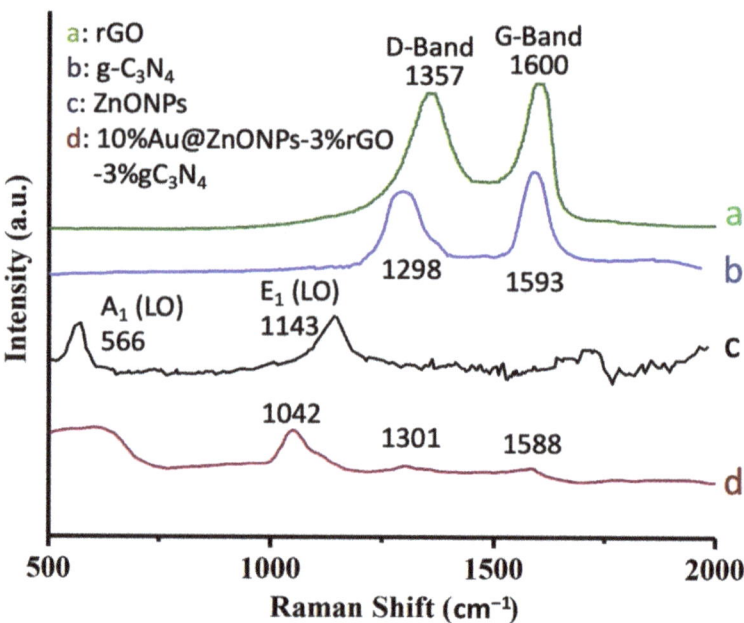

Figure 4. (a) Raman spectra of rGO; (b) gC$_3$N$_4$; (c) ZnONPs; and (d) 10%Au@ZnONPs-3%rGO-3%gC$_3$N$_4$.

The absorption of radiation by catalysts is crucial for their catalytic activity, so the different systems were analyzed using Tauc plots. As shown in Figure 5a, ZnONPs show a bandgap in the UV region (3.23 eV). The incorporation of Au nanoparticles on the surface of the semiconductor (Figure 5b) produces a slight shift at 3.21 eV, still in the UV region. gC$_3$N$_4$ (Figure 5c) and rGO (Figure 5d) show bandgaps at 2.79 and 2.33 eV, respectively, already in the visible range. The addition of these components will be able to displace the bandgap of the catalyst towards the visible region. In fact, Figure 5e shows how the bandgap moves towards 2.77 eV in the 10%Au-ZnONPs-3%rGO composite. The addition of gC$_3$N$_4$ (Figure 5f) still causes a greater displacement, in this case up to 2.73 eV, falling squarely in the visible region. The results obtained justify the activity of the 10%Au@ZnONPs-3%rGO-3%gC$_3$N$_4$ heterostructure under irradiation with visible light, as will be described in the section corresponding to catalytic results.

10%Au@ZnONPs-3%rGO-3%gC$_3$N$_4$ was also characterized by XPS. As shown in Figure 6a, two peaks at 1020.6 eV and 1044.1 eV have been assigned to the binding energies of Zn2p3/2 and Zn2p1/2, respectively, indicating the presence of Zn^{2+} [47]. Furthermore, the spin-orbit splitting of these two peaks at 23.5 eV also confirmed the presence of ZnO [48]. The transition corresponding to O1s (see Figure 6b) showed a major peak at ca. 530.2 eV, which was assigned to O^{2-} species in the ZnO network, and a shoulder at ca. 532.1 eV assigned to O^{2-} in oxygen-deficient regions, respectively [49]. The Au4f peak (Figure 6c) was fitted to two peaks at 83.3 and 86.9 eV, attributed to Au4f$_{7/2}$ and Au4f$_{5/2}$ double peak, respectively, in metallic gold (Au^0) [50]. The C1s transition (Figure 6d) showed a peak at 287.6 eV, which was assigned to C-N-C bonds, and a less intense one at 284.8 eV attributed to the aromatic C atom in the s-triazine ring, respectively [51,52]. Nitrogen from graphitic carbon nitride was evidenced by the N1s transition (see Figure 6e). This clearly asymmetric peak could be deconvolved into two components, showing an intense peak at 398.7 eV, which was assigned to C=N-C, indicating the presence of triazine rings, and a less intense peak at 400.1 eV that was assigned to the presence of tertiary N atoms (N-(C)$_3$) [53,54].

As previously shown, the XPS analysis further confirmed the association of the different components (Au, ZnO, rGO, and gC$_3$N$_4$) in the heterostructure.

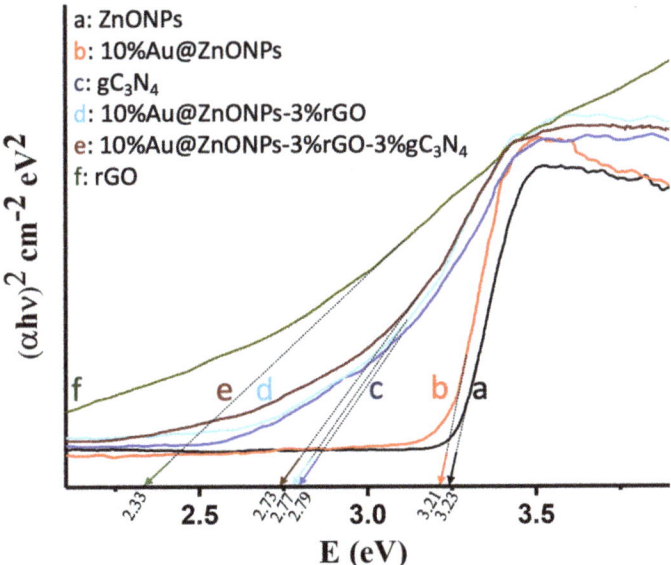

Figure 5. (a) Tauc plots of (αhν)^2 versus energy (eV), and determination of the bandgap energy of ZnONPs; (b) 10%Au@ZnONPs; (c) gC$_3$N$_4$; (d) 10%Au@ZnONPs-3%rGO; (e) 10%Au@ZnONPs-3%rGO-3%gC$_3$N$_4$ and (f) rGO.

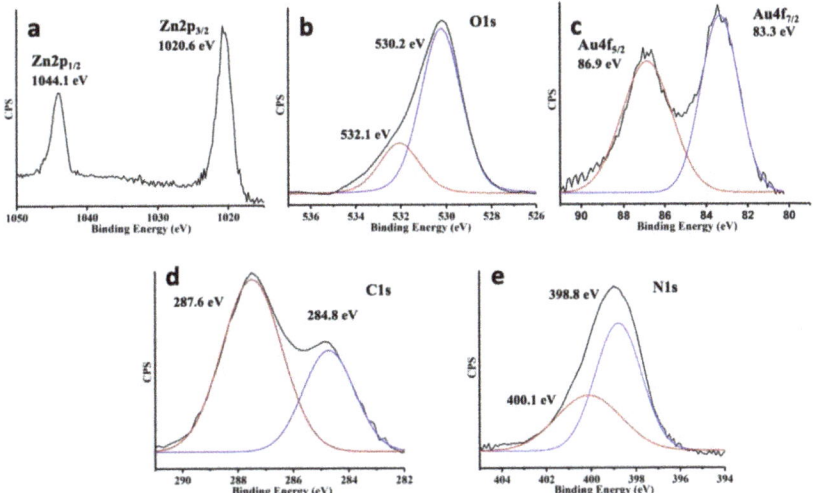

Figure 6. (a) XPS core level spectra for 10%Au@ZnONPs-3%rGO-3%gC$_3$N$_4$: Zn2p; (b) O1s; (c) Au4f; (d) C1s; and (e) N1s.

2.2. Photocatalytic Degradation

Before proceeding to the CFX and LFX photodegradation, several preliminary studies were carried out to establish the optimal reaction conditions. To do this, a study of the optimal pH was initially carried out, and it was established that the most appropriate pH for both antibiotics was to use solutions at pH = 7, which guaranteed the maximum solubility of both contaminants in water. Another parameter that was evaluated was initial antibiotic concentration. For this, the three catalytic systems studied were used using, in all cases, the highest concentration of Au (10%). The results for CFX and LFX shown in Figure S1. As can be seen, there is a clear relationship between the percentage degradation and the initial concentration of the antibiotic. In the case of CFX (Figure S the highest activity in the three catalytic systems studied was obtained at a concentrat of 20 mM. In the case of LFX (see Figure S1b), the optimal concentration was half (10 m possibly because LFX is slightly less soluble in water than CFX. Another of the prelimin studies that were carried out established the optimal amount of catalyst in the react medium. As can be seen in Figure S2, there is a clear correlation between the amount catalyst and the percentage of degradation. For both antibiotics, a sustained increase in efficiency of the process was observed, until reaching a load of 1.1 gL^{-1} in CFX (Figure S and 1 gL^{-1} in LFX (Figure S2b). From these values, a drop in efficiency was observed, wh has been justified on the basis of possible radiation scattering processes that potentia occur from a certain catalyst load. Taking into account the above considerations, the optimi reaction conditions for each antibiotic were established. In the case of CFX, the optimiz conditions were: pH = 7, [CFX]$_i$ = 20 mM, and catalyst loading of 1.1 gL^{-1}, while for L the reaction conditions were: pH = 7, [CFX]$_i$ = 10 mM, and catalyst loading of 1 gL^{-1}.

Figures 7 and 8 show the results of photodegradation of CFX and LFX, respectiv for each of the catalysts studied. The study was carried out for a total time of 180 min a for both contaminants, the behavior was very different depending on the catalyst. In case of CFX (Figure 7), it was observed that after 3 h, the degradation ranged between 7 and 99%. The highest efficiency was obtained with 10%Au@ZnONPs-3%rGO-3%gC$_3$ while the lowest degradation was obtained with 1%Au@ZnONPs. In this last catalyst, percentage of Au on the surface clearly affects the reaction, facilitating absorption a formation of electron-hole pairs, showing greater efficiency with 5%Au@ZnONPs, and maximum efficiency with 10%Au@ZnONPs. In the case of LFX, the behavior is simi showing the highest degradation with 10%Au@ZnONPs-3%rGO-3%gC$_3$N$_4$ (96%), and lowest with 1%Au@ZnONPs (51%). As seen with CFX, the addition of a higher gold lo greatly increased the efficiency of the process. For both CFX and LFX, the incorporati of rGO considerably improved the efficiency of photodegradation, which was maximiz with the addition of gC$_3$N$_4$, especially in catalysts with 10%Au. Another observation consider when comparing the degradation profiles of CFX and LFX is the different r at which the process is carried out. The degradation of CFX is certainly faster than t of LFX, indicating that CFX is more recalcitrant to degradation. The greater degradati observed when incorporating rGO and, additionally, gC$_3$N$_4$ is due to a synergistic effect improvement of the electrical and conductive properties of the material, and to an incre in absorption in the visible region, as evidenced by the determination of the bandgap (Figure 5).

In order to investigate the kinetic behavior of the photodegradation of CFX and L the pseudo-first-order kinetics were studied by representing $-\ln(C/C_0)$ vs. the irradiati time (see Figure S3). The results obtained are shown in Table S1, indicating that the appar rate constant of the catalysts increases with an increasing amount of Au on the surfa and when incorporating rGO and gC$_3$N$_4$. The correlation coefficients R^2 (see Table S1) higher for the linear fits of LFX than CFX although, as derived from the results obtain for both contaminants the degradation clearly follows pseudo-first-order kinetics. T behavior can also be correlated with the BET surface area of these catalysts. Thus, as sho in Table 1, the materials that show the highest BET area values are those that have bet catalytic behavior, with higher apparent rate constants.

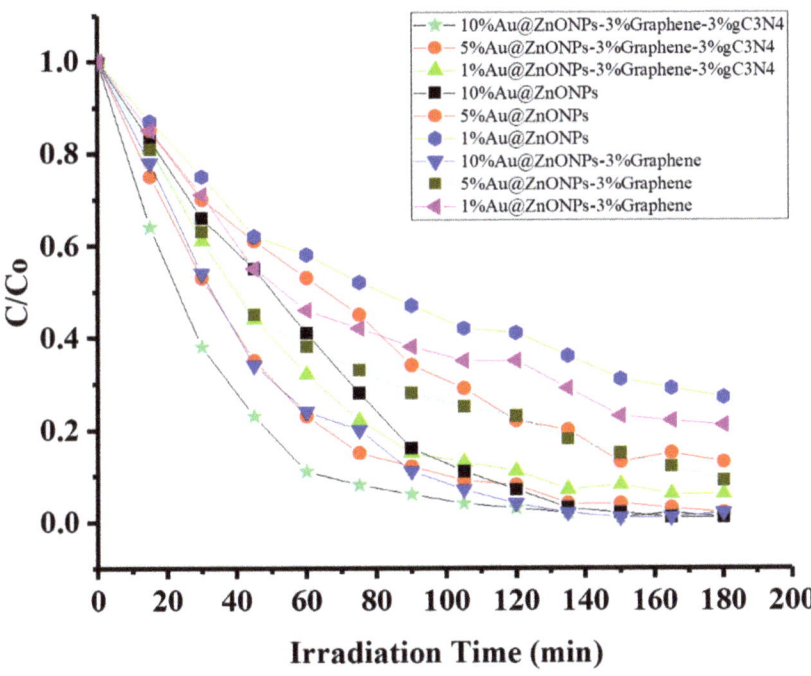

Figure 7. Photodegradation rate of CFX as a function of time, using the different catalysts evaluated.

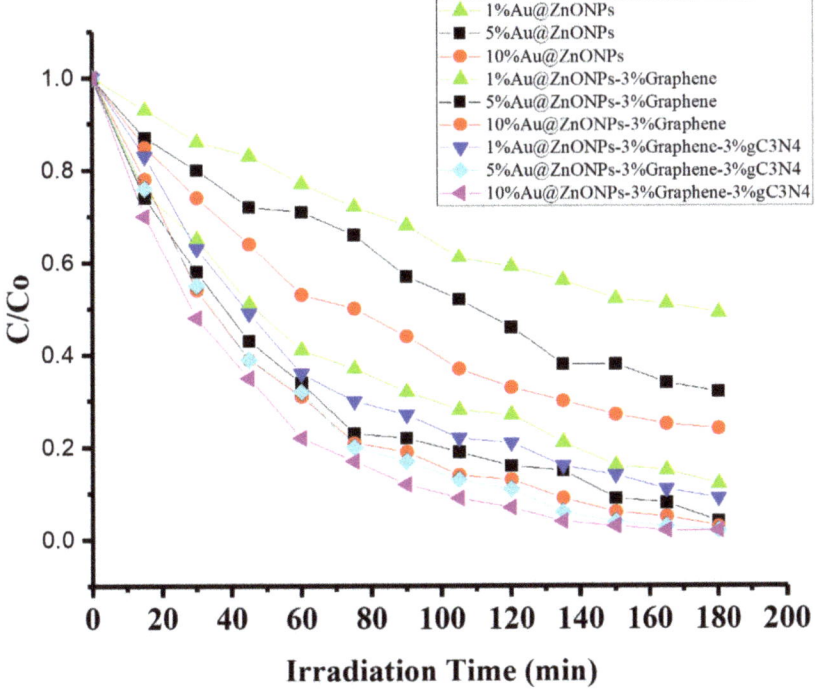

Figure 8. Photodegradation rate of LFX as a function of time, using the different catalysts evaluated.

To further study the photocatalytic activity of 10%Au@ZnONPs-3%rGO-3%gC$_3$N$_4$, control experiments and recyclability tests were carried out. The control experiments for both contaminants were similar (Figure S4). After 3 h of reaction, it was found that CFX and LFX were highly recalcitrant. Under photolysis conditions, that is, in the absence of the catalyst with the rest of the components present, a very slight reduction in the concentration of both CFX and LFX was observed, which indicates that radiation absorption is the only pathway responsible for this change and, as seen in Figure S4, it is clearly inefficient. In the absence of oxygen, a greater degradation of CFX than LFX was observed but, in both cases, the percentage of degradation after 3 h of reaction is not relevant. From control experiments, photocatalysis is shown to be the primary degradation route, indicating that catalysis, even photolysis, is totally insufficient to degrade antibiotics. The results of recyclability 10%Au@ZnONPs-3%rGO-3%gC$_3$N$_4$ for the degradation of CFX and LFX during five cycles are shown in Figure S5. The catalyst was recovered after each cycle by centrifugation (30 rpm for 20 min). After the supernatant was decanted, the catalyst was washed three times with deionized water and once with ethanol (using centrifugation–sonication cycles), and dried for at least 5 h at 60 °C. The parameters used for the degradation of each antibiotic were set to optimal values, as previously described. In CFX (Figure S5a) it was found that the degradation remained practically constant after each cycle, going from 99% after the first cycle to 97% after the fifth cycle. In the case of LFX (Figure S5b), the recyclability was not as good, going from a degradation of 97% in the first cycle to 90% after the fifth cycle. In both cases (CFX and LFX) the photocatalyst recovered after the fifth cycle was analyzed by XRD (not shown) showing a diffraction pattern without notable differences with respect to the XRD of the catalyst before use.

2.3. Proposed Photodegradation Mechanism for CFX and LFX

As previously shown, gold nanoparticles improve catalytic efficiency, which increases appreciably when rGO and gC$_3$N$_4$ are incorporated. The most efficient catalyst for the photodegradation of CFX and LFX (10%Au@ZnONPs-3%rGO-3%gC$_3$N$_4$) incorporates different characteristics that act synergistically on its behavior, that is, increased surface area and smaller bandgap. The considerable increase in the specific area of the catalyst can provide more active sites during the photocatalytic reaction, thus producing more photogenerated electrons, which can also lead to less recombination of photogenerated charge carriers. This decrease in recombination is also favored by the presence of Au nanoparticles, which act as electron sinks. The displacement towards the visible region of the bandgap, with respect to other catalysts studied, is another of the decisive factors for the observed behavior.

In this context, and to evaluate the photodegradation mechanism, some scavengers were added to the reaction. In this photocatalytic study, benzoquinone (BQ), ethylene diaminetetraacetic acid disodium salt (EDTA), and methanol (MetOH), were employed to capture superoxide radicals ($\cdot O^{2-}$), holes (h^+), and hydroxyl radicals ($\cdot OH$), respectively [55,56], determining the percentage of degradation of CFX and LFX after 3 h reaction (see Figure S6). As can be seen, BQ hindered photoactivity noticeably, suggesting the main role of the O^{2-} reactive species in the photodegradation process. EDTA and MetOH hindered the reaction to a lesser extent, which supports the fact that h^+ and $\cdot OH$ do not play a prominent role in the degradation process. This effect is similar for both antibiotics (see Figure S6a,b), although in the case of LFX, the effects of all scavengers were certainly greater.

Taking into account these results, together with the determination of the bandgap (see Figure 5), a photodegradation mechanism of CFX and LFX using 10%Au@ZnONPs-3%rGO-3%gC$_3$N$_4$ (Figure 9) has been proposed. For this, the Mulliken electronegativity theory [57,58] has been used, which allows establishing the band edge position of the different components of the catalyst and, in this way, determining the migration direction of the photogenerated charge carriers in the composite (Equations (1) and (2)).

$$E_{CB} = X - E_C - 0.5E_g \quad (1)$$

$$E_{VB} = E_{CB} + E_g \quad (2)$$

where E_{CB} and E_{VB} are the edge potentials of the valence band (VB) and conduction band (CB), respectively, X is the absolute electronegativity, E_C is the energy of free electrons on the hydrogen scale (4.50 eV) [59,60], and E_g is the bandgap. X values for ZnO and gC_3N_4 are 5.75 [61] and 4.73 eV [62], respectively. The calculated E_{CB} and E_{VB} edge positions for Au@ZnONPs are −0.355 and 2.855 eV, respectively, while for gC_3N_4 the calculated values were −1.165 and 1.625 eV, respectively, being in agreement with values previously determined in other investigations [62,63].

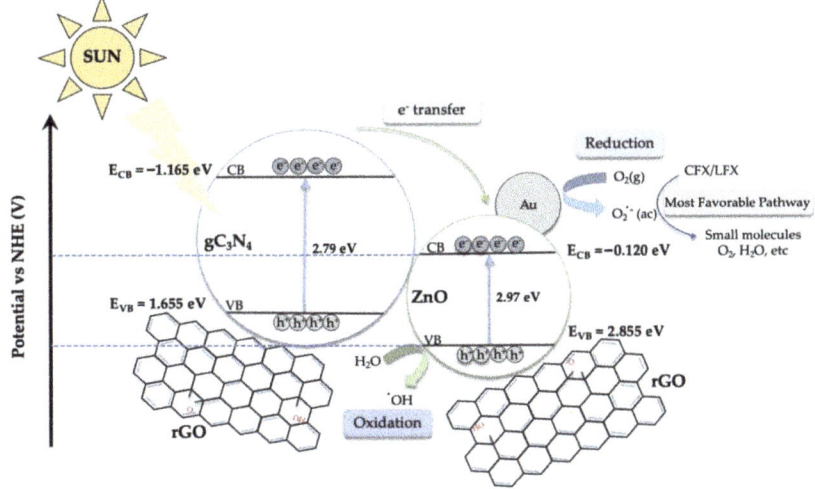

Figure 9. Schematic diagram of the proposed photodegradation mechanism of CFX and LFX under visible radiation.

Under visible radiation (Figure 9), the electrons of the VB of gC_3N_4 are excited towards the conduction band (CB), giving rise to h^+ in the VB. Due to the potential difference with respect to ZnONPs, the electrons of the conduction band of gC_3N_4 move towards the CB of Au@ZnONPs. Au acts as a sink for electrons [64,65], which is why they move towards Au and it is there that they react with the adsorbed molecular oxygen to generate superoxide anions, which in turn can react with water molecules to form hydroxyl radicals. According to studies carried out with different scavengers, the O^{2-} anion is the one that preferentially participates in the photodegradation process of antibiotics, giving rise to small molecules as a by-product of the reaction, CO_2 and water. The holes formed in the VB of gC_3N_4 and, to a lesser extent in ZnO, will promote the oxidation of CFX and LFX also leading to degradation. However, this second pathway, as evidenced above, is not the most favored pathway. AuNPs, in addition to acting as electron sinks and thus reducing recombination processes, also contribute to the system through the surface plasmon resonance mechanism, reacting with molecular oxygen and generating superoxide anions. Photogenerated electrons are also transferred to rGO that, as in the case of gold, acts as an electron acceptor and transport medium in the photocatalytic system, suppressing the recombination of e^--h^+ pairs. Photodegradation reactions therefore also occur on rGO sheets, increasing the specific surface area and active reaction sites [65].

3. Materials and Methods

3.1. Reagents and Materials

All the reagents were used as received, without further purification. All the lutions were prepared using deionized water (Milli-Q water, 18.2 MΩcm^{-1} at 25 °C. Zn(CH$_3$COO)$_2$ (99.99%, trace metal basis), HAuCl$_4$•3H$_2$O (ACS Reagent, 49.0+% Au bas Ethanol (200 proof, anhydrous, ≥99.5%), NaBH$_4$ (99.99% trace metals basis), melam (2,4,6-Triamino-1,3,5-triazine, 99%), reduced graphene oxide (rGO), ethylenediamine traacetic acid, disodium salt dihydrate (EDTA, ACS Reagent, 99.9+%), P-benzoquinc (Spect. Grade, 99.5+%), and methanol (HPLC Plus, 99.9+%) were provided by Sigm Aldrich (Darmstadt, Germany). NaOH (98+%) and HCl (36% w/w aq. soln) were acqui from Alfa Aesar. Whatman® Puradisc 13 disposable syringe filters (Clifton, NJ, USA) w used for the aliquot filtration process.

3.2. Synthesis of Nanomaterials

Zinc oxide nanoparticles (ZnONPs) were synthesized by thermal decomposition anhydrous zinc acetate. In a standard synthesis, 5 g of zinc acetate was introduced in an alumina crucible which was covered by a perforated alumina lid, and subsequen placed in a tube furnace. Next, the sample was subjected to heat treatment in air fl (300 mL/min) using a temperature ramp of 3 °C/min until reaching 450 °C. The sam was kept at this temperature for 2 h then proceeded to cool, maintaining an air flow 100 mL/min and a ramp of −10 °C/min. The sample was recovered the next day. Graph carbon nitride was synthesized using the procedure described by Mo et al. [66]. For t 2 g of melamine were introduced into a crucible, which was subsequently treated at 650 (heating rate of 2 °C/min) in flowing nitrogen (200 mL/min), for 4 h. After treatment, synthesized material was subjected to exfoliation. For this, 200 mg of gC$_3$N$_4$ were pu contact with 20 mL of concentrated H$_2$SO$_4$ (with stirring for 6 h). The solution was th treated with ultrasound for 2 h. The mixture was decanted and centrifuged, washed w water several times until neutral, and finally rinsed with ethanol to facilitate drying. T gC$_3$N$_4$ crystals obtained after the described process showed a characteristic white colo

The deposition of Au NPs has been described elsewhere [49]. In a typical synthe 1g of ZnONPs was dispersed in 100 mL of H$_2$O and the mixture was sonicated for 30 m After that, the desired amount of the gold precursor (HAuCl$_4$·3H$_2$O) was added to reaction mixture and kept stirring for 30 min. Finally, a solution of NaBH$_4$ (10 mg 10 mL of H$_2$O) was added dropwise while maintaining stirring. After the reagent w added, the solution was kept stirring for an additional 30 min. The reaction product w separated by centrifugation, washed 4 times with deionized water, and dried overnig at 80 °C. The different compounds of Au@ZnONPs were identified as 1%Au@ZnON 5%Au@ZnONPs and 10%Au@ZnONPs, where the numbers correspond to the percenta by weight of gold nanoparticles deposited in each sample. The material obtained was us for the following stages of preparation of the catalysts. Thus, 200 mg of ZnONPs containi the gold nanoparticles were dispersed in a solution containing 10 mL of ethanol and 40 r of deionized water, and the mixture was vigorously stirred for 30 min. Subsequently, reduced graphene oxide (rGO) was added, and the suspension was kept under stirring 1 h. Subsequently, the product was separated from the solution by centrifugation, wash 4 times with deionized water and dried overnight at 80 °C. Next, and by same procedu described for rGO, the incorporation of graphitic carbon nitride (gC$_3$N$_4$) was carried c Finally, the product was collected, sealed, and stored at room temperature. The differe catalysts prepared, based on Au@ZnONPs, Au@ZnONPs-rGO, and Au@ZnONPs-rG gC$_3$N$_4$ were identified indicating the percentage of gold incorporated on the surface. In cases, the amount of rGO and gC$_3$N$_4$ used was 3% by weight.

3.3. Characterization of the Catalysts

The morphology of the catalysts was characterized by Scanning Electron Microsco (SEM), using a Hitachi S-3000N instrument (Westford, MA, USA), equipped with a E

Quantax EDS X-Flash 6I30 Analyzer, and by High Resolution Transmission Electron Microscopy (HRTEM), using a JEOL 3000F (Peabody, MA, USA). XPS measurements were performed on an ESCALAB 220i-XL spectrometer (East Grinstead, United Kingdom), using the non-monochromated Mg Kα (1253.6 eV) radiation of a twin-anode, operating at 20 mA and 12 kV in the constant analyzer energy mode, with a PE of 40 eV. Brunauer Emmett Teller (BET) specific areas were measured using a Micromeritics ASAP 2020, according to N_2 adsorption isotherms at 77 K. Raman spectra were collected using a DXR Raman Microscope (Thermo Scientific, Waltham, MA, USA), employing a 532 nm laser source at 5 mW power and a nominal resolution of 5 cm^{-1}. Diffuse reflectance measurements were carried out on a Perkin Elmer Lambda 365 UV-Vis spectrophotometer (Perkin Elmer, Waltham, MA, USA), equipped with an integrating sphere. The bandgap value was obtained from the plot of the Kubelka–Munk function versus the energy of the absorbed light [67]. X-ray powder diffraction patterns (XRD) were collected using a Bruker D8 Discover X-ray Diffractometer (Madison, WI, USA), in Bragg–Brentano goniometer configuration. The X-ray radiation source was a ceramic X-ray diffraction Cu anode tube type Empyrean of 2.2 kW. XRD diffractograms were recorded in angular range of 30–75° at 1° min^{-1}, operating at 40 kV and 40 mA.

3.4. Photocatalytic Experiments

To test the activity of the synthesized catalysts in the photodegradation process of ciprofloxacin and levofloxacin, a solar simulator composed of three white annular bulbs with a total irradiation power of 90 watts was used. In the case of ciprofloxacin, a 2×10^{-5} M solution was prepared, while for levofloxacin the concentration used was 10^{-5} M. Next, the desired catalyst was added, the amount of which also depended on the antibiotic used. After adding the catalyst and adjusting the pH to 7, the system was kept in the dark for 30 min to allow the system to reach adsorption–desorption equilibrium. Then, a small amount of H_2O_2 (2 mL, 0.005%) was added, and additional oxygen was provided to the system by constantly bubbling air into the reaction mixture. The white bulbs were then turned on and the photocatalytic system was maintained with constant agitation and irradiation. The reaction was monitored for a period of 180 min and photodegradation was followed by taking aliquots every 15 min. After filtering the aliquots with Whatman® Puradisc 13 (Clifton, NJ, USA) disposable syringe filters, the samples were immediately analyzed at room temperature with a Perkin Elmer Lambda 35 UV-vis spectrophotometer (San José, CA, USA).

4. Conclusions

The photocatalytic degradation of two quinolone-type antibiotics (CFX and LFX) in aqueous solution was studied, using catalysts based on ZnONPs, which were synthesized by means of a thermal procedure. Subsequently, the efficiency of ZnONPs was optimized by incorporating different cocatalysts of gold, rGO, and gC_3N_4, obtaining a total of nine different catalysts that were used in the photodegradation reaction of CFX and LFX. The most efficient catalyst was 10%Au@ZnONPs-3%rGO-3%gC_3N_4, allowing degradations of both pollutants above 96%. This catalyst has the largest specific area, and its activity has been related to a synergistic effect, involving factors as relevant as the surface of the material and the ability to absorb radiation in the visible region, mainly produced by the incorporation of rGO and gC_3N_4 to the semiconductor. The use of different scavengers during the catalytic process, together with the determination of bandgaps of the different components of the photocatalyst, has made it possible to establish a possible photodegradation mechanism of CFX and LFX in which superoxide radicals ($\cdot O^{2-}$) are the main reactive species involved in the process. The results obtained are certainly relevant because, in less than 3 h, almost complete photodegradation of CFX and LFX occurs, with conversions above 96%. Some catalysts based on TiO_2 doped with boron have shown high degradation percentages (ca. 88%) [68]. In other cases, copper tungstate ($CuWO_4$) catalysts doped with graphene have made it possible to obtain CFX degradations of ca. 97% [69]. Other catalysts based on ZnO doped with silver allowed degradation to 99% of CFX [49], although preliminary studies showed many difficulties in photodegrading LFX. Considering the results published so far, the catalyst developed in this research is cost effective, easy synthesize, and highly effective to for the degradation of both CFX and LFX, opening up wide field of possibilities in environmental decontamination processes. In addition, the developed catalyst could have relevant uses for facing environmental problems generated by other pollutants, from an applied and global point of view.

Supplementary Materials: The following supporting information can be downloaded at: https: www.mdpi.com/article/10.3390/catal12020166/s1, Figure S1: Evaluation of the initial concentration of (a) CFX and (b) LFX on the catalytic efficiency of 10%Au@ZnONPs, 10%Au@ZnONPs-3%rGO, and 10%Au@ZnONPs-3%rGO-3%gC$_3$N$_4$ in the photodegradation reaction; Figure S2: Evaluation of the initial concentration of 10%Au@ZnONPs, 10%Au@ZnONPs-3%rGO, and 10%Au@ZnONPs-3%rGO 3%gC$_3$N$_4$ on the efficiency of the photodegradation reaction of (a) CFX and (b) LFX; Figure Pseudo-first order kinetics of photodegradation of (a) CFX and (b) LFX using different catalysts Figure S4: Control experiments for 10%Au@ZnONPs-3%rGO-3%gC$_3$N$_4$ with (a) CFX and (b) LFX under visible radiation; Figure S5: Recyclability of 10%Au@ZnONPs-3%rGO-3%gC$_3$N$_4$ after five consecutive catalytic cycles of photodegradation of (a) CFX and (b) LFX under visible radiation Figure S6: Photocatalytic activity of 10%Au@ZnONPs-3%rGO-3%gC$_3$N$_4$ on the degradation (a) CFX and (b) LFX in the presence of various scavengers under visible radiation; Table S1: The pseudo-first-order kinetics constants for the photodegradation of CFX and LFX.

Author Contributions: Conceptualization, A.M., F.M.; methodology, F.M., A.M.; formal analysis F.M., M.C.C., J.D.; investigation, A.M., K.F., C.M.; resources, F.M., C.M., F.I.P.; writing—original draft preparation, F.M.; writing—review and editing, F.M., A.M., K.F., C.M., M.C.C., J.D.; supervision A.M., F.M.; project administration, A.M., F.M.; funding acquisition, A.M., F.M., M.C.C., J.D., C.M. F.I.P. All authors have read and agreed to the published version of the manuscript.

Funding: Financial support from the NSF Center for the Advancement of Wearable Technologies CAWT (Grant 1849243), and from the framework of the UE M-ERA.NET 2018 program under the StressLIC Project (Grant PCI2019-103594) are gratefully acknowledged.

Data Availability Statement: The data is contained in the article and is available from the corresponding authors on reasonable request.

Acknowledgments: The authors thank Raúl S. García for his arduous technical help in the procedures for the synthesis and characterization of materials. Technical assistance of I. Poveda from "Servicio Interdepartamental de Investigacion, SIdI" at Autonomous University of Madrid (Spain), gratefully acknowledged. The facilities provided by the National Center for Electron Microscopy Complutense University of Madrid (Spain) is gratefully acknowledged. K.F. thanks PR NASA Space Grant Consortium for a graduate fellowship (#80NSSC20M0052).

Conflicts of Interest: The authors declare no conflict of interest.

References

1. Wang, J.; Svoboda, L.; Němečková, Z.; Sgarzi, M.; Henych, J.; Licciardello, N.; Cuniberti, G. Enhanced Visible-Light Photodegradation of Fluoroquinolone-Based Antibiotics and *E. coli* Growth Inhibition Using Ag–TiO$_2$ Nanoparticles. *RSC Adv.* **2021**, 13980–13991. [CrossRef]
2. Rodríguez-López, L.; Cela-Dablanca, R.; Núñez-Delgado, A.; Álvarez-Rodríguez, E.; Fernández-Calviño, D.; Arias-Estévez, Photodegradation of Ciprofloxacin, Clarithromycin and Trimethoprim: Influence of PH and Humic Acids. *Molecules* **2021**, 3080. [CrossRef] [PubMed]
3. El-Maraghy, C.M.; El-Borady, O.M.; El-Naem, O.A. Effective Removal of Levofloxacin from Pharmaceutical Wastewater Using Synthesized Zinc Oxide, Graphen Oxid Nanoparticles Compared with Their Combination. *Sci. Rep.* **2020**, *10*, 5914. [CrossRef] [PubMed]
4. Larsson, D.G.J.; de Pedro, C.; Paxeus, N. Effluent from Drug Manufactures Contains Extremely High Levels of Pharmaceuticals *J. Hazard. Mater.* **2007**, *148*, 751–755. [CrossRef] [PubMed]
5. Rigos, G.; Bitchava, K.; Nengas, I. Antibacterial Drugs in Products Originating from Aquaculture: Assessing the Risks to Public Welfare. *Medit. Mar. Sci.* **2010**, *11*, 33. [CrossRef]

Hubeny, J.; Harnisz, M.; Korzeniewska, E.; Buta, M.; Zieliński, W.; Rolbiecki, D.; Giebułtowicz, J.; Nałęcz-Jawecki, G.; Płaza, G. Industrialization as a Source of Heavy Metals and Antibiotics Which Can Enhance the Antibiotic Resistance in Wastewater, Sewage Sludge and River Water. *PLoS ONE* **2021**, *16*, e0252691. [CrossRef]

Rozman, U.; Duh, D.; Cimerman, M.; Turk, S.Š. Hospital Wastewater Effluent: Hot Spot for Antibiotic Resistant Bacteria. *J. Water Sanit. Hyg. Dev.* **2020**, *10*, 171–178. [CrossRef]

Church, N.A.; McKillip, J.L. Antibiotic Resistance Crisis: Challenges and Imperatives. *Biologia* **2021**, *76*, 1535–1550. [CrossRef]

Talebi Bezmin Abadi, A.; Rizvanov, A.A.; Haertlé, T.; Blatt, N.L. World Health Organization Report: Current Crisis of Antibiotic Resistance. *BioNanoScience* **2019**, *9*, 778–788. [CrossRef]

Sitara, E.; Ehsan, M.F.; Nasir, H.; Iram, S.; Bukhari, S.A.B. Synthesis, Characterization and Photocatalytic Activity of MoS_2/ZnSe Heterostructures for the Degradation of Levofloxacin. *Catalysts* **2020**, *10*, 1380. [CrossRef]

Dawadi, S.; Thapa, R.; Modi, B.; Bhandari, S.; Timilsina, A.P.; Yadav, R.P.; Aryal, B.; Gautam, S.; Sharma, P.; Thapa, B.B.; et al. Technological Advancements for the Detection of Antibiotics in Food Products. *Processes* **2021**, *9*, 1500. [CrossRef]

Hassan, M.; Zhu, G.; Lu, Y.; AL-Falahi, A.H.; Lu, Y.; Huang, S.; Wan, Z. Removal of Antibiotics from Wastewater and Its Problematic Effects on Microbial Communities by Bioelectrochemical Technology: Current Knowledge and Future Perspectives. *Environ. Eng. Res.* **2020**, *26*, 16–30. [CrossRef]

Huang, A.; Yan, M.; Lin, J.; Xu, L.; Gong, H.; Gong, H. A Review of Processes for Removing Antibiotics from Breeding Wastewater. *Int. J. Environ. Res. Public Health* **2021**, *18*, 4909. [CrossRef] [PubMed]

Rodriguez-Mozaz, S.; Vaz-Moreira, I.; Varela Della Giustina, S.; Llorca, M.; Barceló, D.; Schubert, S.; Berendonk, T.U.; Michael-Kordatou, I.; Fatta-Kassinos, D.; Martinez, J.L.; et al. Antibiotic Residues in Final Effluents of European Wastewater Treatment Plants and Their Impact on the Aquatic Environment. *Environ. Int.* **2020**, *140*, 105733. [CrossRef]

Mahdi, M.H.; Mohammed, T.J.; Al-Najar, J.A. Advanced Oxidation Processes (AOPs) for Treatment of Antibiotics in Wastewater: A Review. *IOP Conf. Ser. Earth Environ. Sci.* **2021**, *779*, 012109. [CrossRef]

Cuerda-Correa, E.M.; Alexandre-Franco, M.F.; Fernández-González, C. Advanced Oxidation Processes for the Removal of Antibiotics from Water. An Overview. *Water* **2019**, *12*, 102. [CrossRef]

Akbari, M.Z.; Xu, Y.; Lu, Z.; Peng, L. Review of Antibiotics Treatment by Advance Oxidation Processes. *Environ. Adv.* **2021**, *5*, 100111. [CrossRef]

Li, H.; Zhang, W.; Liu, Y. HZSM-5 Zeolite Supported Boron-Doped TiO_2 for Photocatalytic Degradation of Ofloxacin. *J. Mater. Res. Technol.* **2020**, *9*, 2557–2567. [CrossRef]

Lops, C.; Ancona, A.; Di Cesare, K.; Dumontel, B.; Garino, N.; Canavese, G.; Hérnandez, S.; Cauda, V. Sonophotocatalytic Degradation Mechanisms of Rhodamine B Dye via Radicals Generation by Micro- and Nano-Particles of ZnO. *Appl. Catal. B Environ.* **2019**, *243*, 629–640. [CrossRef]

Fujishima, A.; Zhang, X.; Tryk, D. TiO_2 Photocatalysis and Related Surface Phenomena. *Surf. Sci. Rep.* **2008**, *63*, 515–582. [CrossRef]

Molinari, R.; Lavorato, C.; Argurio, P. Visible-Light Photocatalysts and Their Perspectives for Building Photocatalytic Membrane Reactors for Various Liquid Phase Chemical Conversions. *Catalysts* **2020**, *10*, 1334. [CrossRef]

Zhang, F.; Wang, X.; Liu, H.; Liu, C.; Wan, Y.; Long, Y.; Cai, Z. Recent Advances and Applications of Semiconductor Photocatalytic Technology. *App. Sci.* **2019**, *9*, 2489. [CrossRef]

Li, H.; Li, Z.; Yu, Y.; Ma, Y.; Yang, W.; Wang, F.; Yin, X.; Wang, X. Surface-Plasmon-Resonance-Enhanced Photoelectrochemical Water Splitting from Au-Nanoparticle-Decorated 3D TiO_2 Nanorod Architectures. *J. Phys. Chem. C* **2017**, *121*, 12071–12079. [CrossRef]

Kholikov, B.; Hussain, J.; Hayat, S.; Zeng, H. Surface Plasmon Resonance Improved Photocatalytic Activity of Au/$TIO2$ Nanocomposite under Visible Light for Degradation of Pollutants. *J. Chin. Chem. Soc.* **2021**, *68*, 1908–1915. [CrossRef]

Ahmadi, M.; Mistry, H.; Roldan Cuenya, B. Tailoring the Catalytic Properties of Metal Nanoparticles via Support Interactions. *J. Phys. Chem. Lett.* **2016**, *7*, 3519–3533. [CrossRef]

Smith, A.T.; LaChance, A.M.; Zeng, S.; Liu, B.; Sun, L. Synthesis, Properties, and Applications of Graphene Oxide/Reduced Graphene Oxide and Their Nanocomposites. *Nano Mater. Sci.* **2019**, *1*, 31–47. [CrossRef]

Siklitskaya, A.; Gacka, E.; Larowska, D.; Mazurkiewicz-Pawlicka, M.; Malolepszy, A.; Stobiński, L.; Marciniak, B.; Lewandowska-Andrałojć, A.; Kubas, A. Lerf–Klinowski-Type Models of Graphene Oxide and Reduced Graphene Oxide Are Robust in Analyzing Non-Covalent Functionalization with Porphyrins. *Sci. Rep.* **2021**, *11*, 7977. [CrossRef]

Fatima, S.; Irfan Ali, S.; Younas, D.; Islam, A.; Akinwande, D.; Rizwan, S. Graphene Nanohybrids for Enhanced Catalytic Activity and Large Surface Area. *MRS Commun.* **2019**, *9*, 27–36. [CrossRef]

Yaqoob, A.A.; Mohd Noor, N.H.b.; Serrà, A.; Mohamad Ibrahim, M.N. Advances and Challenges in Developing Efficient Graphene Oxide-Based ZnO Photocatalysts for Dye Photo-Oxidation. *Nanomaterials* **2020**, *10*, 932. [CrossRef]

Lu, K.-Q.; Li, Y.-H.; Tang, Z.-R.; Xu, Y.-J. Roles of Graphene Oxide in Heterogeneous Photocatalysis. *ACS Mater. Au* **2021**, *1*, 37–54. [CrossRef]

Inagaki, M.; Tsumura, T.; Kinumoto, T.; Toyoda, M. Graphitic Carbon Nitrides (g-C_3N_4) with Comparative Discussion to Carbon Materials. *Carbon* **2019**, *141*, 580–607. [CrossRef]

Vu, M.H.; Nguyen, C.C.; Do, T. Graphitic Carbon Nitride (g-C_3N_4) Nanosheets as a Multipurpose Material for Detection of Amines and Solar-Driven Hydrogen Production. *ChemPhotoChem* **2021**, *5*, 466–475. [CrossRef]

33. Darkwah, W.K.; Ao, Y. Mini Review on the Structure and Properties (Photocatalysis), and Preparation Techniques of Graphitic Carbon Nitride Nano-Based Particle, and Its Applications. *Nanoscale Res. Lett.* **2018**, *13*, 388. [CrossRef] [PubMed]
34. Wang, J.; Wang, S. A Critical Review on Graphitic Carbon Nitride (g-C_3N_4)-Based Materials: Preparation, Modification and Environmental Application. *Coord. Chem. Rev.* **2022**, *453*, 214338. [CrossRef]
35. Mezni, A.; Ben Saber, N.; Ibrahim, M.M.; Shaltout, A.A.; Mersal, G.A.M.; Mostafa, N.Y.; Alharthi, S.; Boukherroub, R.; Altalhi, T. Pt–ZnO/M (M = Fe, Co, Ni or Cu): A New Promising Hybrid-Doped Noble Metal/Semiconductor Photocatalysts. *J. Inorg. Organomet. Polym.* **2020**, *30*, 4627–4636. [CrossRef]
36. Niu, J.; Albero, J.; Atienzar, P.; García, H. Porous Single-Crystal-Based Inorganic Semiconductor Photocatalysts for Energy Production and Environmental Remediation: Preparation, Modification, and Applications. *Adv. Funct. Mater.* **2020**, *30*, 1908984. [CrossRef]
37. Prabavathi, S.L.; Saravanakumar, K.; Park, C.M.; Muthuraj, V. Photocatalytic Degradation of Levofloxacin by a Novel Sm_6WO_{12}/C_3N_4 Heterojunction: Performance, Mechanism and Degradation Pathways. *Sep. Pur. Technol.* **2021**, *257*, 117985. [CrossRef]
38. Han, X.-G.; He, H.-Z.; Kuang, Q.; Zhou, X.; Zhang, X.-H.; Xu, T.; Xie, Z.-X.; Zheng, L.-S. Controlling Morphologies and Tuning Related Properties of Nano/Microstructured ZnO Crystallites. *J. Phys. Chem. C* **2009**, *113*, 584–589. [CrossRef]
39. Ou, H.; Lin, L.; Zheng, Y.; Yang, P.; Fang, Y.; Wang, X. Tri-s-Triazine-Based Crystalline Carbon Nitride Nanosheets for an Improved Hydrogen Evolution. *Adv. Mater.* **2017**, *29*, 1700008. [CrossRef]
40. Xing, W.; Tu, W.; Han, Z.; Hu, Y.; Meng, Q.; Chen, G. Template-Induced High-Crystalline g-C_3N_4 Nanosheets for Enhanced Photocatalytic H_2 Evolution. *ACS Energy Lett.* **2018**, *3*, 514–519. [CrossRef]
41. The International Center for Diffraction Data (ICDD) No. 00-004-0784. Available online: https://www.icdd.com/pdfsearch (accessed on 20 December 2021).
42. Xu, Y.; Sheng, K.; Li, C.; Shi, G. Self-Assembled Graphene Hydrogel via a One-Step Hydrothermal Process. *ACS Nano* **2010**, *4*, 4324–4330. [CrossRef] [PubMed]
43. Stankovich, S.; Dikin, D.A.; Piner, R.D.; Kohlhaas, K.A.; Kleinhammes, A.; Jia, Y.; Wu, Y.; Nguyen, S.T.; Ruoff, R.S. Synthesis of Graphene-Based Nanosheets via Chemical Reduction of Exfoliated Graphite Oxide. *Carbon* **2007**, *45*, 1558–1565. [CrossRef]
44. Sumathi, M.; Prakasam, A.; Anbarasan, P.M. High Capable Visible Light Driven Photocatalytic Activity of WO_3/g-C_3N_4 Hetrostructure Catalysts Synthesized by a Novel One Step Microwave Irradiation Route. *J. Mater. Sci. Mater. Electron.* **2019**, *30*, 3294–3304. [CrossRef]
45. Yan, X.; Li, W.; Aberle, A.G.; Venkataraj, S. Investigation of the Thickness Effect on Material and Surface Texturing Properties of Sputtered ZnO:Al Films for Thin-Film Si Solar Cell Applications. *Vacuum* **2016**, *123*, 151–159. [CrossRef]
46. Murmu, P.P.; Kennedy, J.; Ruck, B.J.; Leveneur, J. Structural, Electronic and Magnetic Properties of Er Implanted ZnO Thin Films. *Nucl. Instrum. Methods Phys. Res. Sect. B Beam Interact. Mater. At.* **2015**, *359*, 1–4. [CrossRef]
47. Naseri, A.; Samadi, M.; Mahmoodi, N.M.; Pourjavadi, A.; Mehdipour, H.; Moshfegh, A.Z. Tuning Composition of Electrospun ZnO/CuO Nanofibers: Toward Controllable and Efficient Solar Photocatalytic Degradation of Organic Pollutants. *J. Phys. Chem. C* **2017**, *121*, 3327–3338. [CrossRef]
48. Qiao, Y.; Li, J.; Li, H.; Fang, H.; Fan, D.; Wang, W. A Label-Free Photoelectrochemical Aptasensor for Bisphenol A Based on Surface Plasmon Resonance of Gold Nanoparticle-Sensitized ZnO Nanopencils. *Biosens. Bioelectron.* **2016**, *86*, 315–320. [CrossRef]
49. Machín, A.; Arango, J.C.; Fontánez, K.; Cotto, M.; Duconge, J.; Soto-Vázquez, L.; Resto, E.; Petrescu, F.I.T.; Morant, C.; Márquez, F. Biomimetic Catalysts Based on Au@ZnO–Graphene Composites for the Generation of Hydrogen by Water Splitting. *Biomimetics* **2020**, *5*, 39. [CrossRef]
50. Briggs, D.; Seah, M. *Practical Surface Analysis*; Wiley: New York, NY, USA, 1994.
51. Yue, B.; Li, Q.; Iwai, H.; Kako, T.; Ye, J. Hydrogen Production Using Zinc-Doped Carbon Nitride Catalyst Irradiated with Visible Light. *Sci. Technol. Adv. Mater.* **2011**, *12*, 034401. [CrossRef]
52. Gu, Q.; Gao, Z.; Zhao, H.; Lou, Z.; Liao, Y.; Xue, C. Temperature-Controlled Morphology Evolution of Graphitic Carbon Nitride Nanostructures and Their Photocatalytic Activities under Visible Light. *RSC Adv.* **2015**, *5*, 49317–49325. [CrossRef]
53. Zhu, Y.-P.; Li, M.; Liu, Y.-L.; Ren, T.-Z.; Yuan, Z.-Y. Carbon-Doped ZnO Hybridized Homogeneously with Graphitic Carbon Nitride Nanocomposites for Photocatalysis. *J. Phys. Chem. C* **2014**, *118*, 10963–10971. [CrossRef]
54. Li, N.; Tian, Y.; Zhao, J.; Zhang, J.; Zuo, W.; Kong, L.; Cui, H. Z-Scheme 2D/3D g-C_3N_4@ZnO with Enhanced Photocatalytic Activity for Cephalexin Oxidation under Solar Light. *Chem. Eng. J.* **2018**, *352*, 412–422. [CrossRef]
55. Zhu, Y.; Xue, J.; Xu, T.; He, G.; Chen, H. Enhanced Photocatalytic Activity of Magnetic Core–Shell Fe_3O_4@Bi_2O_3–RGO Heterojunctions for Quinolone Antibiotics Degradation under Visible Light. *J. Mater. Sci. Mater. Electron.* **2017**, *28*, 8519–8528. [CrossRef]
56. Ye, L.; Liu, J.; Gong, C.; Tian, L.; Peng, T.; Zan, L. Two Different Roles of Metallic Ag on Ag/AgX/BiOX (X = Cl, Br) Visible Light Photocatalysts: Surface Plasmon Resonance and Z-Scheme Bridge. *ACS Catal.* **2012**, *2*, 1677–1683. [CrossRef]
57. Jourshabani, M.; Shariatinia, Z.; Badiei, A. Synthesis and Characterization of Novel Sm_2O_3/S-Doped g-C_3N_4 Nanocomposites with Enhanced Photocatalytic Activities under Visible Light Irradiation. *Appl. Surf. Sci.* **2018**, *427*, 375–387. [CrossRef]
58. Prabavathi, S.L.; Saravanakumar, K.; Nkambule, T.T.I.; Muthuraj, V.; Mamba, G. Enhanced Photoactivity of Cerium Tungstate-Modified Graphitic Carbon Nitride Heterojunction Photocatalyst for the Photodegradation of Moxifloxacin. *J. Mater. Sci. Mater. Electron.* **2020**, *31*, 11434–11447. [CrossRef]

Cao, J.; Li, X.; Lin, H.; Chen, S.; Fu, X. In Situ Preparation of Novel p–n Junction Photocatalyst BiOI/(BiO)$_2$CO$_3$ with Enhanced Visible Light Photocatalytic Activity. *J. Hazard. Mater.* **2012**, *239–240*, 316–324. [CrossRef]

Nethercot, A.H. Prediction of Fermi Energies and Photoelectric Thresholds Based on Electronegativity Concepts. *Phys. Rev. Lett.* **1974**, *33*, 1088–1091. [CrossRef]

Chen, C.; Bi, W.; Xia, Z.; Yuan, W.; Li, L. Hydrothermal Synthesis of the CuWO$_4$/ZnO Composites with Enhanced Photocatalytic Performance. *ACS Omega* **2020**, *5*, 13185–13195. [CrossRef]

Jiménez-Salcedo, M.; Monge, M.; Tena, M.T. The Photocatalytic Degradation of Sodium Diclofenac in Different Water Matrices Using g-C$_3$N$_4$ Nanosheets: A Study of the Intermediate by-Products and Mechanism. *J. Environ. Chem. Eng.* **2021**, *9*, 105827. [CrossRef]

Raizada, P.; Sudhaik, A.; Singh, P. Photocatalytic Water Decontamination Using Graphene and ZnO Coupled Photocatalysts: A Review. *Mater. Sci. Energy Technol.* **2019**, *2*, 509–525. [CrossRef]

Gomes Silva, C.; Juárez, R.; Marino, T.; Molinari, R.; García, H. Influence of Excitation Wavelength (UV or Visible Light) on the Photocatalytic Activity of Titania Containing Gold Nanoparticles for the Generation of Hydrogen or Oxygen from Water. *J. Am. Chem. Soc.* **2011**, *133*, 595–602. [CrossRef] [PubMed]

Liu, H.; Cao, W.-R.; Su, Y.; Chen, Z.; Wang, Y. Bismuth Oxyiodide–Graphene Nanocomposites with High Visible Light Photocatalytic Activity. *J. Col. Interf. Sci.* **2013**, *398*, 161–167. [CrossRef] [PubMed]

Mo, Z.; She, X.; Li, Y.; Liu, L.; Huang, L.; Chen, Z.; Zhang, Q.; Xu, H.; Li, H. Synthesis of g-C$_3$N$_4$ at Different Temperatures for Superior Visible/UV Photocatalytic Performance and Photoelectrochemical Sensing of MB Solution. *RSC Adv.* **2015**, *5*, 101552–101562. [CrossRef]

Makuła, P.; Pacia, M.; Macyk, W. How To Correctly Determine the Band Gap Energy of Modified Semiconductor Photocatalysts Based on UV–Vis Spectra. *J. Phys. Chem. Lett.* **2018**, *9*, 6814–6817. [CrossRef]

Şimşek, E.B. Doping of boron in TiO$_2$ catalyst: Enhanced photocatalytic degradation of antibiotic under visible light irradiation. *J. Boron* **2017**, *2*, 18–27.

Thiruppathi, M.; Kumar, J.V.; Vahini, M.; Ramalingan, C.; Nagarajan, E. A study on divergent functional properties of sphere-like CuWO$_4$ anchored on 2D graphene oxide sheets towards the photocatalysis of ciprofloxacin and electrocatalysis of methanol. *J. Mater. Sci. Mater. Electron.* **2019**, *30*, 10172–10182. [CrossRef]

Article

Comparative Study of ZnO Thin Films Doped with Transition Metals (Cu and Co) for Methylene Blue Photodegradation under Visible Irradiation

William Vallejo [1,*], Alvaro Cantillo [1], Briggitte Salazar [1], Carlos Diaz-Uribe [1], Wilkendry Ramos [1], Eduard Romero [2] and Mikel Hurtado [3,4]

1. Photochemistry and Photobiology Research Group, College of Basic Sciences, Universidad del Atlántico, Puerto Colombia 81007, Colombia; acantilloguzman@mail.uniatlantico.edu.co (A.C.); bssalazar@mail.uniatlantico.edu.co (B.S.); carlosdiaz@mail.uniatlantico.edu.co (C.D.-U.); wramos@mail.uniatlantico.edu.co (W.R.)
2. Deparment of Chemistry, College of Science, Universidad Nacional de Colombia, Bogotá 111321, Colombia; erromerom@unal.edu.co
3. Electronic Engineering College, Cluster NBIC, Universidad Central, Bogotá 111021, Colombia; mhurtadom1@ucentral.edu.co or mikel.hurtado@uniminuto.edu
4. Civil Engineering College, Universidad Minuto de Dios, Bogotá 110111, Colombia
* Correspondence: williamvallejo@mail.uniatlantico.edu.co; Tel.: +57-53599484

Received: 14 March 2020; Accepted: 6 May 2020; Published: 11 May 2020

Abstract: We synthesized and characterized both Co-doped ZnO (ZnO:Co) and Cu-doped ZnO (ZnO:Cu) thin films. The catalysts' synthesis was carried out by the sol–gel method while the doctor blade technique was used for thin film deposition. The physicochemical characterization of the catalysts was carried out by Raman spectroscopy, scanning electron microscopy (SEM), X-ray diffraction, and diffuse reflectance measurements. The photocatalytic activity was studied under visible irradiation in aqueous solution, and kinetic parameters were determined by pseudo-first-order fitting. The Raman spectra results evinced the doping process and suggested the formation of heterojunctions for both dopants. The structural diffraction patterns indicated that the catalysts were polycrystalline and demonstrated the presence of a ZnO wurtzite crystalline phase. The SEM analysis showed that the morphological properties changed significantly, the micro-aggregates disappeared, and agglomeration was reduced after modification of ZnO. The ZnO optical bandgap (3.22 eV) reduced after the doping process, these being ZnO:Co (2.39 eV) and ZnO:Co (3.01 eV). Finally, the kinetic results of methylene blue photodegradation reached 62.6% for ZnO:Co thin films and 42.5% for ZnO:Cu thin films.

Keywords: thin films; ZnO; doping; heterogeneous photocatalysis

1. Introduction

Synthetic dyes are commonly used by various industries, especially textile ones. These physically and chemically stable compounds are harmful to the environment. Synthetic dyes are recalcitrant compounds that exhibit high solubility in water and accumulate in both wastewater and industrial effluents [1,2]. Currently, water pollution is one of the major challenges of the modern world, and the recovery of wastewater by conventional methods is not suitable for emergent pollutants [3,4]. Heterogeneous photocatalysis can be satisfactorily applied for the decontamination of natural samples through the photocatalytic degradation of toxic pollutants from complex matrices, such as river water and wastewater [5]. All renewable technologies have become a promising alternative for both energy generation and wastewater treatments. Solar photocatalysis is a suitable option to degrade recalcitrant pollutants from water [6,7]. Different catalysts have been reported to exhibit photocatalytic activity (e.g., TiO_2 [8], Fe_2O_3 [9], ZnO [10], CuO [11], CdS [12], WO_3, and SnO_2 [13]). ZnO has been used as a photocatalyst, but its high bandgap value (~3.3 eV) is one of its major drawbacks. ZnO is not photocatalytically active at longer wavelengths of the electromagnetic spectrum. As a consequence, using solar irradiation as the main energy source to develop practical applications is a challenge for ZnO [14]. Some alternative strategies to extend ZnO photoresponse in the visible light region are: (a) ZnO doping [15–18], (b) co-doping [19,20], (c) coupling with lower band gap semiconductors [21,22], (d) surface plasmon resonance [23–26], (e) quantum dots [27], and (f) sensitization with natural and synthetic dyes [28–31]. With the doping process, optical and catalytic properties can be tuned by doping [32], and the ZnO bandgap generates intragap electronic states inside the semiconductor [33]. Some transition metals (e.g., Co^{2+}, Ag^+, Cu^{2+}, Mn^{2+}) have been used to enhance the properties of ZnO [34–37]. Among these, copper is an economical option. Because the ionic size of Cu^2 is close to Zn^{2+}, Cu^{2+} ions can replace Zn^{2+} ions to modify the absorption spectrum [38]. Kuriakose et al. reported that Cu-doped ZnO nanostructures photodegraded organic dyes. Their analysis associated the enhanced photocatalytic activity to the combined effects of: (i) the separation of charge carriers and (ii) the optimal Cu doping load [39]. Another transition element is cobalt, as the ionic size of the Co^{2+} ion is close to that of Zn^{2+}. Co^{2+} ions can replace Zn^{2+} and generate minimal distortion in the crystalline lattice [35,40]. Yongchun et al. reported the synthesis of cobalt-doped ZnO nanorods and reported an improvement in the performance of alizarin red photocatalytic degradation [41]. In addition, Kuriakose et al. synthesized Co-doped ZnO nanodisks and nanorods, and reported photocatalytic activity improvement as the by-effect of the doping process due to: (i) the charge separation efficiency and (ii) the surface area [42]. Poornaprakash et al. reported 66.5% efficiency in photocatalytic degradation of Rhodamine B under artificial solar light illumination on Co-doped ZnO nanorods [43]. The incorporation of these kinds of metals inside the ZnO structure can modify optical properties by extending the photodegradation ability towards the largest wavelength of the electromagnetic spectrum [44,45]. In this work, we report a facile wet chemical method for the synthesis of Cu and Co-doped ZnO thin films with highly enhanced photocatalytic activity. The metal doping process leads to highly efficient visible light photocatalytic degradation of methylene blue.

2. Results and Discussion

2.1. Structural Study

Figure 1 shows the XRD pattern for the catalysts synthesized in this study. The hexagonal wurtzite phase (JCPDS No. 36–1451) is identified as a crystalline structure for ZnO thin films, with the signals of the diffraction pattern corresponding to those reported by other authors [46]. The doping process did not affect the main signals in the diffraction patterns, as the XRD patterns for ZnO:Cu and ZnO:Co showed signals of a wurtzite ZnO structure. However, the XRD patterns showed a change in the intensity of the signals, suggesting that metal ions could substitute Zn^{2+} ions after the doping process. Lima et al. suggested that the change in the intensities could be associated with changes in both (i) grain size due to network defects and (ii) oxygen vacancies [47,48].

We used the Debye–Scherrer equation to calculate the crystalline domain size of the catalysts, using the full width at half maximum (FWHM) for the highest peak (101), with θ being the Scherrer diffraction angle [49]. Although there is no clear tendency between the intensity of the signal and metal doping load (see Figure 1b,c), all samples reduced the grain size of catalysis after the doping process (see Table 1). This could be explained by the incorporation of Co^{2+} and Cu^{2+} ions as dopants into the ZnO after the doping process [48,49]. Finally, the structural results suggest that ZnO films incorporated metallic ions. This observation was verified by Raman spectroscopy and diffuse reflectance, as described in the next sections.

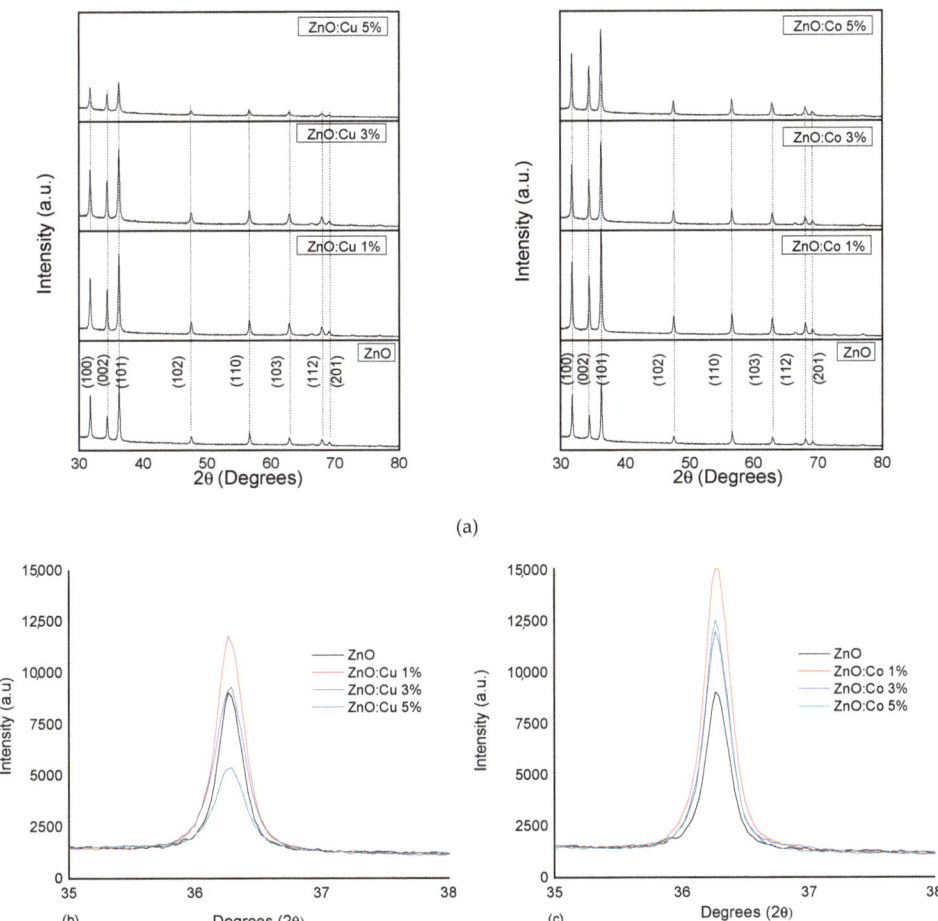

Figure 1. (**a**) X-ray diffraction patterns for the catalysts synthesized in this study. (**b**) Comparison of the highest peak (101) for ZnO:Cu. (**c**) Comparison of the highest peak (101) for ZnO:Co.

Table 1. Crystallographic results for the X-ray characterization of both undoped ZnO and metal-doped ZnO thin films.

Thin Film	FWHM * (101)	Intensity (101) Peak	Grain Size (nm)
ZnO	0.2396	8949	34.9
ZnO:Co 1%	0.2615	14,997	32.0
ZnO:Co 3%	0.2552	11,836	32.3
ZnO:Co 5%	0.2583	12,495	32.4
ZnO:Cu 1%	0.2601	11,649	32.1
ZnO:Cu 3%	0.2803	9202	29.8
ZnO:Cu 5%	0.3075	5300	27.2

* FWHM: full width at half maximum.

2.2. Raman Study

The Raman spectra of the catalysts are shown in Figure 2. All the peaks correspond with wurtzite–ZnO (C^4_{6v}): (i) 97.4 cm^{-1} (vibrational mode E_{2L}), (ii) 340 cm (E_{2H}–E_{2L}), and (iii) 437.0 cm^{-1} and

581 cm^{-1} (A$_1$ vibrational mode) [50,51]. Figure 2a shows the Raman spectra for Cu-doped ZnO thin films. Signals E$_{2L}$ (~99 cm^{-1}) and E$_{2H}$ (~437 cm^{-1}) widen and decrease after the doping process—a behavior that can be explained by the incorporation of Cu^{2+} into the ZnO lattice. Additionally, this behavior has been associated with reduction of ZnO crystallinity by the formation of nanocomposites [52]. For greater Cu loads, Figure 2a shows two new signals, the first one located at 298 cm^{-1} and a second weak one at 614 cm^{-1}. These two signals can be attributed to modes A$_{1g}$ and B$_{2g}$ for CuO, respectively, and this result suggests the formation of a ZnO–CuO heterojunction during the synthesis process. The hydrodynamic stability of the suspension is affected by the concentration of reagents; so for obtaining greater Cu loads, the CuO generation is feasible. This result is in line with previous reports [39].

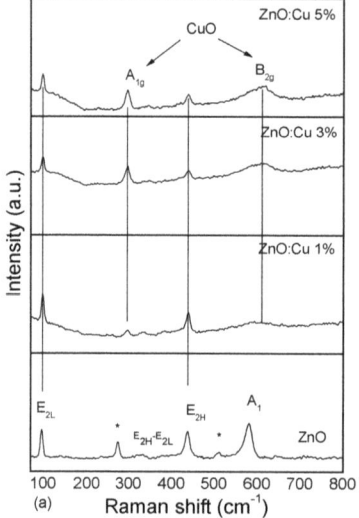 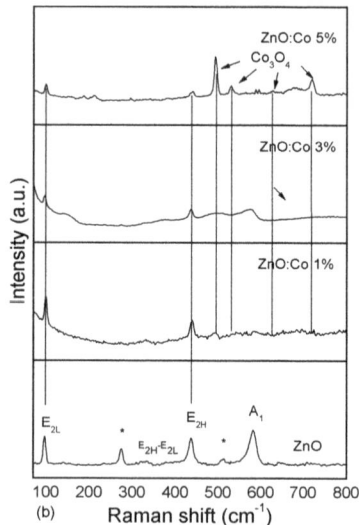

Figure 2. Raman spectra for: (**a**) ZnO:Cu and (**b**) ZnO:Co thin films. Inside the figures are the Raman vibration modes, where (*) corresponds to defects inside the ZnO structure.

Figure 2b shows the Raman spectrum for ZnO:Co; the intensity of A$_{1LO}$, E$_{2L}$, and E$_{2H}$ modes decreases for these films. The ZnO:Co thin films (doping load 5%) show four new signals at 490 cm^{-1}, 526 cm^{-1}, 626 cm^{-1}, and 725 cm^{-1}. These signals could be attributed to the possible presence of Co$_3$O$_4$, and these results confirm both the doping process and the formation of a heterojunction of ZnO–Co$_3$O$_4$ [53,54].

2.3. Morphological Study

Figure 3 shows SEM images for the catalysts. Figure 3a shows that the ZnO films formed microaggregates (~220 nm) composed of quasi-spherical ZnO nanoparticles (around 40 nm in diameter), and this is a typical result for this material sensitized by the sol–gel method. Figure 3b,c shows that the morphological properties changed significantly after the doping process. Regarding the ZnO:Cu thin films, Figure 3b shows the formation of nanorods. Meanwhile, Figure 3c shows the formation of nanosized elongated particles of various shapes (~100 nm) from the ZnO:Co thin films. Likewise, Figure 3c shows that the agglomeration on the catalyst surface reduced and the microaggregates disappeared. Different nanostructures have been reported for ZnO (e.g., nanorods, nanotubes, nanobelts, nanosprings, nanospirals, nanorings) [55]. It is known that ZnO's morphological properties rely on synthesis conditions, and in our case, it is clear that the metal ions used during synthesis reduced the agglomeration and changed the thin films' morphology [56].

Figure 3. SEM images: (**a**) ZnO, (**b**) ZnO:Cu 5% and (**c**) ZnO:Co 5% thin films.

2.4. Optical Study

The diffuse reflectance spectra for the catalysts are shown in Figure 4. We used the Kubelka–Munk (KM) remission function for determining the bandgap energy value of the catalysts [57]. The use of the KM remission function makes it possible to obtain an analog to Tauc plots [58,59]:

$$(F(R_\infty)h\nu)^{\frac{1}{2}} = A(h\nu - E_g) \tag{1}$$

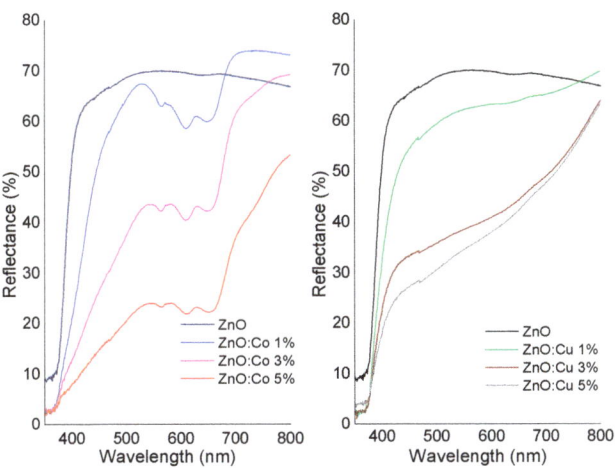

Figure 4. Reflectance diffuse spectra for both catalysts.

Figure 5 shows plots for $(F(R_\infty)h\nu)^{\frac{1}{2}}$ versus $(h\nu)$ and Table 2 lists the optical properties of the catalysts. Figure 5 shows that ZnO had a bandgap value (E_g) of 3.22 eV, a value that corresponds with that reported by Srikant el al. (3.1 eV and 3.2 eV) [60,61]. For doped ZnO catalysts, the E_g value was lower, and this behavior is associated with the reduction of the Fermi level of ZnO by the generation of intragap states. For ZnO:Cu, the modification of the bandgap can be attributed to the induction of 3d states of Cu located inside the bandgap of ZnO [37]. Additionally, the visible light absorption observed for doped ZnO can be attributed to intragap transitions between Cu 3d and Zn 4s states. Furthermore, the ZnO:Co 5% catalyst has a lower bandgap value compared to other catalysts. This reduction is attributed to s-d and p-d exchange interactions between ZnO and Co^{2+} ions [62]. The 3d levels of Co^{2+} are located within the bandgap of ZnO, which can create new bands at larger wavelengths [63]. Some photoluminescence studies of the transition metal doping ZnO nanoparticles suggest that this important reduction in E_g value is due to oxygen deficiency [64]. Finally, the formation of nanoheterojunctions in the catalyst surface leads to an enhanced separation of charge carriers, increasing photocatalytic efficiency in addition to the doping process. The generation of these heterostructures has been reported for photocatalytic applications [65].

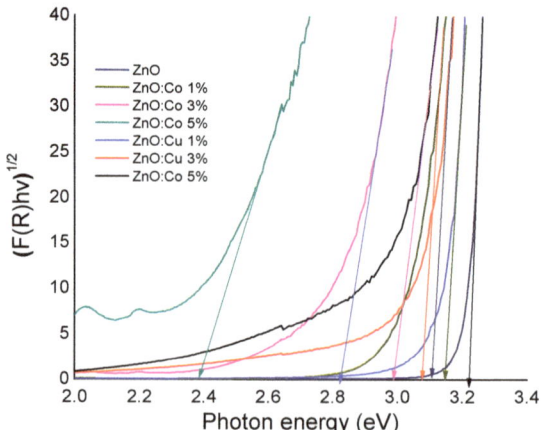

Figure 5. Kubelka–Munk (KM) fitting for the catalysts synthesized in this study.

Table 2. Band gap and results of pseudo-first-order model fitting.

Catalyst	$k_{ap} \times 10^{-3}$ (min^{-1})	Degradation (%)	Band Gap (eV)
ZnO	0.2	2.7	3.22
ZnO:Co 1%	2.6	30.4	3.17
ZnO:Co 3%	4.2	45.7	2.83
ZnO:Co 5%	7.2	62.6	2.39
ZnO:Cu 1%	3.4	36.2	3.12
ZnO:Cu 3%	3.4	37.7	3.07
ZnO:Cu 5%	4.0	42.5	3.01

2.5. Photocatalytic Study

Figure 6 shows the decrease of MB as a function of time for all tests performed under visible irradiation. The MB concentration did not change after 140 min under visible irradiation, verifying the stability of MB dye. Furthermore, ZnO films did not show photocatalytic activity under visible irradiation (<3%). This result is in accordance with the ZnO bandgap energy value, and this photocatalyst is active only under UV irradiation. The ZnO:Co 5% catalyst reported the highest photocatalytic activity. This result can be explained by the lower bandgap value of the ZnO:Co 5% catalyst compared to other catalysts. The ZnO:Cu catalysts showed less photocatalytic activity than the Co-doped ZnO films. Compared to the Co-doped ZnO films, the bandgap values of this catalyst did not change; however, the best photodegradation result for ZnO:Cu was 42.5%, a value greater than that obtained for the ZnO thin films. The combined effect of the doping process and the heterojunction can explain this behavior. The photodegradation kinetics of MB on catalysts were studied by using the pseudo-first-order model [66]:

$$v_{[MB]} = [MB]_o e^{-k_{ap}t} \tag{2}$$

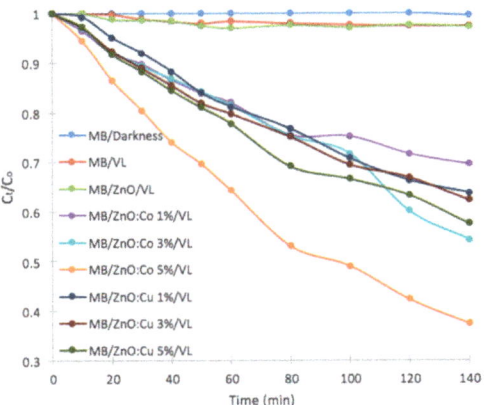

Figure 6. Methylene blue (MB) concentration vs. time of visible irradiation on the synthesized catalysts.

Time (t) is expressed in minutes and k_{app} is the apparent reaction rate constant (min^{-1}). Table 2 lists the kinetic parameters for the studied catalysts. Among all the catalysts, the ZnO thin films ($k_{app} = 0.2 \times 10^{-4}$ min^{-1}) showed the lowest k_{ap} value, while the best results were obtained for ZnO:Co 5% ($k_{app} = 7.2 \times 10^{-3}$ min^{-1}) and ZnO:Cu 5% ($k_{app} = 4 \times 10^{-3}$ min^{-1}). In the best case, the kinetic rate constant was 36 times higher than the ZnO thin films. A combined effect could be present: (i) Cu doping in ZnO and (ii) the formation of a nanoheterojunction (ZnO–CuO). This synergic effect could be a reason for the increase in photocatalytic yields. The heterostructure generation for the methyl orange photodegradation under visible light irradiation has been reported before [67]. Table 3 lists other reports for the use of doped ZnO with different metals as catalysts. Our results indicate that the catalysts produced in this study are suitable options for solar photocatalytic applications.

Table 3. k_{app} values for different catalysts (ZnO doped with different metals) under visible irradiation.

Catalysts/Reference	Pollutant/Molar Concentration	Degradation (%)/Time Test	$k_{ap} \times 10^{-3}$ (min^{-1})
SnS/ZnO [67]	Rhodamine B/5 ppm	99%/175 min	21.2
	Methyl Orange/5 ppm	82%/125 min	13.9
Carbon-ZnO [68]	2,4-dinitrophenol/25 ppm	92%/140 min	18.3
ZnO:Co [41]	Alizarin Red/20 ppm	93%/60 min	—
ZnO:Cu [39]	Methyl Orange/5 ppm	80%/30 min	23
ZnO:Ag [69]	Methylene Blue/10 ppm	65%/140 min	4.1
ZnO:Co/this work	Methylene Blue/10 ppm	63%/140 min	7.2
ZnO:Cu/this work	Methylene Blue/10 ppm	43%/140 min	4.0

The ZnO films did not show photocatalytic activity under visible irradiation. However, two different processes can contribute to photocatalytic degradation under visible irradiation: (i) the intraband transitions as dopants allow the doped ZnO thin films to absorb visible light, generating charge pairs; (ii) CuO and Co$_3$O$_4$ can absorb visible light after the formation of ZnO–CuO and ZnO–Co$_3$O$_4$ heterojunctions, generating charge pairs. In this case, the electron can be transferred to the conduction band of ZnO. After electrons are located at the conduction band of ZnO, different reactive oxygen species (ROS) can be generated (e.g., O_2^-; OH), starting the MB photodegradation. Scheme 1 shows the general scheme of energetic levels for doped ZnO thin films and the ROS generation.

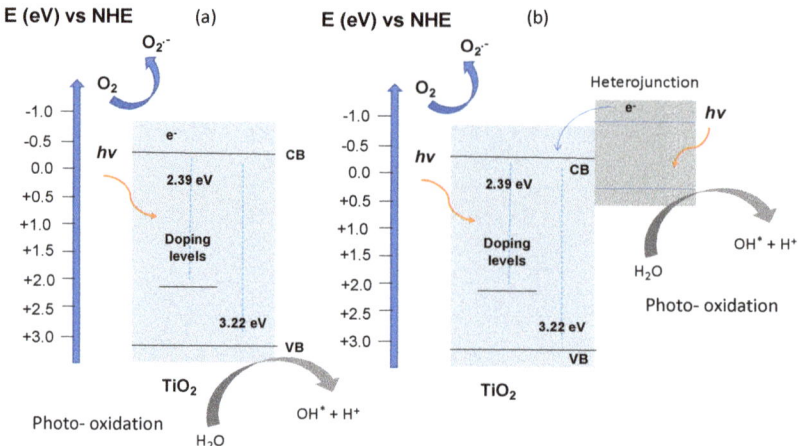

Scheme 1. Hypothetical scheme of energetic levels for metal-doped ZnO thin films: (**a**) Metal doping process. (**b**) Metal doping process and the generation of a heterojunction [64,70,71]. After charge pairs generation, the ROS can be yielded on the catalyst surface and MB degradation starts.

3. Materials and Methods

3.1. Synthesis and Characterization

The ZnO synthesis was carried out according to a previous report [69]: Twenty five mL of ammonium hydroxide (NH_4OH) (25–35% w/w) reactive grade (Merck) was placed in a 250 mL glass beaker, then 0.500 M ($Zn(CH_3COO)_2 \cdot 2H_2O$) (Merck) was added dropwise at a rate of 1.7 mL.min^{-1} for 1 h, at a temperature of 85 °C under constant agitation at 300 rpm. After that, the suspension stood for three days at room temperature then the solid was filtered and dried for 5 h at 100 °C [72,73].

For the ZnO doping process, we used a similar procedure as previously described. While adding Zn^{2+} ions, we also added salts of the doping metals ($CuSO_4 \cdot 5H_2O$) (Merck) for copper doping and ($Co(CH_3COO)_2 \cdot 4H_2O$) (JT Baker) for the cobalt doping processes. The synthesis of doped ZnO powder was performed at 1.0%, 3.0%, and 5.0%. The thin films were deposited using the doctor blade method, and the suspension was placed in a glass measuring 2 cm high and 2 cm wide. The thin films were heated for 30 min at 90 °C to evaporate the solvent, and finally, the sintering process was performed at 500 °C for 2 h [73,74]. Using this procedure, we obtained thin films (6 µm thickness). The film thickness was measured through a Veeco Dektak 150 profilometer. The physical chemistry properties of the films were studied by X-ray diffraction, diffuse reflectance spectrophotometry, and Raman spectroscopy assay. X-ray diffraction patterns were obtained using a Shimadzu 6000 diffractometer using Cu Kα radiation (λ = 0.15406 nm) as an X-ray source with a diffraction angle in the 2θ range (20–80°). Diffuse reflectance spectra were obtained with a Lambda 4 Perkin–Elmer spectrophotometer equipped with an integration sphere. The compositional properties of the materials were studied by Raman spectroscopy in a DXR device equipped with a 780 nm laser. The morphological properties were studied by scanning electron microscopy, under an excitation energy of 5 and 1 kV. The metallic content of the films was determined by plasma emission spectroscopy using the SM 3120 B technique, EPA 3015A modified for solids (see Supporting Information).

3.2. Photocatalytic Test

Methylene blue (MB) was chosen as the pollutant model in this study. The experiments were carried out in a batch reactor using an LED tape as a source of visible radiation (cold white light 17 watts), and the incident photon flow per unit volume I_o was 5.8 × 10^{-7} Einstein*L^{-1}s^{-1}. Before irradiation, the MB solution was kept in the dark for 90 min at 250 rpm to reach adsorption–desorption equilibrium on the catalysts' surface. Photodegradation was carried out using 50 ± 0.025 mL of an MB solution (10 mg·L^{-1}) saturated with oxygen at pH 7.0. The concentration of dye was determined through the spectrophotometric method (Thermo Scientific–Genesys 10S) using 665 nm as a fixed wavelength, with a calibration curve (correlation coefficient R = 0.997) for the use of the Lambert–Beer equation.

4. Conclusions

We synthesized and characterized ZnO thin films doped with Co and Cu. Raman results corroborated the doping process, which suggested the generation of a heterostructure. For the doped ZnO catalysts, the results show a reduction in the E_g values (from 3.22 to 2.42 eV for the best catalyst). This behavior is associated with a reduction of the Fermi level of ZnO by the presence of intragap states. The photocatalytic test indicated that doped catalysts had greater photocatalytic activity than unmodified catalysts, which could be attributed to (i) the generation of intraband states for the insertion of Co and Cu into ZnO, and (ii) the generation of ZnO–CuO and ZnO–Co$_3$O$_4$ heterojunctions. Furthermore, the rate of the photodegradation process (ZnO:Co 5%) was 36 times greater than the rate for unmodified ZnO, and the k_{app} values for the catalysts synthesized in this study had a suitable photocatalytic activity compared to other reports.

Supplementary Materials: The following are available online at http://www.mdpi.com/2073-4344/10/5/528/s1. Figure S1: ICP calibration curve for: (a) Co and (b) Cu content. Table S1: Results ICP for catalysts. Fitting curves of kinetic model.

Author Contributions: Conceptualization, W.V.; methodology, W.V., C.D.-U., B.S., A.C., W.R., E.R. and M.H.; validation, W.V., C.D.-U., B.S., A.C.; formal analysis, W.V., C.D.-U., B.S., A.C.; investigation, W.V., C.D.-U., B.S., A.C.; resources, W.V., C.D.-U., B.S., A.C., W.R., E.R. and M.H.; data curation, W.V., C.D.-U., B.S., A.C.; writing—original draft preparation, W.V., C.D.-U., B.S., A.C., writing—review and editing, W.V., C.D.-U., B.S., A.C., W.R., E.R., M.H.; visualization, W.V., C.D.-U.; supervision, W.V., C.D.-U.; project administration, W.V., C.D.-U.; funding acquisition, W.V., C.D.-U. All authors have read and agreed to the published version of the manuscript.

Funding: This research was funded by Universidad del Atlántico.

Acknowledgments: W.V., C.D.-U., B.S., and A.C. would like to thank Universidad del Atlántico.

Conflicts of Interest: The authors declare that there is no conflict of interest.

References

1. Lellis, B.; Fávaro-Polonio, C.Z.; Pamphile, J.A.; Polonio, J.C. Effects of textile dyes on health and the environment and bioremediation potential of living organisms. *Biotechnol. Res. Innov.* **2019**, *3*, 275–290. [CrossRef]
2. Hassan, M.M.; Carr, C.M. A critical review on recent advancements of the removal of reactive dyes from dyehouse effluent by ion-exchange adsorbents. *Chemosphere* **2018**, *209*, 201–219. [CrossRef]
3. Fabbri, D.; López-Muñoz, M.J.; Daniele, A.; Medana, C.; Calza, P. Photocatalytic abatement of emerging pollutants in pure water and wastewater effluent by TiO 2 and Ce-ZnO: Degradation kinetics and assessment of transformation products. *Photochem. Photobiol. Sci.* **2019**, *18*, 845–852. [CrossRef]
4. Zelinski, D.W.; dos Santos, T.P.M.; Takashina, T.A.; Leifeld, V.; Igarashi-Mafra, L. Photocatalytic Degradation of Emerging Contaminants: Artificial Sweeteners. *Water. Air. Soil Pollut.* **2018**, *229*, 1–12. [CrossRef]
5. Regulska, E.; Rivera-Nazario, D.; Karpinska, J.; Plonska-Brzezinska, M.; Echegoyen, L. Zinc Porphyrin-Functionalized Fullerenes for the Sensitization of Titania as a Visible-Light Active Photocatalyst: River Waters and Wastewaters Remediation. *Molecules* **2019**, *24*, 1118. [CrossRef] [PubMed]
6. Ansari, S.A.; Ansari, S.G.; Foaud, H.; Cho, M.H. Facile and sustainable synthesis of carbon-doped ZnO nanostructures towards the superior visible light photocatalytic performance. *New J. Chem.* **2017**, *41*, 9314–9320. [CrossRef]
7. Naldoni, A.; Riboni, F.; Guler, U.; Boltasseva, A.; Shalaev, V.M.; Kildishev, A.V. Solar-Powered Plasmon-Enhanced Heterogeneous Catalysis. *Nanophotonics* **2016**, *5*, 112–133. [CrossRef]
8. Sedghi, R.; Heidari, F. A novel & effective visible light-driven TiO$_2$/magnetic porous graphene oxide nanocomposite for the degradation of dye pollutants. *RSC Adv.* **2016**, *6*, 49459–49468.

9. Balu, S.; Uma, K.; Pan, G.T.; Yang, T.C.K.; Ramaraj, S.K. Degradation of methylene blue dye in the presence of visible light using SiO2@α-Fe2O3 nanocomposites deposited on SnS2 flowers. *Materials* **2018**, *11*, 1030. [CrossRef]
10. Loh, K.; Gaylarde, C.C.; Shirakawa, M.A. Photocatalytic activity of ZnO and TiO2 'nanoparticles' for use in cement mixes. *Constr. Build. Mater.* **2018**, *167*, 853–859. [CrossRef]
11. Saravanakkumar, D.; Oualid, H.A.; Brahmi, Y.; Ayeshamariam, A.; Karunanaithy, M.; Saleem, A.M.; Kaviyarasu, K.; Sivaranjani, S.; Jayachandran, M. Synthesis and characterization of CuO/ZnO/CNTs thin films on copper substrate and its photocatalytic applications. *OpenNano* **2019**, *4*, 100025. [CrossRef]
12. Zyoud, A.; Zaatar, N.; Saadeddin, I.; Helal, M.H.; Campet, G.; Hakim, M.; Park, D.; Hilal, H.S. Alternative natural dyes in water purification: Anthocyanin as TiO2-sensitizer in methyl orange photo-degradation. *Solid State Sci.* **2011**, *13*, 1268–1275. [CrossRef]
13. Elango, G.; Roopan, S.M. Efficacy of SnO2 nanoparticles toward photocatalytic degradation of methylene blue dye. *J. Photochem. Photobiol. B Biol.* **2016**, *155*, 34–38. [CrossRef]
14. Das, A.; Malakar, P.; Nair, R.G. Engineering of ZnO nanostructures for efficient solar photocatalysis. *Mater. Lett.* **2018**, *219*, 76–80. [CrossRef]
15. Schumann, J.; Eichelbaum, M.; Lunkenbein, T.; Thomas, N.; Álvarez Galván, M.C.; Schlögl, R.; Behrens, M. Promoting Strong Metal Support Interaction: Doping ZnO for Enhanced Activity of Cu/ZnO:M (M = Al, Ga, Mg) Catalysts. *ACS Catal.* **2015**, *5*, 3260–3270. [CrossRef]
16. Kumari, V.; Mittal, A.; Jindal, J.; Yadav, S.; Kumar, N. S-, N- and C-doped ZnO as semiconductor photocatalysts: A review. *Front. Mater. Sci.* **2019**, *13*, 1–22. [CrossRef]
17. Bharat, T.C.; Mondal, S.; Gupta, H.S.; Singh, P.K.; Das, A.K. Synthesis of Doped Zinc Oxide Nanoparticles: A Review. *Mater. Today Proc.* **2019**, *11*, 767–775. [CrossRef]
18. Poornaprakash, U.; Chalapathi, K.; Subramanyam, S.V.; Prabhakar Vattikuti, Y.; Shun, S.P. Effects of Ce incorporation on the structural, morphological, optical, magnetic, and photocatalytic characteristics of ZnO nanoparticles. *Mater. Res. Express* **2019**, *6*, 105356. [CrossRef]
19. Poornaprakash, B.; Subramanyam, K.; Vattikuti, S.V.P.; Pratap Reddy, M.S. Achieving enhanced ferromagnetism in ZnTbO nanoparticles through Cu co-doping. *Ceram. Int.* **2019**, *45*, 16347–16352. [CrossRef]
20. Poornaprakash, B.; Chalapathi, U.; Poojitha, P.T.; Vattikuti, S.V.P.; Reddy, M.S.P. (Al, Cu) Co-doped ZnS nanoparticles: Structural, chemical, optical, and photocatalytic properties. *J. Mater. Sci. Mater. Electron.* **2019**, *30*, 9897–9902. [CrossRef]
21. Kaur, M.; Umar, A.; Mehta, S.K.; Singh, S.; Kansal, S.K.; Fouad, H.; Alothman, O.Y. Rapid Solar-Light Driven Superior Photocatalytic Degradation of Methylene Blue Using MoS2-ZnO Heterostructure Nanorods Photocatalyst. *Material* **2018**, *11*, 2254.
22. Vallejo, W.; Díaz-Uribe, C.; Rios, K. Methylene Blue Photocatalytic Degradation under Visible Irradiation on In2S3 Synthesized by Chemical Bath Deposition. *Adv. Phys. Chem.* **2017**, *2017*, 1–5. [CrossRef]
23. Subash, B.; Krishnakumar, B.; Swaminathan, M.; Shanthi, M. Highly Efficient, Solar Active, and Reusable Photocatalyst: Zr-Loaded Ag–ZnO for Reactive Red 120 Dye Degradation with Synergistic Effect and Dye-Sensitized Mechanism. *Langmuir* **2013**, *29*, 939–949. [CrossRef] [PubMed]
24. Aby, H.; Kshirsagar, A.; Pk, K. Plasmon Mediated Photocatalysis by Solar Active Ag/ZnO Nanostructures: Degradation of Organic Pollutants in Aqueous Conditions. *J. Mater Sci Nanotechnol* **2016**, *4*. [CrossRef]
25. Díaz-Uribe, C.; Viloria, J.; Cervantes, L.; Vallejo, W.; Navarro, K.; Romero, E.; Quiñones, C. Photocatalytic Activity of Ag-TiO2 Composites Deposited by Photoreduction under UV Irradiation. *Int. J. Photoenergy* **2018**, *2018*, 1–8. [CrossRef]
26. Chen, L.; Tran, T.T.; Huang, C.; Li, J.; Yuan, L.; Cai, Q. Synthesis and photocatalytic application of Au/Ag nanoparticle-sensitized ZnO films. *Appl. Surf. Sci.* **2013**, *273*, 82–88. [CrossRef]
27. Poornaprakash, B.; Chalapathi, U.; Poojitha, P.T.; Vattikuti, S.V.P.; Park, S.H. Co-Doped ZnS Quantum Dots: Structural, Optical, Photoluminescence, Magnetic, and Photocatalytic Properties. *J. Supercond. Nov. Magn.* **2020**, *33*, 539–544. [CrossRef]
28. Youssef, Z.; Colombeau, L.; Yesmurzayeva, N.; Baros, F.; Vanderesse, R.; Hamieh, T.; Toufaily, J.; Frochot, C.; Roques-Carmes, T.; Acherar, S. Dye-sensitized nanoparticles for heterogeneous photocatalysis: Cases studies with TiO2, ZnO, fullerene and graphene for water purification. *Dye. Pigment.* **2018**, *159*, 49–71. [CrossRef]

29. Vallejo, W.; Diaz-Uribe, C.; Cantillo, Á. Methylene blue photocatalytic degradation under visible irradiation on TiO_2 thin films sensitized with Cu and Zn tetracarboxy-phthalocyanines. *J. Photochem. Photobiol. A Chem.* **2015**, *299*, 80–86. [CrossRef]
30. Vallejo, W.; Rueda, A.; Díaz-Uribe, C.; Grande, C.; Quintana, P. Photocatalytic activity of graphene oxide–TiO_2 thin films sensitized by natural dyes extracted from Bactris guineensis. *R. Soc. Open Sci.* **2019**, *6*, 181824. [CrossRef]
31. Diaz-Uribe, C.; Vallejo, W.; Camargo, G.; Muñoz-Acevedo, A.; Quiñones, C.; Schott, E.; Zarate, X. Potential use of an anthocyanin-rich extract from berries of Vaccinium meridionale Swartz as sensitizer for TiO_2 thin films—An experimental and theoretical study. *J. Photochem. Photobiol. A Chem.* **2019**, *384*, 112050. [CrossRef]
32. Hamid, S.B.A.; Teh, S.J.; Lai, C.W. Photocatalytic Water Oxidation on ZnO: A Review. *Catalysts* **2017**, *7*, 93. [CrossRef]
33. Türkyılmaz, Ş.Ş.; Güy, N.; Özacar, M. Photocatalytic efficiencies of Ni, Mn, Fe and Ag doped ZnO nanostructures synthesized by hydrothermal method: The synergistic/antagonistic effect between ZnO and metals. *J. Photochem. Photobiol. A Chem.* **2017**, *341*, 39–50. [CrossRef]
34. Bouzid, H.; Faisal, M.; Harraz, F.A.; Al-Sayari, S.A.; Ismail, A.A. Synthesis of mesoporous Ag/ZnO nanocrystals with enhanced photocatalytic activity. *Catal. Today* **2015**, *252*, 20–26. [CrossRef]
35. Altintas Yildirim, O.; Arslan, H.; Sönmezoğlu, S. Facile synthesis of cobalt-doped zinc oxide thin films for highly efficient visible light photocatalysts. *Appl. Surf. Sci.* **2016**, *390*, 111–121. [CrossRef]
36. Ahmad, M.; Ahmed, E.; Ahmed, W.; Elhissi, A.; Hong, Z.L.; Khalid, N.R. Enhancing visible light responsive photocatalytic activity by decorating Mn-doped ZnO nanoparticles on graphene. *Ceram. Int.* **2014**, *40*, 10085–10097. [CrossRef]
37. Polat, İ.; Yılmaz, S.; Altın, İ.; Bacaksız, E.; Sökmen, M. The influence of Cu-doping on structural, optical and photocatalytic properties of ZnO nanorods. *Mater. Chem. Phys.* **2014**, *148*, 528–532. [CrossRef]
38. Mittal, M.; Sharma, M.; Pandey, O.P. UV–Visible light induced photocatalytic studies of Cu doped ZnO nanoparticles prepared by co-precipitation method. *Sol. Energy* **2014**, *110*, 386–397. [CrossRef]
39. Kuriakose, S.; Satpati, B.; Mohapatra, S. Highly efficient photocatalytic degradation of organic dyes by Cu doped ZnO nanostructures. *Phys. Chem. Chem. Phys.* **2015**, *17*, 25172–25181. [CrossRef]
40. Thennarasu, G.; Sivasamy, A. Metal ion doped semiconductor metal oxide nanosphere particles prepared by soft chemical method and its visible light photocatalytic activity in degradation of phenol. *Powder Technol.* **2013**, *250*, 1–12. [CrossRef]
41. Lu, Y.; Lin, Y.; Wang, D.; Wang, L.; Xie, T.; Jiang, T. A high performance cobalt-doped ZnO visible light photocatalyst and its photogenerated charge transfer properties. *Nano Res.* **2011**, *4*, 1144–1152. [CrossRef]
42. Kuriakose, S.; Satpati, B.; Mohapatra, S. Enhanced photocatalytic activity of Co doped ZnO nanodisks and nanorods prepared by a facile wet chemical method. *Phys. Chem. Chem. Phys.* **2014**, *16*, 12741. [CrossRef]
43. Poornaprakash, B.; Chalapathi, U.; Subramanyam, K.; Vattikuti, S.V.P.; Park, S.H. Wurtzite phase Co-doped ZnO nanorods: Morphological, structural, optical, magnetic, and enhanced photocatalytic characteristics. *Ceram. Int.* **2020**, *46*, 2931–2939. [CrossRef]
44. Rajbongshi, B.M.; Samdarshi, S.K. Cobalt-doped zincblende–wurtzite mixed-phase ZnO photocatalyst nanoparticles with high activity in visible spectrum. *Appl. Catal. B Environ.* **2014**, *144*, 435–441. [CrossRef]
45. Rajbongshi, B.M.; Samdarshi, S.K. ZnO and Co-ZnO nanorods—Complementary role of oxygen vacancy in photocatalytic activity of under UV and visible radiation flux. *Mater. Sci. Eng. B* **2014**, *182*, 21–28. [CrossRef]
46. Muchuweni, E.; Sathiaraj, T.S.; Nyakotyo, H. Synthesis and characterization of zinc oxide thin films for optoelectronic applications. *Heliyon* **2017**, *3*, e00285. [CrossRef]
47. Yuhas, B.D.; Zitoun, D.O.; Pauzauskie, P.J.; He, R.; Yang, P. Transition-Metal Doped Zinc Oxide Nanowires. *Angew. Chemie* **2006**, *118*, 434–437. [CrossRef]
48. Wang, X.; Sø, L.; Su, R.; Wendt, S.; Hald, P.; Mamakhel, A.; Yang, C.; Huang, Y.; Iversen, B.B.; Besenbacher, F. The influence of crystallite size and crystallinity of anatase nanoparticles on the photo-degradation of phenol. *J. Catal.* **2014**, *310*, 100–108. [CrossRef]
49. Lima, M.K.; Fernandes, D.M.; Silva, M.F.; Baesso, M.L.; Neto, A.M.; de Morais, G.R.; Nakamura, C.V.; de Oliveira Caleare, A.; Hechenleitner, A.A.W.; Pineda, E.A.G. Co-doped ZnO nanoparticles synthesized by an adapted sol-gel method: Effects on the structural, optical, photocatalytic and antibacterial properties. *J. Sol-Gel Sci. Technol.* **2014**, *72*, 301–309. [CrossRef]
50. Calleja, J.M.; Cardona, M. Resonant Raman scattering in ZnO. *Phys. Rev. B* **1977**, *16*, 3753–3761. [CrossRef]

51. Cuscó, R.; Alarcón-Lladó, E.; Ibáñez, J.; Artús, L.; Jiménez, J.; Wang, B.; Callahan, M.J. Temperature dependence of Raman scattering in ZnO. *Phys. Rev. B* **2007**, *75*, 165202. [CrossRef]
52. Wang, W.; Zhou, Q.; Fei, X.; He, Y.; Zhang, P.; Zhang, G.; Peng, L.; Xie, W. Synthesis of CuO nano- and micro-structures and their Raman spectroscopic studies. *CrystEngComm* **2010**, *12*, 2232. [CrossRef]
53. Winiarski, J.; Tylus, W.; Szczygieł, B. EIS and XPS investigations on the corrosion mechanism of ternary Zn–Co–Mo alloy coatings in NaCl solution. *Appl. Surf. Sci.* **2016**, *364*, 455–466. [CrossRef]
54. Xuan, H.; Yao, C.; Hao, X.; Liu, C.; Ren, J.; Zhu, Y.; Xu, C.; Ge, L. Fluorescence enhancement with one-dimensional photonic crystals/nanoscaled ZnO composite thin films. *Colloids Surfaces A Physicochem. Eng. Asp.* **2016**, *497*, 251–256. [CrossRef]
55. Hasnidawani, J.N.; Azlina, H.N.; Norita, H.; Bonnia, N.N.; Ratim, S.; Ali, E.S. Synthesis of ZnO Nanostructures Using Sol-Gel Method. *Procedia Chem.* **2016**, *19*, 211–216. [CrossRef]
56. Pourrahimi, A.M.; Liu, D.; Pallon, L.K.H.; Andersson, R.L.; Martínez Abad, A.; Lagarón, J.-M.; Hedenqvist, M.S.; Ström, V.; Gedde, U.W.; Olsson, R.T. Water-based synthesis and cleaning methods for high purity ZnO nanoparticles–comparing acetate, chloride, sulphate and nitrate zinc salt precursors. *RSC Adv.* **2014**, *4*, 35568–35577. [CrossRef]
57. Simmons, E.L. Relation of the Diffuse Reflectance Remission Function to the Fundamental Optical Parameters. *Opt. Acta Int. J. Opt.* **1972**, *19*, 845–851. [CrossRef]
58. Pal, M.; Pal, U.; Jiménez, J.M.G.Y.; Pérez-Rodríguez, F. Effects of crystallization and dopant concentration on the emission behavior of TiO_2:Eu nanophosphors. *Nanoscale Res. Lett.* **2012**, *7*, 1. [CrossRef]
59. Viezbicke, B.D.; Patel, S.; Davis, B.E.; Birnie, D.P. Evaluation of the Tauc method for optical absorption edge determination: ZnO thin films as a model system. *Phys. Status Solidi* **2015**, *252*, 1700–1710. [CrossRef]
60. Srikant, V.; Clarke, D.R. On the optical band gap of zinc oxide. *J. Appl. Phys.* **1998**, *83*, 5447–5451. [CrossRef]
61. El-Atab, N.; Chowdhury, F.; Ulusoy, T.G.; Ghobadi, A.; Nazirzadeh, A.; Okyay, A.K.; Nayfeh, A. ~3-nm ZnO Nanoislands Deposition and Application in Charge Trapping Memory Grown by Single ALD Step. *Sci. Rep.* **2016**, *6*, 38712. [CrossRef] [PubMed]
62. Liu, X.-C.; Shi, E.-W.; Chen, Z.-Z.; Zhang, H.-W.; Song, L.-X.; Wang, H.; Yao, S.-D. Structural, optical and magnetic properties of Co-doped ZnO films. *J. Cryst. Growth* **2006**, *296*, 135–140. [CrossRef]
63. Qiu, X.; Li, G.; Sun, X.; Li, L.; Fu, X. Doping effects of Co^{2+} ions on ZnO nanorods and their photocatalytic properties. *Nanotechnology* **2008**, *19*, 215703. [CrossRef] [PubMed]
64. Ramya, E.; Rao, M.V.; Jyothi, L.; Rao, D.N. Photoluminescence and Nonlinear Optical Properties of Transition Metal (Ag, Ni, Mn) Doped ZnO Nanoparticles. *J. Nanosci. Nanotechnol.* **2018**, *18*, 7072–7077. [CrossRef] [PubMed]
65. Xu, H.; Shi, M.; Liang, C.; Wang, S.; Xia, C.; Xue, C.; Hai, Z.; Zhuiykov, S. Effect of Zinc Acetate Concentration on Optimization of Photocatalytic Activity of p-Co_3O_4/n-ZnO Heterostructures. *Nanoscale Res. Lett.* **2018**, *13*, 195. [CrossRef]
66. Konstantinou, I.K.; Albanis, T.A. TiO_2—Assisted photocatalytic degradation of azo dyes in aqueous solution: Kinetic and mechanistic investigations A review. *Appl. Catal. B Environ.* **2004**, *49*, 1–14. [CrossRef]
67. Jayswal, S.; Moirangthem, R.S. Construction of a solar spectrum active SnS/ZnO p–n heterojunction as a highly efficient photocatalyst: The effect of the sensitization process on its performance. *New J. Chem.* **2018**, *42*, 13689–13701. [CrossRef]
68. Jun Park, S.; Sankar Das, G.; Schütt, F.; Adelung, R.; Kumar Mishra, Y.; Malika Tripathi, K.; Kim, T. Visible-light photocatalysis by carbon-nano-onion-functionalized ZnO tetrapods: Degradation of 2,4-dinitrophenol and a plant-model-based ecological assessment. *Asia Mater.* **2019**, *11*, 1–13.
69. Vallejo, W.; Cantillo, A.; Dias-Uribe, C. Methylene Blue Photodegradation under Visible Irradiation on Ag-Doped ZnO Thin Films. *Int. J. Photoenergy* **2020**, *2020*, 112. [CrossRef]
70. Prasad, C.; Tang, H.; Liu, Q.Q.; Zulfiqar, S.; Shah, S.; Bahadur, I. An overview of semiconductors/layered double hydroxides composites: Properties, synthesis, photocatalytic and photoelectrochemical applications. *J. Mol. Liq.* **2019**, *289*, 111114. [CrossRef]
71. Hernández-Alonso, M.D.; Fresno, F.; Suárez, S.; Coronado, J.M. Development of alternative photocatalysts to TiO_2: Challenges and opportunities. *Energy Environ. Sci.* **2009**, *2*, 1231–1257. [CrossRef]
72. Pérez, J.A.; Gallego, J.L.; Wilson Stiven Roman, H.R.L. Zinc Oxide Nanostructured Thin Films. *Sci. Tech.* **2008**, *39*, 416–421.

73. Ramírez Vinasco, D.; Vera, L.; Patricia, L.; Riascos Landázuri, H. Zn1-xMnxO Thin Films. *Sci. Tech.* **2009**, *41*, 273–278.
74. Quiñones, C.; Ayala, J.; Vallejo, W. Methylene blue photoelectrodegradation under UV irradiation on Au/Pd-modified TiO2 films. *Appl. Surf. Sci.* **2010**, *257*, 367–371. [CrossRef]

© 2020 by the authors. Licensee MDPI, Basel, Switzerland. This article is an open access article distributed under the terms and conditions of the Creative Commons Attribution (CC BY) license (http://creativecommons.org/licenses/by/4.0/).

Magnetite, Hematite and Zero-Valent Iron as Co-Catalysts in Advanced Oxidation Processes Application for Cosmetic Wastewater Treatment

Jan Bogacki *, Piotr Marcinowski, Dominika Bury, Monika Krupa, Dominika Ścieżyńska and Prasanth Prabhu

Faculty of Building Services, Hydro and Environmental Engineering, Warsaw University of Technology, 00-653 Warsaw, Poland; piotr.marcinowski@pw.edu.pl (P.M.); dominika2609@o2.pl (D.B.); monika.krupa@onet.eu (M.K.); domino233@wp.pl (D.Ś.); lytic.sam@gmail.com (P.P.)
* Correspondence: jan.bogacki@pw.edu.pl

Abstract: Background: There is a need for more effective methods of industrial wastewater treatment. Methods: Cosmetic wastewater was collected and subjected to $H_2O_2/Fe_3O_4/Fe_2O_3/Fe^0$ and $UV/H_2O_2/Fe_3O_4/Fe_2O_3/Fe^0$ process treatment. Results: Total organic carbon (TOC) was decreased from an initial 306.3 to 134.1 mg/L, 56.2% TOC removal, after 120 min of treatment for 1:1 H_2O_2/COD mass ratio and 500/500/1000 mg/L $Fe_3O_4/Fe_2O_3/Fe^0$ catalyst doses. The application chromatographic analysis allowed for the detection and identification of pollutants present in the wastewater. Identified pollutants were removed during the treatment processes. Processes carried out at a pH greater than 3.0 were ineffective. The UV process was more effective than the lightless process. Conclusions: The applied processes are effective methods for wastewater treatment. Chromatographic results confirmed the effectiveness of the treatment method. The kinetics of the process were described by the modified second-order model. On the basis of ANOVA results, the hypothesis regarding the accuracy and reproducibility of the research was confirmed.

Keywords: industrial wastewater; advanced oxidation processes; zero valent iron; magnetite; hematite

1. Introduction

The cosmetics market is booming and it is one of the fastest growing consumer markets. It globally generated EUR 474.2 billion in 2019. The coronavirus pandemic resulted in a decrease in industry revenues in 2020 by only 1.2%, to EUR 468.3 billion [1].

The constantly increasing production of cosmetics is accompanied by the side effect of producing increasing amounts of waste and wastewater. Cosmetic wastewater (CW) is created by washing production lines with water with surfactants and disinfectants, so CW contains the same compounds as those that are present in cosmetics. A typical industrial-scale CW treatment method is coagulation coupled with dissolved air flotation (C/DAF) followed by biological treatment [2,3]. This method is highly effective, but not enough [4] to remove micropollutants considered to be particularly harmful, such as polycyclic musk, UV filters, heavy metals, and microplastics [5–11]. Fragrances and UV filters are contaminants of emerging concern (CEC) [5]. The most commonly used and thus detected in the polycyclic musk environment are galaxolide (1,3,4,6,7,8-hexahydro-4,6,6,7,8,8,-hexamethyl-cyclopenta[g]benzopyran, HHCB) and tonalide (6-acetyl-1,1,2,4,4,7-hexamethyltetraline, AHTN), while the most important UV filters are benzophenone-3 (2-hydroxy-4-methoxyphenyl)-phenylmethanone, BP-3) and 4-MBC (4-methylbenzylidene camphor). These compounds often have the potential for bioaccumulation and also show estrogenic activity [6]. Heavy metals such as Zn, Cu, and Fe are typically used in cosmetics as physical UV filters, dyes, or enzyme components. However, even metals such as silver or bismuth are used as bactericides or mask ingredients [7]. Plastics are usually chemically inert, but under environmental conditions, they are broken down into microscopic grai(ns) that penetrate the body even at the cellular level. Their content in organisms increases (as) they move up the food chain [8]. Due to their persistence to decomposition, they for(m) layers or even islands floating on the water or they accumulate in the soil or bottom se(di)ments, depending on their density [9]. There is a need to develop a policy for dealing w(ith) substances that are components of cosmetics [10]. Cosmetic micropollutants during trea(t)ment in a biological treatment plant do not decompose but pass into the sludge phase [ ]. Their presence is detected in globally collected environmental samples, in concentratio(ns) usually below 100 µg/L or 100 µg/kg, depending on sample type [11].

In order to increase the effectiveness of CW treatment, the possibilities of improvi(ng) classically used coagulation and DAF processes were investigated [12–15]. Advanced (ox)idation processes (AOPs) [15–19] and the improvement of biological treatment [20–2(3)] were also tested. Attempts were also made to improve the entire treatment, including bo(th) chemical and biological methods [24–26].

Many alternatives to classical coagulation and DAF for CW treatment technologi(es) are being developed. Promising ones are AOPs, consisting of the effective generation (of) strong oxidants, namely radicals. In the case of AOPs in which the production of radica(ls) is catalyzed by the presence of Fe^{2+} ions (Fenton's process and its modifications), a maj(or) problem [27] is to ensure the appropriate quantity and availability of Fe^{2+} ions. The amou(nt) of Fe^{2+} ions in a solution is influenced by many factors, including pH, the efficiency of Fe^{2+} ion recovery from Fe^{3+}, and the rate of Fe^{2+} ion release from the carrier. This problem (is) solved in two ways: by controlling the Fe^{2+}/Fe^{3+} ions ratio or by the controlled continuo(us) introduction of Fe^{2+} ions into the solution. Both strategies pose numerous technical diffic(ul)ties when applied in practice; therefore, iron-based heterogeneous cocatalysts are gaini(ng) interest. Among them are Fe^0 (metallic iron, zero-valent iron, ZVI), Fe_2O_3 (hematit(e) and Fe_3O_4 (magnetite) [28–31]. Oxides act through coordinating surface sites of Fe^{2+} th(at) form complexes with contaminants and reduce them [28].

The aim of this study is to determine the effectiveness of the joint use of Fe^0, $Fe_2O(_3)$ and Fe_3O_4 as mutually supportive catalysts using synergy effects in the AOP treatment (of) industrial wastewater. This is the first article where Fe_2O_3, Fe_3O_4, and Fe^0 were mutua(lly) used in one process as co-catalysts supporting modified Fenton processes to treat cosme(tic) wastewater.

2. Results

2.1. Raw Wastewater

CW parameters used in the experiments are shown in Table 1. Low values of param(e)ters indicating the content of organic compounds (total organic carbon, TOC and chemic(al) oxygen demand, COD), and the almost complete absence of suspended solids (TSS) a(nd) nitrogen compounds (total Kjeldahl nitrogen, TNK), indicated the effective operation (of) the preliminary treatment (C/DAF) at the production plant. The very high value of t(he) electrolytic conductivity indicated significant wastewater salinity that cannot be derive(d) from only aluminum coagulants used in wastewater treatment in the factory or reagents f(or) pH correction. Despite the high five day biochemical oxygen demand (BOD_5) value an(d) theoretically high potential susceptibility to biological treatment (described as $BOD_5/CO(D)$ ratio, 0.382), even a small amount of raw cosmetic wastewater has a negative effect o(n) biological wastewater treatment plants, hence the need for further treatment.

Table 1. Cosmetic wastewater parameters.

Parameter	Unit	Value
TOC	mg/L	306.3
COD	mg/L	904
BOD_5	mg/L	345
TSS	mg/L	7
pH	-	8.7
Conductivity	mS/cm	13.8
Surfactants	mg/L	7
TKN	mg/L	<0.1

2.2. Kinetics Matching

In the case of the classical Fenton process, it involves catalytic radical oxidation and final coagulation combined with coprecipitation. The applied modification of the process causes the concentration of iron (II) ions to change due to the dissolution of metallic iron—Fe (II)—amount constantly increasing. In addition, as a result of UV irradiation, there is an increased reduction in iron (III) during the Fenton reaction to iron (II), which results in at least a theoretical decrease in homogeneous catalyst demand. The use of magnetite and hematite as cocatalysts, and metallic iron, leads to the appearance of the surface of solid catalyst heterogeneous processes, including sorption or ion exchange. All these processes are at least partially independent and sometimes even antagonistic. Therefore, describing the kinetics of the treatment process is not easy. Four equations were used to describe the kinetics:

- First-order reaction with respect to the TOC value:

$$TOC = TOC_0 \times e^{-kt} \qquad (1)$$

- Second-order reaction with respect to the TOC value:

$$TOC = (kt + 1/TOC_0)^{-1} \qquad (2)$$

- Modified first-order reaction with respect to the TOC value:

$$TOC = (TOC_0 - b) \times e^{-kt} + b \qquad (3)$$

- Modified second-order reaction with respect to the TOC value:

$$TOC = (kt + (TOC_0 - b)^{-1})^{-1} + b \qquad (4)$$

Equations (1) and (2) are the descriptions of the usual first- and second-order kinetics. Typically, however, the kinetics of a specific chemical reaction is described with clearly defined substrates and products. In the case of the description of wastewater, it is a complex mixture of many chemical compounds present in various concentrations. From a practical point of view, it is not possible to determine the concentrations of all chemical compounds and, most importantly, to predict all chemical reactions taking place. Therefore, collective parameters such as BOD5, COD, and TOC are described. In the case of treatment processes where hydrogen peroxide is used, it may remain after the process. While we ran the process to ensure that it was decomposed (and iodometrically checked), the decomposition of hydrogen peroxide takes time. Hydrogen peroxide is a well-known disruptor in COD measurement. During radical oxidation, wastewater is at least partially sterilized, which also affects BOD determination. Although both disturbing factors (in the determination of BOD and COD) can be eliminated, from a practical point of view, it is easier to use TOC notation, which is considered more reliable and unambiguous in its interpretation. Therefore, it was decided to describe all kinetics in relation to one collective TOC parameter. The idea behind first- and second-order kinetics is that the reaction can be completed, i.e., until the substrate is completely used. Under the conditions of our experiment, this means zeroing the TOC value and complete decomposition of the pollutant. However, it is not possible to obtain complete TOC elimination. There is always a certain amount left, hence the idea to modify the description of kinetics. As such, there was a certain number of compounds that could be removed in our process, but some would be persistent to decomposition. The amounts of these substances can be described as possible and impossible to remove TOC. Value "b" Equations (3) and (4), represents the content of this persistent, so-called "hard" TOC. The remaining amount of nonpersistent TOC can still be decomposed, and the description is related to first- or second-order kinetics. An example of the application of four kinetic models is presented in Figure 1. The best match was obtained for the modified second-order kinetics model.

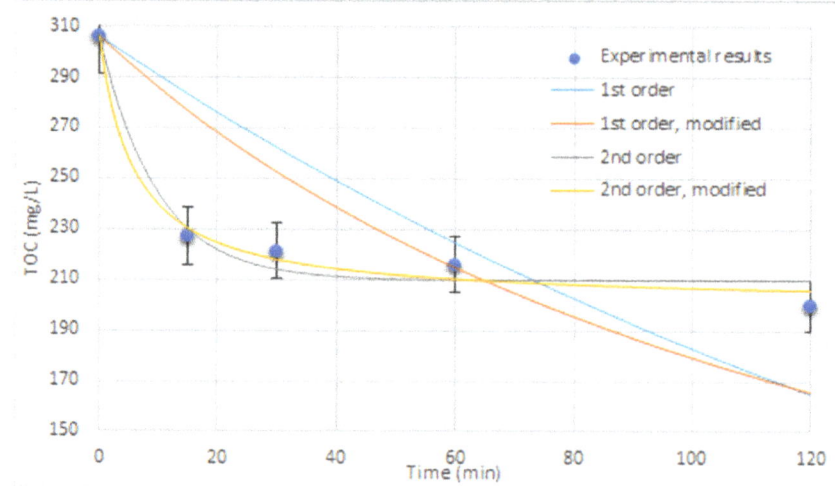

Figure 1. Example kinetic model results: 1500/1500/1000 $Fe_3O_4/Fe_2O_3/Fe^0$ doses (mg/L), H_2O_2/COD mass ratio 1:1, UV irradiation, pH = 3.0.

2.3. Treatment Processes

Detailed doses and proportions of the reagents used during the research on CW catalytic treatment are presented in Table S1, while treatment results are shown in Figures 2–5.

Treatment is more effective as the process takes longer to run. The use of UV light increases the effectiveness of the treatment compared to a non-light-assisted process with the same doses of reagents.

In each of the non-light-assisted experiments, the most intensive TOC removal, around 50 mg/L, was obtained in the first 15 min. Such a situation could be observed for, e.g., 2:1 H_2O_2/COD ratio and 250/250/1500 mg/L $Fe_3O_4/Fe_2O_3/Fe^0$ catalyst doses (Figure 3). In the mentioned sample, after 30 min of the process, TOC was 221.7 mg/L, for 15 min from the first measurement, it was decreased by 27 mg/L. During subsequent measurements made at 15-, 30-, and 60-min time differences, TOC decreased more slowly. A better treatment effect was obtained by using a lower ratio of 1:1 H_2O_2/COD. The lowest TOC, 182.0 mg/L, was obtained for 1:1 H_2O_2/COD ratio and 500/500/3000 mg/L $Fe_3O_4/Fe_2O_3/Fe^0$ catalyst doses after a 120 min process time (Figure 2). The second-lowest TOC, 198.4 mg/L, was for 1:1 H_2O_2/COD ratio and 500/500/1000 mg/L $Fe_3O_4/Fe_2O_3/Fe^0$ catalyst doses. The highest TOC 218.0 mg/L, only 28.8% TOC removal after a 120-min process time, was obtained for 375/375/250 mg/L $Fe_3O_4/Fe_2O_3/Fe^0$ catalyst doses at 2:1 H_2O_2/COD ratio. On the basis of the presented data, for non-light-assisted process with regard to catalyst doses at constant values of hydrogen peroxide, lower TOC values were achieved for higher doses of the catalysts: 4000 mg/L, slightly lower for 2000 mg/L and the lowest for 1000 mg/L. In most cases, lower values of TOC were recorded for the lower 1:1 H_2O_2/COD ratio. The TOC values determined after the treatment process with the 2:1 H_2O_2/COD ratio were higher than the value for 1:1 H_2O_2/COD ratio. The exception was the process involving 1000 mg/L of hematite, 1000 mg/L of magnetite, and 2000 mg/L of metallic iron, in which the difference was 40 mg/L.

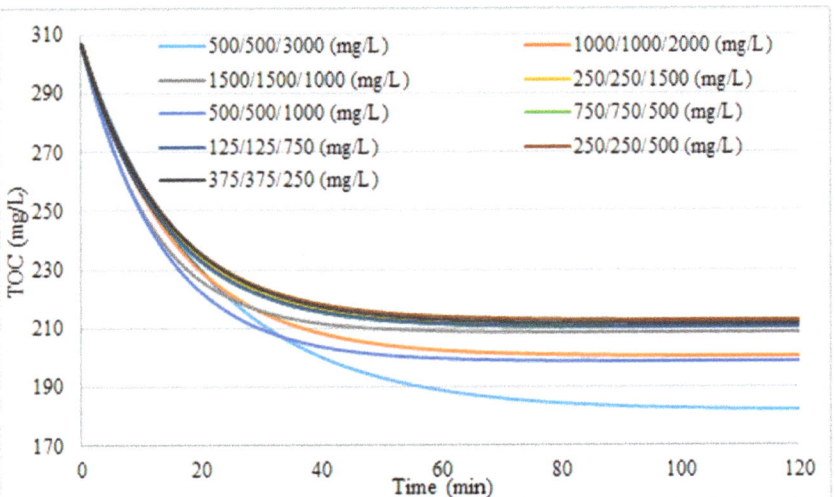

Figure 2. Cosmetic wastewater (CW) treatment results with different Fe_3O_4/Fe_2O_3/Fe^0 doses (mg/L) H_2O_2/COD mass ratio 1:1, without UV irradiation, pH = 3.0.

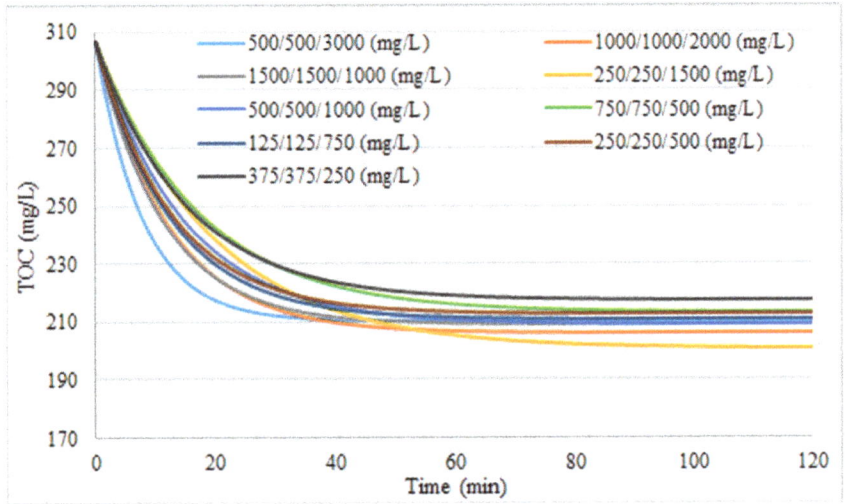

Figure 3. CW treatment results with different Fe_3O_4/Fe_2O_3/Fe^0 doses (mg/L) H_2O_2/COD mass ratio 2:1, without UV irradiation, pH = 3.0.

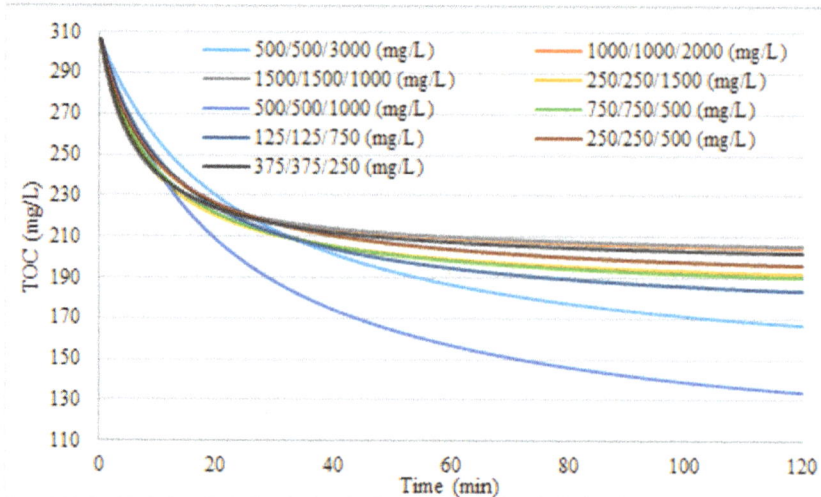

Figure 4. CW treatment results with different $Fe_3O_4/Fe_2O_3/Fe^0$ doses (mg/L) H_2O_2/COD mass ratio 1:1, UV irradiation, pH = 3.0.

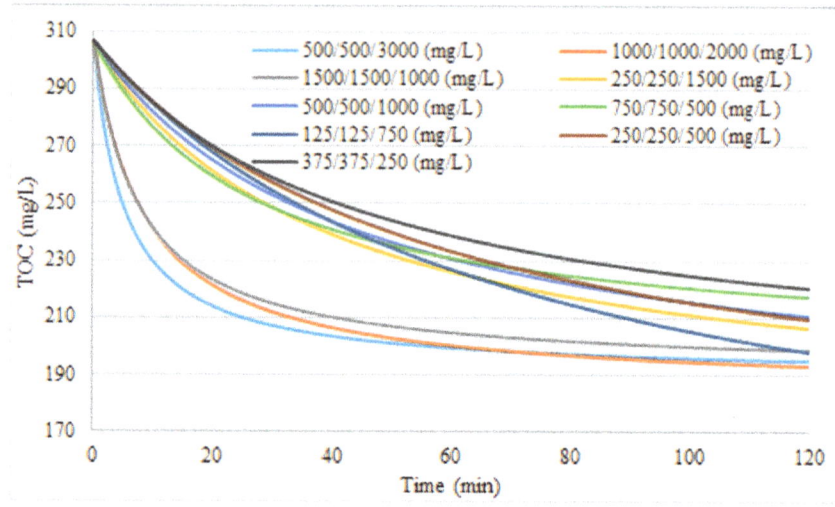

Figure 5. CW treatment results with different $Fe_3O_4/Fe_2O_3/Fe^0$ doses (mg/L) H_2O_2/COD mass ratio 2:1, UV irradiation, pH = 3.0.

In the light-assisted process, the lowest TOC, 134.1 mg/L (56.2% TOC removal) was obtained after 120 min for 1:1 H_2O_2/COD ratio and 500/500/1000 mg/L $Fe_3O_4/Fe_2O_3/$ Fe^0 catalyst doses (Figure 5). In the initial phase of the process, a slower decrease in TOC value was visible for the 2:1 H_2O_2/COD ratio, to 260–270 mg/L. However, a faster decrease was observed in the mixture of iron compounds mass equal to 4000 mg/L. This may indicate that the weight of the catalysts was optimal for the higher dose of hydrogen peroxide and accelerated the process. For smaller total catalyst concentrations, 2:1 H_2O_2/COD ratio led to the inhibition of the reaction. The decrease in TOC value for 4000 mg/L of catalyst 2 H_2O_2/COD ratio was comparable with that in samples with a lower oxidant concentration for up to 30 min. After this time, for the 1:1 H_2O_2/COD ratio, reactions slowed down significantly. An exception may be the sample of 500/500/1000 mg/L $Fe_3O_4/Fe_2O_3/F$

catalyst doses, where TOC decreased throughout the experiment. In this case, the heterocatalytic reaction could have contributed to the steady decline. The optimal selection of reagent doses ensured the decomposition of the organic pollutants on the surface of the catalysts. For a higher concentration of the oxidant, the given doses of the catalysts did not give an outstanding result, and at a lower concentration of H_2O_2, it had a greater effect on TOC decomposition. In the experiments where the concentration of iron compounds was 4000 mg/L, there was no visible difference in the rate of the processes, resulting from the concentration of hydrogen peroxide. For the experiments carried out at the concentration of iron compounds of 2000 and 1000 mg/L, however, there was a difference according to the dose of hydrogen peroxide. Lower concentrations of the oxidant resulted in faster TOC removal in the first few minutes of the experiment. At higher concentrations, removal was slower, and more time was required for the reaction to come to a halt. Even though the reaction took longer, in most cases, efficiency for TOC removal at a higher concentration of H_2O_2 did not exceed the effectiveness for the 1:1 H_2O_2/COD ratio. The exceptions were the concentrations of 1000/1000/3000 mg/L Fe_3O_4/Fe_2O_3/Fe^0, in which efficiency was higher at a higher oxidant concentration.

Additionally, an experiment was performed that demonstrated the influence of pH on the efficiency of the pollutant oxidation process (Figure 6). The process was the most effective at pH 2 and 3. The processes carried out at a pH greater than 3 were ineffective, and the decrease in TOC value from 15 min until the end of the experiment was not significant. At pH 2 and 3, TOC decrease was visible throughout the process. At pH 2, sediment in the sample was swollen and occupied the largest volume compared to in the other samples. Sediment after the process carried out at pH 4 was a reddish color and had a volume comparable to that in the processes at pH 3. Obtained sediment during the experiment at pH 5 was brown, and its amount was the smallest in comparison to that formed during the process at other pH values. In the experiment carried out at pH 6, the sediment after the process was reddish, and its structure was comparable to that of the sediment at pH 3 and 4. The red may have indicated the presence of iron (III) hydroxide, which is formed at a high solution pH.

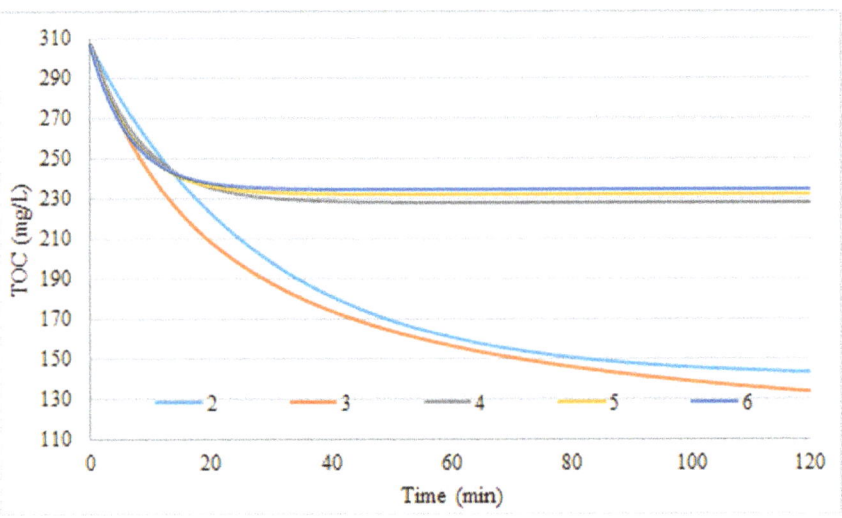

Figure 6. CW treatment results at different pH, 500/500/1000 Fe_3O_4/Fe_2O_3/Fe^0 doses (mg/L) H_2O_2/COD mass ratio 1:1, UV irradiation.

H_2O_2 is a weak acid. Its stability increases with a decrease in pH value, which may impede its catalytic decomposition into a hydroxyl ion and radical. However, these effects were not strongly observed during the experiment. Hydrogen peroxide under alkal conditions decomposes rapidly with the evolution of oxygen. For this reason, no atten was made to operate the process under alkaline conditions. However, the presence the formation of oxygen bubbles hindering the sedimentation of the formed sludge w observed in all experiments at the termination stage of the Fenton reaction. In the case iron (II) and (III), its source in the process was twofold: the surface of stable iron oxic magnetite and hematite, and the dissolution of metallic iron. The solubility (corrosion) metallic iron occurs quickly under strongly acidic conditions; under neutral conditic the process is very slow. Therefore, in neutral conditions, contact between the reacta is difficult to achieve during the entire duration of the process. In an acidic environme due to the constant increase in the content of dissolved iron, its availability, and thus intensity of the Fenton reaction, increases steadily with time. Additionally, the form iron is strongly pH-dependent. At pH 5, iron hydroxides with small solubility begin form, and the coagulation process begins. For iron hydroxide, due to minimal solubil the optimal value for carrying out the coagulation process was around 6.0 and above This was another reason for abandoning the experiment in alkaline conditions, as radic were terminated on the sludge flocs, which resulted in rapidly decreasing process efficier

The statistical analysis was described on the basis of Miller [32]. ANOVA was used determine the magnitude of variability in the average concentrations of TOC and to che whether differences in the average test results for TOC for individual process conditic (for different doses of catalysts) may have been caused by random errors (Tables S2 and Figures S4 and S5).

Variance is estimated using two methods: the method determining the variabil within a given sample, and variability between samples. The difference in the perform tests was the different durations and reagent doses of the process. The above statement i null hypothesis.

If the hypothesis were true, then there would be no large difference between calculat values. If the hypothesis were not true, then the between-group estimate would be grea than the intergroup estimate. This is due to the high variability between samples. To che if the difference was significant, Snedecor's one-sided F test with $\alpha = 0.05$ was perform

The following null hypothesis was made for ANOVA: the tests were perform accurately, and reproducibility was achieved in TOC results. The value of the F parame was lower than that of the critical F, which means that the hypothesis is true. The me values in the samples were similar, and similar conclusions were found during TC analysis. Even the optimal value did not significantly differ from the other values. From perspective of the performed tests, this is a favorable phenomenon, as it proved the accura and repeatability of the performed tests. However, the process itself was not effecti and from the perspective of the conducted process, its effectiveness was not favorable.

2.4. HS-SPME–GC–MS Analysis

Head space-solid phase micro extraction–gas chromatography–mass spectrome (HS-SPME–GC–MS) analysis results are shown in Table 2. GC-MS chromatograms, raw and treated samples (the sample with the lowest TOC after the process was selecte are shown in Figures S1 and S2.

The identified compounds were mainly cosmetic bases (e.g., decamethyltetrasil ane or decamethylcyclopentasiloxane) and fragrances (e.g., 1,3,4,6,7,8-hexahydro-4,6 7,8,8-hexamethyl-cyclopenta(g)-2-benzopyran (galaxolide) or 4-isopropenyl-1-methyl cyclohexene (limonene)).

All compounds detected with HS-SPME-GC-MS were removed during the treatme process. No new compound was detected during the process. HS-SPME-GC-MS is a use tool that can be used to confirm the high efficiency of the treatment process.

Table 2. Raw wastewater head space-solid phase micro extraction–gas chromatography–mass spectrometry (HS-SPME–GC–MS) analysis results.

No.	Retention Time (s)	Peak Area	Compound Name
1	842.56	263936015	2,2,4,6,6-pentamethylheptane
2	914.76	76933033	2-ethyl-1-hexanol
3	922.12	55446786	4-isopropenyl-1-methyl-1-cyclohexene
4	969.37	40480391	decamethyltetrasiloxane
5	995.26	135240755	6-ethyl-2-methyl-6-hepten-2-ol
6	1042.25	407718598	3,7-dimethyl-3-octanol
7	1049.08	376591195	3,7-dimethyl-1,6-octadien-3-ol
8	1126.13	376591195	decamethylcyclopentasiloxane
9	1191.81	87920970	5-methyl-2-(1-methylethyl)-cyclohexanol
10	1220.03	87920970	3,7-dimethyl-1,6-octadien-3-ol
11	128073	342408048	dodecamethylpentasiloxane
12	1344.91	201821493	2,6-dimethyloctane
13	13.92.76	84107649	tridecane
14	1424.61	293329202	dodecamethylcyclohexasiloxane
15	1441.31	523263467	2,2,4,4,6,8,8-heptamethylnonane
16	1473.59	439161977	undecylcyclohexane
17	1667.25	592268444	2-dodecanol
18	1791.80	913690562	hexadecamethylheptasiloxane
19	1842.23	471919073	hexadecane
20	1933.23	175348881	di-n-octyl ether (1,1′-oxybisoctane)
21	1938.56	193030178	cyclopentaneacetic acid, 3-oxo-2-pentyl-,methyl ester
22	1968.92	152991091	7a-isopropenyl-4,5-dimethy octahydro-1H-inden-4-yl)methanol
23	1978.11	93958573	unidentified compound
24	2035.27	1390556189	2-butylooctanol
25	2067.00	474433634	2,4-dimethyl-1-heptanol
26	2074.26	161503287	isobutyl nonyl carbonate
27	2091.65	569595240	2-methyl-1-decanol
28	2115.29	467909425	oxalic acid, cyclohexylmethyl tridecyl ester
29	2150.05	226358082	unidentified compound
30	2206.07	980264063	1,3,4,6,7,8-hexahydro-4,6,6,7,8,8-hexamethyl-cyclopenta(g)-2-benzopyran
31	2210.77	678685873	1-hexadecanol
32	2307.06	327313756	ether, di-n-octyl-(1,1′-oxybis-octane),
33	2374.06	79843231	1-methylethyl hexadecanoate

3. Discussion

An innovative solution was applied that has not yet been used as a cosmetic wastewater treatment technique, nor has it appeared in other industries. After separate analysis of the effectiveness of each catalyst, metallic iron with hematite and metallic iron with magnetite [33], the compounds were combined to create a unique mixture in order to check its properties and effectiveness during the treatment of cosmetic wastewater. The highest efficiency of the process was achieved when using the catalyst proportion in which there was a significant advantage of metallic iron and a comparable lower dose of magnetite and hematite (quantitative ratio of the compounds was 250/250/1500 mg/L). The lowest efficiency of the process was obtained when the catalyst was used, in which a significantly lower dose of metallic iron of 250 mg/L, and slightly higher concentrations of hematite and magnetite were used, 375 mg/L in all cases. When using a total concentration of 2000 mg/L of the mixed catalyst, the process was the most effective. The lowest concentrations were obtained with the use of a lower dose of 1000 mg/L reagent. The lower dose was insufficient to efficiently perform oxidation. When a higher dose of hydrogen peroxide was used, the process was also not as effective as when a higher dose of the mixed catalysts was used.

Analyzing the obtained results, they complied with those of earlier research [33], indicating that the excess of hydrogen peroxide adversely affects the performed process, decreasing its effectiveness.

The use of a higher dose of the oxidant caused a lower efficiency in TOC removal from CW. Excess hydrogen peroxide is an inhibitor of the reaction. Then, a process may take place (reaction Equation (5)) where a hydroperoxide radical is formed (oxidation reduction potential 1.7 V) which is much less reactive than the hydroxyl radical (2.8 V). Hydroperoxide radicals react with hydroxyl radicals (reaction Equation (6)) to form a water molecule and an oxygen molecule. Reactive molecules merge with each other to form substrates that are not strong oxidants.

$$HO^\bullet + H_2O_2 \rightarrow HO_2^\bullet + H_2O$$

$$HO_2^\bullet + HO^\bullet \rightarrow H_2O + O_2$$

$$Fe^{2+} + H_2O_2 \rightarrow FeO^{2+} + H_2O$$

$$FeO^{2+} + H_2O_2 \rightarrow Fe^{2+} + O_2 + H_2O$$

$$FeO^{2+} + RH \rightarrow Fe^{2+} + ROH$$

$$Fe(C_2O_4)] + H_2O_2 \rightarrow [Fe(C_2O_4)]^+ + OH^- + OH^\bullet \quad (1)$$

$$[Fe(C_2O_4)_3]^{3-} + h\nu \rightarrow [Fe(C_2O_4)_2]^{2-} + C_2O_4^{\bullet-} \quad (1)$$

$$Fe^{3+} + C_2O_4^{\bullet-} \rightarrow Fe^{3+} + 2CO_2 \quad (1)$$

$$Fe(RCOO)^{2+} + h\nu \rightarrow Fe^{2+} + R^\bullet + CO_2 \quad (1)$$

Intermediate compounds are formed in the catalytic cycle. One is oxoiron $Fe^{IV}O$ [3] which is formed as a result of the reaction initiating the Fenton process, described reaction Equation (7).

With an excess of hydrogen peroxide, oxoion reacts with it and forms Fe^{2+}, oxygen and water (reaction (8)). This is a reaction that stops the process. Oxoiron is also involved in the oxidation of organics (reaction (9)), which is much slower than the one with the hydroxyl radical.

Factors accelerating the formation of radicals may be ligands present in CW, which form complexes and chelates with iron. The oxalate ligand reacts with hydrogen peroxide form a hydroxyl radical according to reaction Equation (10). Oxalates (diethyl, dimethyl, diisopropyl, diisobutyl, sodium) are present in cosmetics and thereby also in CW. In acid conditions, at a low pH of 3.0, acid hydrolysis of oxalates esters could take place.

The process was supported by UV radiation (full emission spectrum is shown Figure S3), so its influence on reaction kinetics in the presence of oxalates should considered. Absorbed radiation causes the decarboxylation of the ligand with the release of CO_2 and reduction in Fe^{3+} to Fe^{2+} (reactions (11) and (12)). The oxidation of organic compounds occurs at high speed in the presence of oxalates. Complex compounds with carboxylic acids under UV radiation reduce the iron, and alkyl radical and carbon dioxide are produced (reaction (13)). These reactions lead to the fast mineralization of organic compounds in wastewater.

Improperly selected doses of catalysts (too small an amount of iron) have a negative effect on the high-efficiency treatment process. Regardless of the amount of catalyst and hydrogen peroxide, the effectiveness of the treatment processes increases with time.

The longest process time, 120 min, was the most effective. The 15 min process time was the least effective, as this was not enough time to carry out the treatment process. However, the duration and higher costs of the process should be considered, and the optimal time should be selected so as to maintain high efficiency with an appropriate cost of treatment. For lightless processes and 120-min process time, the process is no longer profitable, and equally high efficiency was achieved after 60 min of treatment. Such an observation was not made in the case of light-assisted processes. For them, extending the process time to 120 min is still profitable. The process showed the greatest efficiency in relation to its duration during the first 15 min, and TOC was mostly decreased in a short period of time. Then, sequential TOC measurements showed lower speed of the treatment process.

Most of the studies that used hematite or magnetite were carried out on pollutants present in components. For example, hematite was used as a catalyst by Araujo et al. [35]. In total, 20 g of hematite and a dose of 800 mg/L of hydrogen peroxide allowed for achieving 99% treatment efficiency of the process after 120 min. Our research was carried out for much lower catalyst doses and a different type of contaminant with a much more complex matrix (wastewater from the cosmetic industry).

In samples taken after 15- and 30-min process time, a large amount of evolving gas bubbles was visible because of the decomposition of unreacted hydrogen peroxide. The color of the precipitate depended on the ratio of iron to magnetite and hematite. Dependence was visible; the greater the amount of metallic iron in the sample was, the more orange the sediment was. No greenish sediment was observed, which would indicate the presence of Fe^{2+} ions. On this basis, it was concluded that the oxidant doses were correctly selected.

There was a separation in the sediment phases into an upper rusty one (oxidized iron) and lower black one (magnetite, metallic iron). The exact mechanism of the separation process is unknown, but this could be due to the difference between densities, as iron hydroxide has a lower density than that of magnetite and metallic iron.

4. Materials and Methods

4.1. Wastewater

Samples of real cosmetic wastewater, pre-treated by coagulation coupled with dissolved air flotation, were taken for the tests. The samples were taken from an industrial plant located in Poland.

4.2. Treatment Process

Zero-valent iron (Ferox Target, 325 mesh) was supplied by Hepure (Hillsborough, NJ, USA). Hematite (10 μm) was supplied by Kremer (Aichstetten, Germany); magnetite (10 μm) was supplied by Kremer (Aichstetten, Germany); 30% H_2O_2 solution was supplied by Stanlab (Lublin, Poland).

Doses of H_2O_2, Fe^0, magnetite, and hematite were selected in preliminary tests. Treatment processes were carried out in a 1.5 L reactor filled with 1 L of the sample, stirred at 300 rpm (Heidolph MR3000, Schwabach, Germany). pH in the treatment processes was set to 3.0, unless otherwise stated. The experiment on the influence of pH on the efficiency of the process was carried out at pH 2.0, 3.0, 4.0, 5.0 and 6.0. Samples were taken after 15, 30, 60 and 120 min of the process. The process was stopped as a result of alkalization to pH 9.0 with 3M NaOH (POCh, Gliwice, Poland).

The source of radiation was medium-pressure Fe/Co 400W lamp type HPA 400/30 SDC with 94W UVA power (Philips, Amsterdam, The Netherlands). The lamp spectrum is shown in Figure S3. Details of the experiment setup are shown in Table S1.

4.3. Analytical Methods

Total organic carbon (TOC), five-day biochemical oxygen demand (BOD_5), chemical oxygen demand (COD), total Kjeldahl nitrogen (TKN), total suspended solids (TSS), ammonia, surfactants, pH, and conductivity were determined according to the standard methods.

TOC was determined using a TOC-L analyzer (Shimadzu, Kyoto, Japan) with an OCT-L8-port sampler (Shimadzu, Kyoto, Japan). Chromatographic analysis was conducted with an Agilent 7890A (Santa Clara, CA, USA) gas chromatograph coupled with a Leco TruTOF (St. Joseph, MI, USA) mass spectrometer. The detailed methodology is described elsewhere [33].

5. Conclusions

Due to the increasing consumption of cosmetics, an effective and inexpensive method of CW treatment is needed. The effective treatment of CW in accordance with applicable legal standards is difficult. The results of this research confirmed the effectiveness the pretreatment of wastewater currently used by most industrial plants with the use C/DAF.

CW can be effectively treated with both the UV/H_2O_2/Fe_3O_4/Fe_2O_3/Fe^0 and H_2O_2/Fe_3O_4/Fe_2O_3/Fe^0 process. The condition for the application of an effective C treatment by catalytic oxidation is the use of an appropriate dose of hydrogen peroxide a catalyst.

The best match of the results to kinetic models was obtained by second-order equatio with a modification, taking into account the amount of undegraded compounds during treatment process.

On the basis of the ANOVA results, the hypothesis regarding the accuracy and rep ducibility of the research was confirmed.

Supplementary Materials: The following are available online at https://www.mdpi.com/207 344/11/1/9/s1, Table S1: Experimental setup, Table S2: One-way analysis of variance for me values of TOC for 1:1 H_2O_2/COD ratio, Table S3: One-way analysis of variance for mean values TOC for 2:1 H_2O_2/COD ratio. Figure S1: GC-MS chromatogram: Raw wastewater, Figure S2: C MS chromatogram: Treated wastewater, 500/500/1000 mg/L Fe_3O_4/Fe_2O_3/Fe^0 catalyst do 1:1 H_2O_2/COD ratio, pH = 3.0, 120 min, Figure S3: The emission spectrum of the lamp, ure S4: Anova plots of the UV/H_2O_2/Fe_3O_4/Fe_2O_3/Fe^0 process, Figure S5: Anova plots of H_2O_2/Fe_3O_4/Fe_2O_3/Fe^0 process.

Author Contributions: Conceptualization, methodology, writing—original draft preparation, view and editing, supervision: J.B. and P.M.; laboratory work and data processing: J.B., P.M., D M.K., D.Ś. and P.P. All authors have read and agreed to the published version of the manuscript.

Funding: This research received no external funding.

Acknowledgments: The authors thank Hepure for providing ZVI Ferox Target samples as resea material.

Conflicts of Interest: The authors declare no conflict of interest.

References

1. Wpływ Światowej Pandemii Koronowirusa na Branżę Kosmetyczną. Biuro Strategii i Analiz Międzynarodowych, Pekao July 2020. Available online: https://wspieramyeksport.pl/api/public/files/1948/Kosmetyczna_lipiec_2020.pdf (accessed 25 September 2020).
2. Ma, G.; Chen, J. Nitrogen and phosphorus pollutants in cosmetics wastewater and its treatment process of a certain bra *IOP Conf. Ser. Earth Environ. Sci.* **2018**, *113*, 012051. [CrossRef]
3. Bello, L.A.; Omoboye, A.J.; Abiola, T.O.; Oyetade, J.A.; Udorah, D.O.; Ayeola, E.R. Treatment technologies for wastewater fr cosmetic industry—A review. *Int. J. Chem. Biol. Sci.* **2018**, *4*, 69–80.
4. De Melo, E.D.; Mounteer, A.H.; de Souza Leão, L.H.; Bahia, R.C.B.; Campos, I.M.F. Toxicity identification evaluation of cosme industry wastewater. *J. Hazard. Mater.* **2013**, *244–245*, 329–334. [CrossRef] [PubMed]
5. Biel-Maeso, M.; Corada-Fernández, C.; Lara-Martín, P.A. Removal of personal care products (PCPs) in wastewater and slud treatment and their occurrence in receiving soils. *Water Res.* **2019**, *150*, 129–139. [CrossRef]
6. Juliano, C.; Magrini, G.A. Cosmetic ingredients as emerging pollutants of environmental and health concern. A mini-revi *Cosmetics* **2017**, *4*, 11. [CrossRef]
7. Amneklev, J.; Augustsson, A.; Sorme, L.; Bergback, B. Bismuth and silver in cosmetic products a source of environmental a resource concern? *J. Ind. Ecol.* **2015**, *20*, 99–106. [CrossRef]
8. Guerranti, C.; Martellini, T.; Perra, G.; Scopetani, C.; Cincinelli, A. Microplastics in cosmetics: Environmental issues and needs global bans. *Environ. Toxicol. Pharmacol.* **2019**, *68*, 75–79. [CrossRef]
9. Kalcíkova, G.; Alic, B.; Skalar, T.; Bundschuh, M.; Gotvajn, Z.A. Wastewater treatment plant effluents as source of cosme polyethylene microbeads to freshwater. *Chemosphere* **2017**, *188*, 25–31. [CrossRef]
10. Vita, N.A.; Brohem, C.A.; Canavez, A.D.P.M.; Oliveira, C.F.S.; Kruger, O.; Lorencini, M.; Carvalho, C.M. Parameters for assessi the aquatic environmental impact of cosmetic products. *Toxicol. Lett.* **2018**, *287*, 70–82. [CrossRef]
11. Hopkins, Z.R.; Blaney, L. An aggregate analysis of personal care products in the environment: Identifying the distribution environmentally-relevant concentrations. *Environ. Int.* **2016**, *92–93*, 301–316. [CrossRef]
12. El-Gohary, F.; Tawfik, A.; Mahmoud, U. Comparative study between chemical coagulation/precipitation (C/P) versus coa lation/dissolved air flotation (C/DAF) for pre-treatment of personal care products (PCPs) wastewater. *Desalination* **2010**, *2* 106–112. [CrossRef]

Michel, M.M.; Tytkowska, M.; Reczek, L.; Trach, Y.; Siwiec, T. Technological conditions for the coagulation of wastewater from cosmetic industry. *Ecol. Eng.* **2019**, *20*, 78–85. [CrossRef]

Michel, M.M.; Siwiec, T.; Tytkowska, M.; Reczek, L. Analysis of flotation unit operation in coagulation of wastewater from a cosmetic factory. *Przem. Chem.* **2015**, *11*, 2000–2005. (In Polish)

Perdigon-Melon, J.; Carbajo, J.; Petre, A.; Rosal, R.; García-Calvo, E. Coagulation-Fenton coupled treatment for ecotoxicity reduction in highly polluted industrial wastewater. *J. Hazard. Mater.* **2010**, *181*, 127–132. [CrossRef] [PubMed]

Martins, R.C.; Nunes, M.; Gando-Ferreira, L.M.; Quinta-Ferreira, R.M. Nanofiltration and Fenton's process over iron shavings for surfactants removal. *Environ. Technol.* **2014**, *35*, 2380–2388. [CrossRef] [PubMed]

Bautista, P.; Casas, J.A.; Zazo, J.A.; Rodriguez, J.J.; Mohedano, A.F. Comparison of Fenton and Fenton-like oxidation for the treatment of cosmetic wastewater. *Water Sci. Technol.* **2014**, *70*, 472–478. [CrossRef] [PubMed]

Martins de Andrade, P.; Dufrayer, C.R.; de Brito, N.N. Treatment of real cosmetic effluent resulting from the manufacture of hair conditioners by reduction degradation, adsorption and the fenton reaction. *Ozone Sci. Eng.* **2019**, *41*, 221–230. [CrossRef]

Ebrahiem, E.E.; Al-Maghrabi, M.N.; Mobarki, A.R. Removal of organic pollutants from industrial wastewater by applying photo-Fenton oxidation technology. *Arab. J. Chem.* **2017**, *10*, S1674–S1679. [CrossRef]

Friha, I.; Feki, F.; Karray, F.; Sayadi, S. A pilot study for cosmetic wastewater using a submerged flat sheet membrane bioreactor. *Procedia Eng.* **2012**, *44*, 819–820. [CrossRef]

Monsalvo, V.M.; Lopez, J.; Mohedano, A.F.; Rodriguez, J.J. Treatment of cosmetic wastewater by a full-scale membrane bioreactor. *Environ. Sci. Pollut. Res.* **2014**, *21*, 12662–12670. [CrossRef]

Zhang, C.; Ning, K.; Guo, Y.; Chen, J.; Liang, C.; Zhang, X.; Wang, R.; Guo, L. Cosmetic wastewater treatment by a combined anaerobic/aerobic (ABR + UBAF) biological system. *Desal. Water Treat.* **2015**, *53*, 1606–1612. [CrossRef]

Puyol, D.; Monsalvo, V.M.; Mohedano, A.F.; Sanz, J.L.; Rodriguez, J.J. Cosmetic wastewater treatment by upflow anaerobic sludge blanket reactor. *J. Hazard. Mater.* **2011**, *185*, 1059–1065. [CrossRef] [PubMed]

Muszyński, A.; Marcinowski, P.; Maksymiec, J.; Beskowska, K.; Kalwarczyk, E.; Bogacki, J. Cosmetic wastewater treatment with combined light/Fe^0/H_2O_2 process coupled with activated sludge. *J. Hazard. Mater.* **2019**, *378*, 120732. [CrossRef] [PubMed]

Chávez, A.M.; Gimeno, O.; Rey, A.; Pliego, G.; Oropesa, A.L.; Álvarez, P.M.; Beltrán, F.J. Treatment of highly polluted industrial wastewater by means of sequential aerobic biological oxidation-ozone based AOPs. *Chem. Eng. J.* **2019**, *361*, 89–98. [CrossRef]

Banerjee, P.; Dey, T.; Sarkar, S.; Swarnakar, S.; Mukhopadhyay, A.; Ghosh, S. Treatment of cosmetic effluent in different configurations of ceramic UF membrane based bioreactor: Toxicity evaluation of the untreated and treated wastewater using catfish (Heteropneustes fossilis). *Chemosphere* **2016**, *146*, 133–144. [CrossRef]

Babuponnusami, A.; Muthukumar, K. A review on Fenton and improvements to the Fenton process for wastewater treatment. *J. Environ. Chem. Eng.* **2014**, *2*, 557–572. [CrossRef]

Sun, Y.; Li, J.; Huang, T.; Guan, X. The influences of iron characteristics, operating conditions and solution chemistry on contaminants removal by zero-valent iron: A review. *Water Res.* **2016**, *100*, 277–295. [CrossRef]

Pereira, M.C.; Oliveira, L.C.A.; Murad, E. Iron oxide catalysts: Fenton and Fenton-like reactions a review. *Clay Miner.* **2012**, *47*, 285–302. [CrossRef]

He, H.; Zhong, Y.; Liang, X.; Tan, W.; Zhu, J.; Wang, C.Y. Natural Magnetite: An efficient catalyst for the degradation of organic contaminant. *Sci. Rep.* **2015**, *5*, 10139. [CrossRef]

Hadjltaief, H.B.; Sdiri, A.; Gálvez, M.E.; Zidi, H.; Da Costa, P.; Zina, M.B. Natural Hematite and Siderite as Heterogeneous Catalysts for an Effective Degradation of 4-Chlorophenol via Photo-Fenton Process. *ChemEngineering* **2018**, *2*, 29. [CrossRef]

Miller, J. *Statystyka I Chemometria W Chemii Analitycznej*; Wydawnictwo Naukowe PWN: Warsaw, Poland, 2016.

Marcinowski, P.; Bury, D.; Krupa, M.; Ścieżyńska, D.; Prabhu, P.; Bogacki, J. Magnetite and hematite in advanced oxidation processes application for cosmetic wastewater treatment. *Processes* **2020**, *8*, 1343. [CrossRef]

Vorontsov, A.V. Advancing Fenton and photo-Fenton water treatment through the catalyst design. *J. Hazard. Mater.* **2019**, *372*, 103–112. [CrossRef] [PubMed]

Araujo, F.V.F.; Yokoyama, L.; Teixeira, L.A.C.; Campos, J.C. Heterogenous Fenton Process using the mineral hematite for the discolouration of a reactive dye solution. *Braz. J. Chem. Eng.* **2011**, *28*, 605–616. [CrossRef]

Alkali-Activated Materials as Catalysts for Water Purification

Anne Heponiemi *, Janne Pesonen, Tao Hu and Ulla Lassi

Research Unit of Sustainable Chemistry, University of Oulu, P.O. Box 4300, FI-90014 Oulu, Finland; janne.pesonen@oulu.fi (J.P.); tao.hu@oulu.fi (T.H.); ulla.lassi@oulu.fi (U.L.)
* Correspondence: anne.heponiemi@oulu.fi

Abstract: In this study, novel and cost-effective alkali-activated materials (AAMs) for catalytic applications were developed by using an industrial side stream, i.e., blast furnace slag (BFS). AAMs can be prepared from aluminosilicate precursors under mild conditions (room temperature using non-hazardous chemicals). AAMs were synthesized by mixing BFS and a 50 wt % sodium hydroxide (NaOH) solution at different BFS/NaOH ratios. The pastes were poured into molds, followed by consolidation at 20 or 60 °C. As the active metal, Fe was impregnated into the prepared AAMs by ion exchange. The prepared materials were examined as catalysts for the catalytic wet peroxide oxidation (CWPO) of a bisphenol A (BPA) aqueous solution. As-prepared AAMs exhibited a moderate surface area and mesoporous structure, and they exhibited moderate activity for the CWPO of BPA, while the iron ion-exchanged, BFS-based catalyst (Fe/BFS30-60) exhibited the maximum removal of BPA (50%) during 3 h of oxidation at pH 3.5 at 70 °C. Therefore, these new, inexpensive, AAM-based catalysts could be interesting alternatives for catalytic wastewater treatment applications.

Keywords: alkali-activated material; geopolymer; blast furnace slag; catalytic wet peroxide oxidation; Fe-catalyst; bisphenol A

1. Introduction

Alkali-activated materials (AAMs) are inorganic, amorphous compounds that contain $[SiO_4]^{4-}$ and $[AlO_4]^{5-}$, which can be prepared by using aluminosilicates in addition to hydroxides, carbonates, or silicates of alkali and alkaline earth metals. The calcium content affects the AAM structure; therefore, materials with a low Ca content comprise a three-dimensional, highly interconnected aluminosilicate framework (also known as a geopolymer) [1], while those with a high Ca content comprise a cross-linked and non-cross linked structure that resembles that of tobermorite [2]. The use of AAMs, and especially geopolymers, has been investigated in the building industry as a more environment-friendly alternative to Portland cement, due to their chemical and physical stability [3–5], as well as in more advanced applications, such as adsorbents for the removal of impurities from wastewater [6–9] and composite materials [4]. Moreover, owing to the fact that the structure of AAMs is similar to that of zeolites, their use as catalytic materials can also be exploited. Compared with the synthesis of zeolites, that of AAMs can be performed at ambient pressure and room temperature, using cost-effective raw materials (kaolin clay or industrial waste, such as fly ash), making AAMs fascinating, environment-friendly alternatives to commercial zeolite. Only a few years ago, Sazama et al. reported the synthesis of AAM-based catalysts [10] by the modification of metakaolin geopolymers for the catalytic reduction of nitrogen oxides by ammonia, as well as the total oxidation of volatile hydrocarbons. Furthermore, metakaolin-based geopolymers and steel slag-containing AAMs also have been examined for photocatalytic applications [11,12] and biodiesel production [13]. Therefore, AAMs are interesting alternatives as catalysts, as well as for water-phase applications like catalytic wet peroxide oxidation (CWPO).

Bisphenol A (BPA) is an estrogenic compound commonly used in the production of polycarbonate plastic and epoxy resin, which are further utilized in several daily-use plastic products, such as drinking bottles, containers, thermal paper, etc. [14]. BPA can spread into water bodies during the manufacturing process, and also from daily-use plas products. Due to its endocrine disrupting character for humans and environment [15], removal of it from wastewaters is essential. Several techniques, such as activated slud treatment [16], membrane bioreactors [17], and sorption [18,19] have been used for removal of BPA from wastewaters. In addition of these, advanced oxidation process (AOPs) have been effectively used for the oxidation of BPA in wastewaters [20]. Techniqu like photolysis [21] and ozonation [22], as well as hybrid processes like UV/H_2O_2 [2 UV/O_3 [24], UV/TiO_2 [25], and O_2/H_2O_2 [26], have been successfully used for the idation of BPA. In this study, one of AOPs, CWPO, is studied for the removal of B from water. In CWPO, hydrogen peroxide is used as an oxidizing agent to decompo organic compounds from wastewater. Transition metals, typically Fe and Cu, are us as catalysts in the reaction to decompose hydrogen peroxide to active hydroxyl radic (Equation (1)) [27].

$$M^{n+} + H_2O_2 \rightarrow M^{(n+1)+} + HO^- + HO^\bullet$$

The formed hydroxyl radicals can further oxidize organic compounds, according Equation (2):

$$RH + HO^\bullet_2 \rightarrow R^\bullet + H_2O_2$$

To obtain the reduced form of the active metals, they must be dispersed on suital supports [28,29]. Various materials have been used as supports in CWPO. Carbon-bas materials, such as activated carbon [30], graphite, carbon black [31], carbon nanotubes [3 and biomass-based carbons [33], have been successfully used for the degradation of ganics with CWPO. Moreover, zeolites [34,35] and clay materials [36,37] also have be applied as supports for Fe and Cu; therefore, AAMs with a chemical composition similar those of zeolites and clay minerals are interesting alternatives as carriers for use in CWI

However, for the use of industrial side streams as raw materials in a catalyst, prepared material must exhibit stability, especially when the prepared material is used water treatment applications. Typically, catalyst stability for photocatalytic experiments h been evaluated in consecutive tests [38] and by characterization of the used materials af experiments, e.g., by X-ray diffraction (XRD) and Fourier transform infrared spectrosco (FTIR) [39,40].

In this study, the industrial side stream from the steel industry, i.e., blast furna slag (BFS), was applied as a raw material to produce cost-effective catalytic materials water purification. Catalysts were prepared by mixing different ratios of BFS and NaC followed by their consolidation at 20 °C and 60 °C. Moreover, iron was impregnated as active metal in the AAMs via ion exchange. The as-prepared materials were characteriz e.g., by XRD, and their surface area and catalytic activity were examined for the CWI of a bisphenol A (BPA) aqueous solution. Particular attention has been focused on t stability of materials; therefore, the possible leaching of the main elements (such as C Si, Al, Mg, and Na) has been investigated before the CWPO of BPA under 2 MPa and 1 °C. Furthermore, the concentrations of the main elements of the as-prepared materi was analyzed after oxidation experiments from water samples, as well as those from t used catalysts.

2. Results and Discussion

In this section, the stability and characteristics, such as phase composition and speci surface area, of the prepared materials are discussed. In addition, the activity of AAM for the CWPO of BPA is evaluated. The prepared materials were named according to th NaOH concentration and consolidation temperature (Section 3.1).

2.1. Stability of Alkali-Activated Materials

Table 1 lists the conductivity values of aqueous solutions after 4 h at 150 °C under N_2 atmosphere of 2 MPa and an AAM concentration of 4 g/dm^3.

Table 1. Conductivity of aqueous solutions after 4 h at 150 °C under an N_2 atmosphere of 2 MPa and AAM concentration of 4 g/dm^3.

AAM	Conductivity [µS/cm]
BFS	533
BFS17.5-20	252
BFS17.5-60	332
BFS20-20	320
BFS20-60	288
BFS25-20	191
BFS25-60	166
BFS30-20	204
BFS30-60	195

The conductivity values of aqueous solutions after 4 h of experiments were 200–300 µS/cm; according to these values, the alkali activation of BFS stabilized the material. With the increase in the amount of NaOH in the sample, the conductivity of aqueous solutions decreased slightly. In addition, the curing temperature affected the conductivity, i.e., samples that were first cured at 60 °C for 24 h exhibited lower conductivity than those cured at room temperature.

Table 2 lists the concentrations of Ca, Si, and Al in aqueous solutions after 4 h at 150 °C, under an N_2 atmosphere of 2 MPa and an AAM concentration of 4 g/dm^3.

Table 2. Ca, Si, and Al concentrations of aqueous solutions after 4 h at 150 °C, under an N_2 atmosphere of 2 MPa and an AAM concentration of 4 g/dm^3.

AAM	Ca [mg/dm^3]	Si [mg/dm^3]	Al [mg/dm^3]
BFS17.5-20	29	16.0	2.6
BFS17.5-60	15	8.5	1.3
BFS20-20	20	9.6	1.8
BFS20-60	22	8.1	2.3
BFS25-20	34	7.7	3.5
BFS25-60	23	9.9	2.8
BFS30-20	27	12.0	4.9
BFS30-60	31	11.0	4.3

In addition to those of Ca, Si, and Al, Mg and Na concentrations also were analyzed from water samples by inductively coupled plasma–optical emission spectroscopy (ICP-OES). However, the magnesium concentration was less than the detection limit (≤ 0.1 mg/dm^3), and the maximum sodium concentration was 1 mg/dm^3 after 4 h at 150 °C and an N_2 atmosphere of 2 MPa. All of the samples exhibited almost the same Ca and Si concentrations. However, with the increase in the amount of NaOH in the samples, the leaching of aluminum increased. Clearly, alkali activation immobilized Al in the inorganic matrix, but basicity enhanced its dissolution [41]. Furthermore, curing at room temperature led to the enhanced dissolution of Ca and Al from AAMs. A curing temperature of 60 °C has been found to be favorable for geopolymer preparation. Muñiz-Villarreal et al. [42] have reported that the optimum dissolution and formation of hydroxy species and oligomers, as well as further polymerization or condensation, occur at 60 °C. Therefore, with the increase in the curing temperature of the BFS-based AAMs, the leaching of Ca, Si, and Al decreased. Thus, based on these stability tests, AAMs that are first cured at 60 °C for 24 h are further characterized by XRD, diffuse-reflectance infrared Fourier transform spectroscopy (DRIFTS), field emission scanning electron microscope with energy-dispersive X-ray spectroscopy (FESEM-EDS), ICP-OES, and surface area techniques, as well as being examined for the CWPO of a BPA aqueous solution.

2.2. Characterization of AAMs

Table 3 lists the results of the Brunauer–Emmett–Teller (BET) surface areas of the prepared AAMs.

Table 3. BET-specific surface areas (SSA) and pore volumes (PV) of AAMs.

Sample	SSA [m^2/g]	PV [m^2/g]
BFS	1.21	0.003
BFS17.5-20	19.5	0.081
BFS17.5-60	13.2	0.062
BFS20-20	14.6	0.077
BFS20-60	14.8	0.083
BFS25-20	11.0	0.047
BFS25-60	23.7	0.119
BFS30-20	26.5	0.112
BFS30-60	27.3	0.162
Fe/BFS17.5-60	38.0	0.120
Fe/BFS30-60	52.0	0.162

The specific surface area of BFS was negligible, while alkali activation led to the increased surface area of all samples (Table 3). Samples prepared by using the higher amount of NaOH exhibited the highest specific surface area, as well as the highest pore volume. Clearly, a low Si/Na ratio favored the formation of a porous structure in the samples. Sindhunata et al. [43] have reported the highest pore volume for fly-ash-based geopolymers at a SiO$_2$/Na$_2$O ratio < 1. Moreover, the samples cured at 60 °C for 24 exhibited a slightly higher specific surface area, and hence a higher pore volume, than those prepared at room temperature. The higher curing temperature promoted the moval of excess water from the material structure, which in turn increased the porosity samples further. Furthermore, higher curing temperatures (>50 °C) have been reported particularly increase the amount of mesopores in the material [43].

As can be observed from the surface area results, no significant differences between the AAMs were observed. Therefore, to examine the effect of the Na concentration samples on the catalytic behavior, samples with the lowest and highest Na concentration (BFS17.5-60 and BFS30-60) were selected as support materials for Fe catalysts. Surprising the surface areas of the Fe catalysts were greater than those of the BFS17.5-60 and BFS30- pure supports (Table 3). This result was related to the calcination performed for Fe catalyst During heat treatment, excess water and carbon dioxide of the support material, and w as traces of Fe salt, evaporated from the AAM structure, enabling the increase in the speci surface area [44]. In addition, the calcination of Fe catalysts led to the decomposition hydrotalcite (Figure 1, XRD results), which also affected the surface area of materials [4 Furthermore, as-prepared AAMs mainly exhibited a mesoporous structure (i.e., po diameter between 2 and 50 nm), with ~10% of pores exhibiting a diameter of less than 2 n However, by the addition of Fe to BFS17.5-60 and BFS30-60 via ion exchange, the numb of mesopores decreased to 80%, while micropores accounted for only a small percentage the total pore volume. Moreover, macropores accounted for only ~15% of the total po volume in Fe/BFS17.5-60 and Fe/BFS30-60, while before Fe ion exchange, pores great than 50 nm were not detected (i.e., heat treatment enhanced the formation of large pore

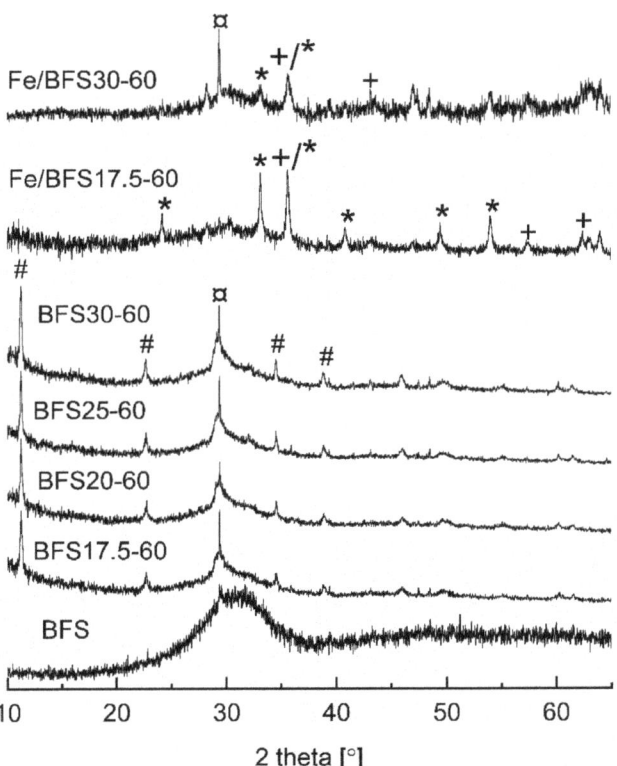

Figure 1. X-ray diffractograms of BFS raw material, supports, and Fe catalysts. (#) ICDD file 00-022-0700 ($Mg_6Al_2CO_3(OH)_{16}\cdot 4H_2O$, hydrotalcite); (¤) ICDD file 01-083-4609 ($CaCO_3$); (*) ICDD file 04-015-7029 (Fe_2O_3); (+) ICDD file 04-008-8146 (Fe_3O_4).

Figure 1 shows the X-ray diffractograms of the BFS raw material; BFS17.5-60, BFS20-60, BFS25-60, and BFS30-60 supports; and Fe/BFS17.5-60 and Fe/BFS30-60 catalysts. In the X-ray diffractogram of BFS, crystal peaks were not observed, but only one wide halo at 2θ between 22° and 40° was observed, which is characteristic of an amorphous material. After the alkali activation of BFS with NaOH, peaks were observed at 2θ values of 11.3°, 22.8°, 34.5°, and 38.6° (denoted with #), corresponding to hydrotalcite ($Mg_6Al_2CO_3(OH)_{16}\cdot 4H_2O$ (ICDD file 00-022-0700)), and the high-intensity peak at ~29.4° corresponded to $CaCO_3$ (ICDD file 01-083-4609). However, the broad "hump" observed at 2θ of 28–35° was still present in the X-ray diffractograms of all supports, indicative of a partly amorphous structure. After the ion exchange of BFS17.5-60 and BFS30-60 with the Fe solution, the peaks observed at 2θ values of 24.1°, 33.2°, 35.6°, 40.9°, 49.5°, and 54.1° (denoted by *) and at 35.7°, 43.4°, 57.4°, and 63.0° (denoted by +) revealed the presence of Fe_2O_3 (ICDD file 04-015-7029) and Fe_3O_4 (ICDD file 04-008-8146) phases, respectively. Owing to the heat treatment of Fe catalysts, hydrotalcite was decomposed [44], and peaks corresponding to hydrotalcite were not observed in the X-ray diffractograms of Fe/BFS17.5-60 and Fe/BFS30-60.

Figure 2 shows the DRIFT spectra of BFS, BFS17.5-60, BFS30-60, Fe/BFS17.5-60, and Fe/BFS30-60. In the DRIFT spectrum of the BFS raw material, only a few peaks were observed. The band at ~1420 cm^{-1} corresponded to Na_2CO_3 [45], and the strong peak at ~1110 cm^{-1} corresponded to pure silica [46].

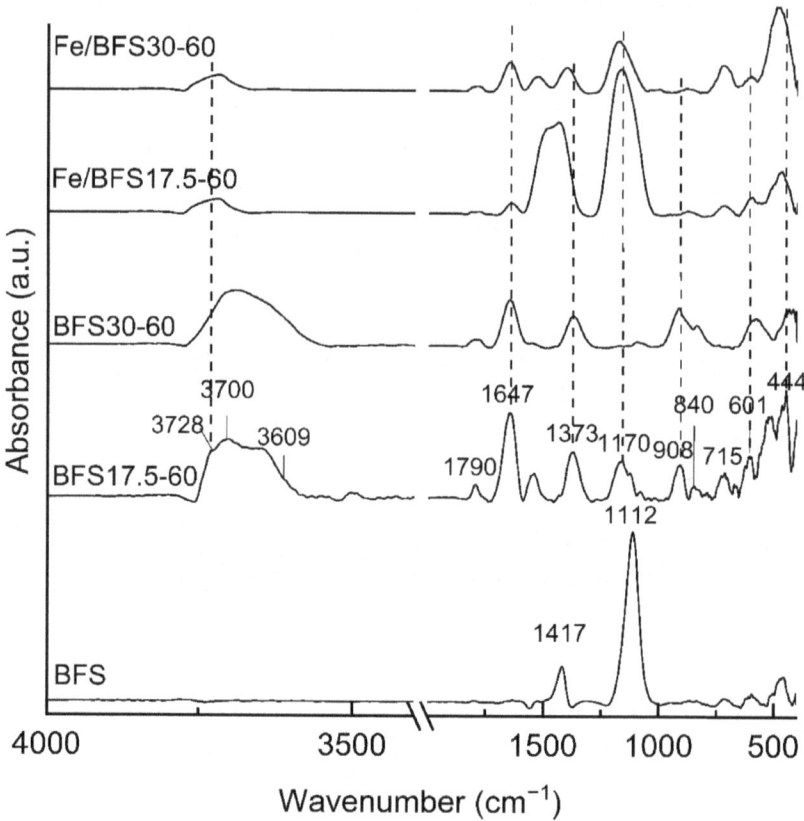

Figure 2. DRIFT spectra of BFS, BFS17.5-60, BFS30-60, Fe/BFS17.5-60, and Fe/BFS30-60 samples.

Alkali-activated samples exhibited several peaks in the analyzed region. The pe at 3730 cm^{-1} for BFS17.5-60 corresponded to silanol groups, which interact with oth atoms—for example, in silanol nests [47]—and the absorption peak at ~3700 cm^{-1} reveal the presence of four coordinated Al [48]. Moreover, the band at 3610 cm^{-1} for BFS17.5- corresponded to the bridging hydroxyl groups [49]. In the DRIFT spectra of Fe/BFS17.5- and Fe/BFS30-60, the peak centers were shifted to higher wavenumbers than those for t samples without iron, probably due to calcination, and the absorption band correspondi to the silanol groups (3730 cm^{-1}) disappeared by the introduction of iron into AAMs [4

The absorbance bands for BFS17.5-60 and BFS30-60 were observed at 715, 840, 13 and 1790 cm^{-1}, corresponding to CO_3^{2-}-containing compounds [46]. Bands at 840 a 1790 cm^{-1} connected to Na_2CO_3, and the band at 715 cm^{-1} corresponded to $CaCO_3$ [4 while that observed at 1373 cm^{-1} corresponded to hydrotalcite [50], which was al detected in the X-ray diffractograms of these samples (Figure 1). In the DRIFT spectra Fe/BFS17.5-60 and Fe/BFS30-60, these peaks were slightly shifted to higher wavenumbe especially for the band corresponding to hydrotalcite, indicative of its decomposition a result of heat treatment. Furthermore, the peak at ~1650 cm^{-1} observed in all samp corresponded to the H–OH stretching vibrations characteristic of absorbed water [51], t intensity of which slightly decreased due to the heat treatment of iron-containing samp

All AAMs exhibited several bands corresponding to the Al and Si bonds. The ban at 435–483 cm^{-1} corresponded to the Si–O–Si and O–Si–O bending vibrations [52], wh the absorption peak at ~600 cm^{-1} revealed the presence of Si–O–Si and Al–O–Si symmet stretching vibrations [53]. The band at ~900 cm^{-1} in the spectra of BFS17.5-60 and BFS30-

corresponded to the Si–O stretching and Si–OH bending modes [53]. Moreover, the band at ~1170 cm^{-1} corresponded to the Si–O–Si and Al–O–Si asymmetric stretching vibrations [53], and this band was broadened in the spectra of Fe/BFS17.5-60 and Fe/BFS30-60, due to the calcination of these samples [45]. According to [54–56], Fe_2O_3 and Fe_3O_4 species should exhibit IR vibrations at 550 and 780 cm^{-1} and 571 and 590 cm^{-1}, respectively. However, owing to the overlap of the Si and Al vibrations in this wavenumber region, peaks corresponding to Fe cannot be observed in the DRIFT spectra of the prepared samples.

Table 4 lists the concentrations (as wt %) of Ca, Si, Al, Mg, Fe, and Na of BFS17.5-60, BFS30-60, Fe/BFS17.5-60, and Fe/BFS30-60, as determined by ICP-OES analysis.

Table 4. Metal concentrations (as wt %) of selected samples, as determined by ICP-OES analysis.

Sample	Ca	Si	Al (wt %)	Mg	Fe	Na
BFS17.5-60	23.3	14.7	4.34	5.32	0.70	0.24
BFS30-60	21.9	14.4	4.17	5.07	1.86	0.29
Fe/BFS17.5-60	20.3	16.2	4.90	5.77	6.99	0.19
Fe/BFS30-60	20.0	15.1	4.58	5.49	4.95	0.19

The Ca concentrations of the prepared samples were several times lower than those in BFS, while the Si, Al, Mg, and Na concentrations were about the same as those in the raw material (Table 5, experimental). The leaching of Ca probably occurred during the washing of the AAMs using deionized water. BFS contained ~0.5 wt % iron, and ion exchange led to the increase in the iron concentration to 5–7 wt % for Fe/BFS30-60 and Fe/BFS17.5-60, respectively. The theoretical amount of iron by the employed impregnation method was 5.3 wt %, indicating that ion exchange between BFS17.5-60 and the iron salt is slightly better than that between BFS30-60 and the iron salt.

Table 5. Elemental composition of the blast furnace slag as determined by ICP-OES analysis [1].

	[wt %]										
	Ca	Si	Al	Mg	S	Ti	K	Fe	Na	Mn	Ba	V
BFS	28.70	16.30	5.00	4.87	1.44	0.60	0.58	0.53	0.50	0.28	0.06	0.04

[1] Elements with wt % > 0.01 were reported.

Figure 3 shows the FESEM images of BFS, Fe/BFS17.5-60, and Fe/BFS30-60. AAMs clearly exhibited an irregular, non-crystalline shape (Figure 3b,c). According to EDS analysis, the Al and Mg concentrations were ~5 wt %, while on the Fe catalyst surface, the Si and Ca concentrations were a few percent less than those in the bulk, as determined by ICP-OES (Table 4).

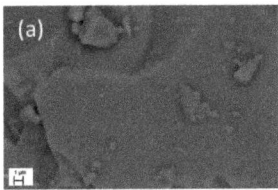 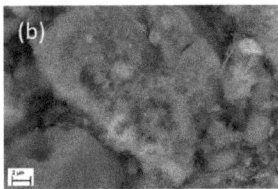 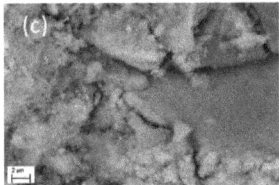

Figure 3. FESEM micrographs of BFS (a), Fe/BFS17.5-60 (b), and Fe/BFS30-60 (c). Dimensions in figures: 1 μm (a) and 2 μm (b,c).

2.3. Oxidation Experiments with AAMs

The prepared AAMs, namely BFS17.5-60, BFS20-60, BFS25-60, and BFS30-60, whi(ch) were first cured at 60 °C for 24 h, were examined for the CWPO of a BPA aqueous soluti(on). Figure 4 shows the results of these experiments.

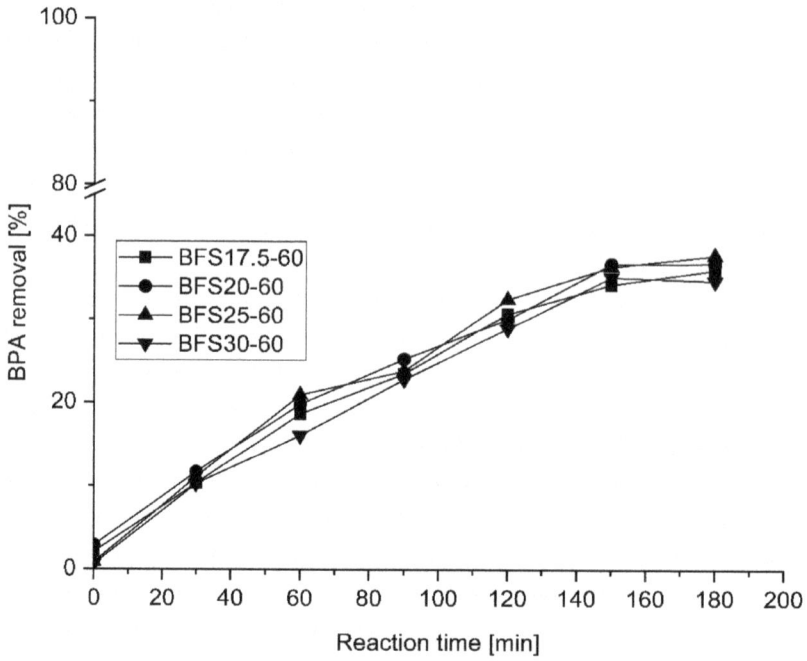

Figure 4. Removal of bisphenol A over AAMs as a function of the reaction time. Reaction conditions: concentration [c], c[BPA] = 60 mg/dm^3, c[H$_2$O$_2$] = 1.5 g/dm^3, c[catalyst] = 4 g/dm^3, temperature [T] = 50 °C, initial pH (6–7).

Oxidation reactions were performed at 50 °C at an initial pH of 6–7, a catalyst conce(n)tration of 4 g/dm^3, and a H$_2$O$_2$ concentration of 1.5 g/dm^3. In the absence of a catalyst (r(ot) shown), only ~10% of BPA removal was observed, while in the presence of AAMs, B(PA) removal of 35–39% after 180 min oxidation was observed. Oxidation proceeded duri(ng) 2.5 h for all samples and stabilized for 3 h. The oxidant H$_2$O$_2$ was added in batches; hen(ce) the final addition was performed at 2 h sampling. The total organic carbon (TOC) w(as) measured from the initial and final samples, and 27–31% of organics were removed. T(he) dissolved oxygen (DO) concentration of the BPA samples changed from ~9.5 mg O$_2$/d(m^3) to 8.1 mg O$_2$/dm^3 during 180 min oxidation, revealing that at the end of the run, oxygen(was) still present in the samples. Probably, the used reaction temperature (50 °C) was not su(ffi)ciently high for the effective decomposition of H$_2$O$_2$ to form active ·OH radicals. In seve(ral) studies, a higher reaction temperature has been reported to enhance the degradation (of) H$_2$O$_2$, thereby enhancing pollutant removal [57–59].

As all of the AAMs exhibited similar activities for the removal of BPA, samples wi(th) the lowest and highest NaOH concentration were selected for further research. Iron w(as) impregnated onto BFS17.5-60 and BFS30-60 samples by ion exchange (Section 3.1), and t(he) prepared Fe catalysts were examined under different reaction conditions.

First, the effect of the addition of the active metal on BFS17.5-60 and BFS30-60 w(as) examined at 50 °C at the initial pH, and a catalyst loading of 4 g/dm^3. After 3 h oxidati(on) BPA removal of 42% and 45% for Fe/BFS17.5-60 and Fe/BFS30-60 were observed, resp(ec)tively (Figure 5). Using the comparison of BPA removal over AAMs without the acti(ve) metal (Figure 4), the addition of Fe led to the increased activity of both catalysts, name(ly)

BPA removal of 6% and 10% for BFS17.5-60 and BFS30-60, respectively. TOC removal after 3 h oxidation was at the same level for both catalysts compared to that over the pure supports (30% and 33% for Fe/BFS17.5-60 and Fe/BFS30-60, respectively). During oxidation, the DO concentration decreased slightly from ~9 mg/O_2 dm^3 to 6.2–6.6 mg/O_2 dm^3, indicating that hydrogen peroxide is not consumed completely in the runs. Therefore, Fe/BFS17.5-60 and Fe/BFS30-60 were further examined at higher reaction temperatures.

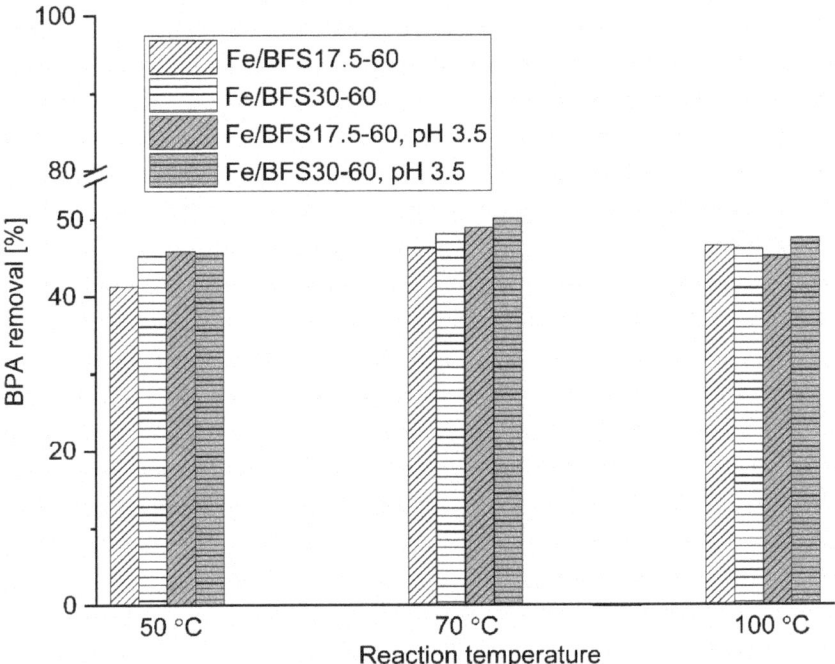

Figure 5. Bisphenol A removal at reaction temperatures of 50 °C, 70 °C, and 100 °C with the Fe/BFS17.5-60 and Fe/BFS30-60 catalysts, at an initial pH of 3.5, a reaction time of 3 h, [BPA] = 60 mg/dm^3, $c[H_2O_2]$ = 1.5 g/dm^3, and c[catalyst] = 4 g/dm^3.

To investigate the effect of temperature on the CWPO of BPA over Fe/BFS17.5-60 and Fe/BFS30-60, oxidation experiments were performed at 70 °C and 100 °C at the initial pH of the BPA aqueous solution. Typically, with the increase in the reaction temperature, the oxidation rate increases. Furthermore, the decomposition rate of H_2O_2 to active hydroxyl radicals also increases. A higher reaction temperature led to the improved degradation of BPA during 3 h oxidation, with the maximum of 5% over Fe/BFS17.5-60 at 70 °C (Figure 5). The increase in the reaction temperature to 100 °C did not affect BPA removal. During oxidation, the DO concentration decreased from 8.0 mg/O_2 dm^3 to 5.7 mg/O_2 dm^3 and from ~10.0 mg/O_2 dm^3 to 4.2 mg/O_2 dm^3 at 70 °C and 100 °C, respectively, revealing that hydrogen peroxide is consumed in the reaction. However, owing to the low degradation level of BPA at 100 °C, hydrogen peroxide was probably decomposed directly to H_2O without the formation of hydroxyl radicals.

Typically, homogeneous iron catalysts for CWPO (Fenton process) are used at a pH of ~3, which is known to be optimum for the decomposition of organic compounds [60]. The effect of pH on the degradation level of BPA was investigated at pH 3.5, in addition to the initial pH (6–7) by using Fe/BFS17.5-60 and Fe/BFS30-60 catalysts. The effect of pH was examined at 50 °C, 70 °C, and 100 °C. The pH of the BPA solution was adjusted to 3.5 using 2.0 M HNO_3 before oxidation. At 50 °C and pH 3.5, BPA removal increased by 5% over Fe/BFS17.5-60, while over Fe/BFS30-60, it was almost the same after 3 h oxidation compared to experiments performed at the initial pH of BPA (Figure 5). T DO concentration of the liquid samples was considerably higher (at the end of the run Fe/BFS17.5-60, 14 mg/O_2 dm^3) than that after oxidation at the initial pH. Therefore, aci pH promotes the formation of hydroxyl radicals during the reaction. However, Fe cataly did not exhibit considerably higher activity for BPA removal than that at the initial p probably due to the basic surfaces of Fe/BFS17.5-60 and Fe/BFS30-60.

At pH 3.5 and 70 °C (Figure 5), the DO concentration was the same during te compared to that in experiments at the initial pH, and the pH change of the BPA soluti led to an increase in BPA removal by only 3% and 2% over Fe/BFS17.5-60 and Fe/BFS30- respectively. At 100 °C and pH 3.5, BPA removal after 3 h was around the same for both catalysts compared with that observed at 100°C and at the initial pH. However, notal owing to the basicity of Fe catalysts, the pH of the BPA solution changed to basic duri runs in all experiments. The decomposition of H_2O_2 to ·OH radicals is the key step CWPO. However, under a basic reaction pH, the generation of hydroxyl radicals w restricted, thereby further decreasing the degradation of BPA [61]. Thus, the change in j marginally affects BPA removal.

The adsorption capacity of the Fe catalysts was examined under the severest reacti conditions in this study, i.e., pH of 3.5, a reaction temperature of 100 °C, in the absence the oxidant, and a catalyst concentration of 4 g/dm^3. For Fe/BFS17.5-60 and Fe/BFS30- during the 3 h experiment, 12% and 17% of BPA was adsorbed, respectively, reveali that Fe/BFS17.5-60 is catalytically more active than Fe/BFS30-60. The higher adsorpti capacity of Fe/BFS30-60 was related to the higher specific surface area of this samp (Table 3).

2.4. Stability of the Used Catalysts

The possible leaching of the elements from the prepared AAMs was examined by IC OES in detail, in addition to the leaching tests (Section 2.1) after oxidation. The oxidiz water samples were immediately filtered after 3 h CWPO using a 0.45 μm cellulose nitr filter to remove the solid catalysts. The Al, Si, and Ca concentrations were determined fro the oxidized water samples catalyzed by BFS17.5-60, BFS20-60, BFS25-60, and BFS30- and in addition to these elements, Fe was analyzed from the filtered samples catalyz by Fe/BFS17.5-60 and Fe/BFS30-60. According to the results, the leaching of Ca and was observed under all of the utilized reaction conditions with all catalysts. In all of t oxidized water samples, the Ca concentration was 25–50 mg/dm^3, and the Si concentrati was 11–19 mg/dm^3. However, notably, the Ca concentration was slightly lower in the wa samples catalyzed by AAMs without iron. Therefore, the heat treatment of Fe cataly (Section 3.1) led to the increased dissolution of Ca in the water phase during oxidati treatment. The leaching of Ca was around the same level as that detected in stabil tests (Section 2.1) with BFS17.5-60, BFS20-60, BFS25-60, and BFS30-60, revealing that the samples also can be used at reaction temperatures > 100 °C and pressures ≥ 2 MPa.

The Al concentration of aqueous BPA samples oxidized at the initial pH was 1.0–1.4 mg/d The dissolution of Al was slightly higher at 150 °C and 2.0 MPa (Table 1), i.e., under conditic of the stability test, than that under CWPO reaction conditions. However, in the case of oxidati experiments performed at a pH of 3.5, and at temperatures 50 °C, 70 °C, and 100 °C over that Fe/BFS17.5-60 and Fe/BFS30-60, 0.6–1.2 mg/dm^3 of Al was leached from the catalysts in t obtained effluents. Therefore, the dissolution of Al from the prepared AAMs is more domina in the CWPO of BPA, which is conducted at the initial pH. Onisei et al. [41] have investigated t leaching behavior of several elements (e.g., Si, Pb, Ca, Zn, Al) from fly ash-based geopolyme The study was performed in the pH range of 6–13. According to their results, the leaching Al increased slightly in the pH range of 10.5–13.0. In the CWPO of BPA, the initial pH of t BPA solution was 6–7. However, at the end of the run, the effluent pH was ~11, due to the ba character of the Fe catalysts. Moreover, in CWPO experiments, which were started at a pH of the pH of the BPA solution was ~10 in the oxidized water sample. Therefore, the adjustment the pH at the start of the CWPO of BPA did not considerably affect the removal of BPA, but it decreased the leaching of Al from the Fe catalysts.

However, the leaching of Ca, Si, and Al was not related to the removal of BPA, because those elements were not active in CWPO. The stability of the material is a key characteristic of the catalyst; therefore, the preparation method of AAM-based catalysts should be carefully considered. Moreover, the leaching of iron was rather negligible (maximum of 0.2 mg/dm^3 at 70 °C and at the initial pH) using Fe/BFS17.5-60 and Fe/BFS30-60 at the employed reaction temperatures and pH. Therefore, the CWPO of BPA with these catalysts proceeded via a heterogeneous reaction.

The activity and durability in consecutive tests and the effect of heat treatment as a regeneration method were examined using Fe/BFS7.5-60 at 50 °C, at the initial pH, and at H$_2$O$_2$ and catalyst concentrations of 1.5 g/dm^3 and 4.0 g/dm^3, respectively. To have sufficient material for consecutive tests and regeneration, 12 runs were performed in total, and the catalysts used in these experiments were collected and combined. Between consecutive experiments, the used catalyst was filtered from the effluent and dried at 105 °C for the subsequent runs.

After the first oxidation reaction, BPA removal of 41% was observed, which decreased to ~6% after the second run using the same catalyst (Figure 6). Furthermore, BPA removal after the third experiment, which used Fe/BFS17.5-60 twice, was practically the same (34%) as that observed in the second run, indicative of the catalyst's stability for multiple cycles in the CWPO of BPA. However, the removal of BPA after three cycles using Fe/BFS17.5-60 was around the same as that using BFS17.5-60 for one cycle, revealing that the addition of Fe does not significantly affect catalytic activity due to the basic reaction pH. Moreover, TOC results confirmed the reusability of Fe/BFS17.5-60, while BPA removal was the same during the three consecutive tests.

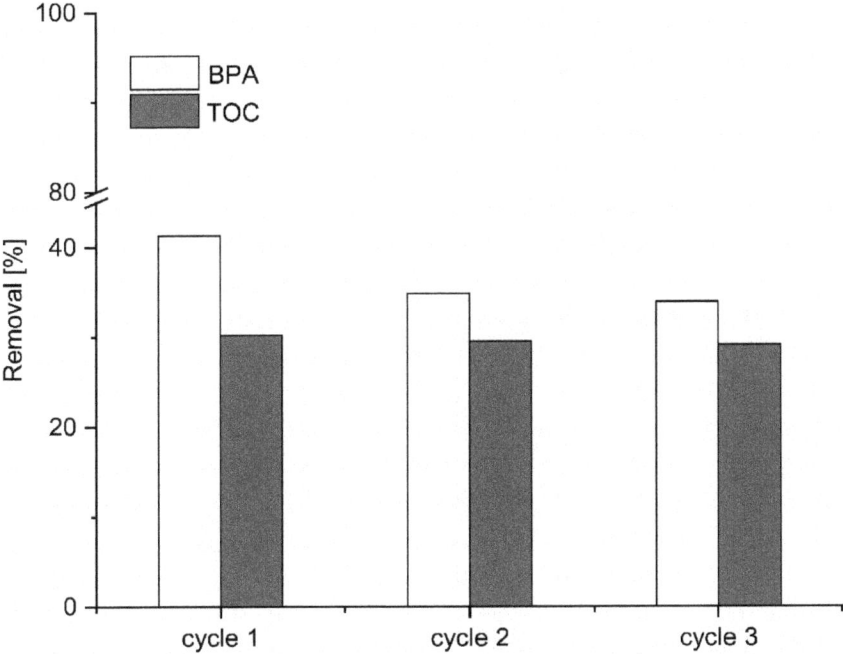

Figure 6. Consecutive tests using Fe/BFS17.5-60 for the CWPO of BPA. Reaction conditions: c[BPA] = 60 mg/dm^3, c[H$_2$O$_2$] = 1.5 g/dm^3, c[catalyst] = 4 g/dm^3, T = 50 °C, initial pH (6–7).

The regeneration of once-used Fe/BFS17.5-60 was examined by heat treatment 250 °C and 500 °C. The procedure was performed by increasing the temperature at a ra of 1 °C/min to the reaction temperature, at which the catalyst was kept for 2 h. Af oxidation, 34% and 32% BPA removal was observed at 250 °C and 500 °C, respective using Fe/BFS17.5-60. Therefore, the regeneration procedure is not effective at returning t activity of the catalysts to the original level. In addition, carbon deposition is confirm to not be responsible for the activity decrease of Fe/BFS17.5-60, because heat treatme at elevated temperatures is a typical regeneration procedure for catalysts with carb deactivation [62].

2.5. Characterization of the Used Catalysts

Figure 7 shows the X-ray diffractograms of BFS17.5-60, BFS30-60, Fe/BFS17.5- and Fe/BFS30-60 after oxidation at the initial pH and at a reaction temperature of 50 ° According to XRD analysis, the hydrotalcite phase (denoted by #, ICDD file 00-022-07 was still present in BFS17.5-60 and BFS30-60, and the $CaCO_3$ phase (o, ICDD file 01-0 4609) was observed in all samples. Moreover, in the X-ray diffractograms of Fe/BFS17.5- and Fe/BFS30-60, the Fe_3O_4 and Fe_2O_3 iron phases (denoted by +: ICDD file 04-008-81 and *: ICDD file 04-015-7029, respectively) were still present, but the high Ca concentrati of samples led to the overlap of the $CaCO_3$ peaks with those of Fe_3O_4 and Fe_2O_3 at 2θ 36° and Fe_3O_4 at 2θ of 43°.

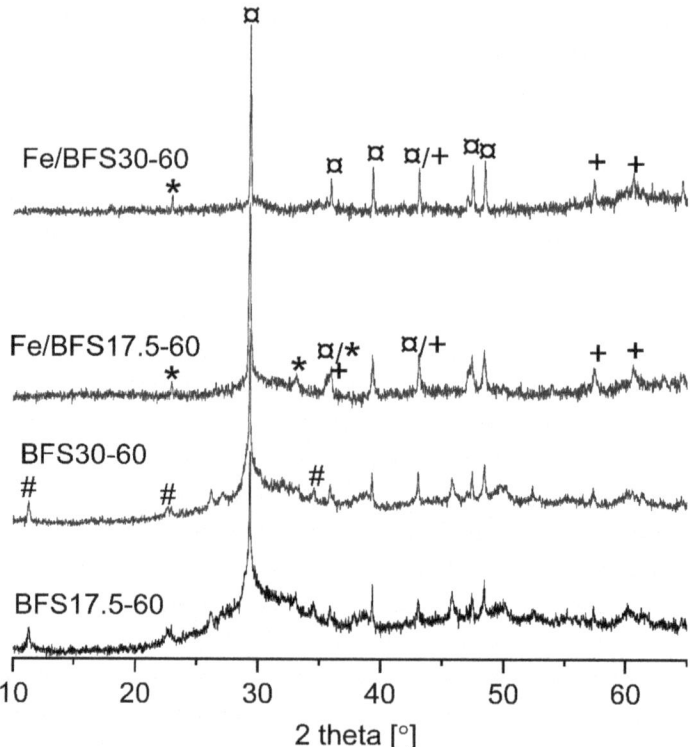

Figure 7. X-ray diffractograms of used BFS17.5-60, BFS30-60, Fe/BFS17.5-60, and Fe/BFS30-60. ICDD file 00-022-0700 ($Mg_6Al_2CO_3(OH)_{16}\cdot 4H_2O$, hydrotalcite); (o) ICDD file 01-083-4609 (CaCC (*) ICDD file 04-015-7029 (Fe_2O_3); (+) ICDD file 04-008-8146 (Fe_3O_4).

Furthermore, acidic pH and a higher reaction temperature did not affect the phase structure, and hydrotalcite was still observed in the X-ray diffractograms of BFS30-60 and BFS17.5-60 (results not shown).

The specific surface area of samples was analyzed after the oxidation of BPA at the initial pH and at reaction temperatures of 50 °C for BFS17.5-60 and BFS30-60, as well as reaction temperatures of 50 °C and 70 °C for Fe/BFS17.5-60 and Fe/BFS30-60. The BET results for used BFS17.5-60 and BFS30-60 revealed that the specific surface areas were 10.4 and 18.5 m^2/g, respectively, revealing that the surface area of BFS30-60 decreases to ~30%, while a rather negligible change in the surface area of BFS17.5-60 was observed (Table 3). For Fe catalysts, the specific surface areas increased after oxidation. In case of Fe/BFS17.5-60, the surface area was ~45% higher, and in case of Fe/BFS30-60, it doubled compared to that of the fresh catalyst (Table 3). Clearly, during oxidation, the Fe catalyst surface is refined by H_2O_2. For example, Han et al. [63] and Liu et al. [64] have used hydrogen peroxide to modify surface properties, i.e., to increase the surface area and porosity of materials. However, the larger specific surface area did not improve the removal of BPA in consecutive tests using Fe/BFS17.5-60 (Figure 6); therefore, the CWPO of BPA is not a surface area-specific reaction, as is the case for the catalytic wet air oxidation of BPA [65].

3. Materials and Methods

In this section, the preparation method and characterization techniques of catalysts are described. In addition, reaction conditions for the CWPO of the bisphenol A aqueous solution are presented.

3.1. Preparation of Alkali-Activated Materials and Fe Catalysts

AAMs were synthesized using powdered blast furnace slag (BFS) obtained from the Finnish steel industry. Table 5 lists the elemental composition of the slag, as determined by ICP-OES analysis.

AAMs were prepared by mixing 40 g of BFS with 17.5, 20.0, 25.0, and 30.0 g of 50 wt % sodium hydroxide (NaOH ≥ 97%, Merck, Darmsdtadt, Germany). The formed pastes were poured into molds, followed by consolidation in plastic bags at room temperature for 168 h. For comparison, consolidation was first performed at 60 °C for 24 h and then for 144 h at room temperature. The prepared samples were named according to their NaOH concentration and consolidation temperature (Table 6). Before use, the materials were crushed using a jaw crusher, sieved to a particle size of 0.5–2.0 mm, and washed with de-ionized water. As the active metal, Fe was impregnated on the AAMs by ion exchange. First, 5 g of AAM and 0.5 dm^3 of 0.01 M $(NH_4)_2Fe(SO_4)_2$ (99.0–101.5%, Merck, Darmsdtadt, Germany) were mixed in a sand bath and left overnight at 80 °C. The solids were collected by filtration, washed with deionized water, and dried overnight at 120 °C. Finally, the prepared Fe catalysts were subjected to calcination at 500 °C for 2 h, increasing the temperature from room temperature to the target temperature at a rate of 1 °C/min.

Table 6. Abbreviations of the prepared AAMs and Fe catalysts. Samples were named according to their NaOH concentrations and curing temperatures.

Studied Samples	50 wt % NaOH [g]	Consolidation T [°C]	Si/Na Ratio
BFS17.5-20	17.5	20	3.24
BFS17.5-60	17.5	60	3.24
BFS20-20	20.0	20	2.84
BFS20-60	20.0	60	2.84
BFS25-20	25.0	20	2.27
BFS25-60	25.0	60	2.27
BFS30-20	30.0	20	1.89
BFS30-60	30.0	60	1.89
Fe/BFS17.5-60	17.0	60	3.24
Fe/BFS30-60	30.0	60	2.84

3.2. Stability of AAMs

The stability of as-prepared AAMs was examined by measuring the possible leaching of the main elements of samples (i.e., Al, Ca, Mg, Na, and Si) to the water phase. Before the test, samples were crushed using a jaw crusher, sieved to a particle size of 1–2 mm, washed with deionized water, and dried. Stability tests were performed in a pressurized reactor at 150 °C and under a nitrogen atmosphere of 2 MPa. The crushed AAMs at concentration of 4 g/dm^3 were continuously stirred for 4 h with 0.16 dm^3 of deionized water. The ready leaching of elements was detected by conductivity measurement during and after the test, while the possible dissolution of Al, Ca, Mg, Na, and Si was analyzed by inductively coupled plasma–optical emission spectroscopy (ICP-OES, Thermo Electron iCAP 6500 Duo, Thermo Fisher Scientific, Waltham, MA, USA).

3.3. Characterization of Samples

The surface morphology and chemical composition of the prepared AAMs were analyzed by field emission scanning electron microscopy (FESEM; Carl Zeiss Microscopy GMbH, Jena, Germany) combined with energy-dispersive X-ray spectroscopy (EDS; analyzer at the Centre for Material Analysis, University of Oulu, Finland). The phase composition of AAMs was determined by powder X-ray diffraction (XRD) with a PANalytical X'Pert Pro X-ray diffractometer (Malvern PANalytical, Almelo, The Netherlands). XRD analysis was performed by scanning two theta values between 10° and 70° with monochromatic Cu Kα1 (λ = 1.5406 Å) at 45 kV and 40 mA at a scan speed of 0.021 °/s. Crystalline phases were identified by HighScore Plus software using the Powder Diffraction File standards from the International Centre for Diffraction DATA ICDD (PDF-4+ 2020 RDB). Diffuse-reflectance infrared Fourier transform spectroscopy (DRIFTS) was employed to investigate the degree of polymerization of the prepared samples. DRIFT spectra were recorded on a Bruker PMA 50 Vertex 80 V (Bruker, Billerica, MA, USA), equipped with Harrick Praying Mantis diffuse reflection accessory and a high-temperature reaction chamber, by baseline measurement using KBr. Before analysis, the sample chamber was flushed with nitrogen (100 cm^3/min), heated at a rate of 10 °C/min to the target temperature 120 °C, and maintained at that temperature for 30 min. Measurements were conducted 400–4000 cm^{-1} with a resolution of 4 cm^{-1} and 500 scans per minute. The specific surface areas and porosity were obtained from nitrogen adsorption–desorption isotherms at the liquid nitrogen temperature (−196 °C) by the Brunauer–Emmett–Teller (BET) method on Micromeritics ASAP 2020 system (Micromeritics Instrument Corporation, Norcross, GA, USA). The pore size distribution was calculated by density functional theory (DFT) [6]. Furthermore, the main elements of the prepared samples (Al, Ca, Mg, Na, and Si) and the active metal Fe were analyzed by ICP-OES analysis (Thermo Electron iCAP 6500 Duo, Thermo Fisher Scientific, Waltham, MA, USA).

3.4. Catalytic Wet Peroxide Oxidation Experiments

Oxidation experiments with a BPA aqueous solution (60 mg/L) were performed in a three-necked flask equipped with a reflux condenser. BFS30-60, BFS25-60, BFS20-60, BFS17.5-60, Fe/BFS30-60, and Fe/BFS17.5-60 were examined at a reaction temperature of 50 °C, a H$_2$O$_2$ concentration of 1.5 g/dm^3 (stoichiometric amount to total oxidation of BPA), and a catalyst loading of 4 g/dm^3, with a reaction volume of 0.16 dm^3. Oxidation was started while the reaction temperature was reached by the addition of H$_2$O$_2$, which was added in batches to maintain a stable oxidation agent concentration during the test. Water samples were taken as a function of reaction time, which were filtered using a 0.45 µm filter paper. The pH and dissolved oxygen (DO) were measured from water samples during the experiment. Furthermore, the effects of pH and temperature on oxidation were examined using Fe/BFS30-60 and Fe/BFS17.5-60. For evaluating the stability and reusability of the prepared materials, Fe/BFS17.5-60 was examined in three consecutive oxidation reactions. In addition, the regeneration of used Fe/BFS17.5-60 was performed by heating the catalyst for 2 h at two temperatures (i.e., 250 °C and 500 °C) to examine the effect of heat treatment on the activity of the used catalyst.

3.5. Water Sample Analysis

The BPA concentration of the water samples was determined by high-pressure liquid chromatography (HPLC) equipped with a Waters 996 photodiode array (PDA) detector (Waters Corp., Milford, MA, USA) at a wavelength of 226 nm. A mixture of 0.1% trifluoracetic acid (TFA) in methanol and 0.1% TFA in water at a flow rate of 0.4 cm^3/min was used as the eluent mixture to separate compounds on a SunFireTM C18 5-m 2.1 × 100 mm column (Waters Corp., Milford, MA, USA) operated at 30 °C. The total organic carbon (TOC) concentration of water samples was determined from the initial and final samples on a Skalar FormacsHT Total Organic Carbon/total nitrogen analyzer (Breda, The Netherlands). Possible leaching of Al, Ca, Mg, Na, and Si was analyzed from the final samples after oxidation by ICP-OES analysis (Thermo Electron iCAP 6500 Duo, Thermo Fisher Scientific, Waltham, MA, USA).

4. Conclusions

In this study, novel, eco-efficient, BFS-based alkali-activated materials were prepared and examined as catalysts for the CWPO of a BPA aqueous solution. AAMs consolidated at 60 °C were selected for catalytic studies, as they were more stable in the aqueous phase, and the phase structure was more porous than that of the samples cured at room temperature. BFSXX-60 samples exhibited moderate activity for the CWPO of BPA at 50°C and at the initial pH. The catalytic activities of Fe/BFS17.5-60 and Fe/BFS30-60 were examined at reaction temperatures of 50 °C, 70 °C, and 100 °C, and at the initial pH and a pH of 3.5. The addition of iron to the BFS-based materials led to the increased removal of BPA, with the highest BPA removal (50%) achieved using Fe/BFS30-60 at a pH of 3.5 at 70 °C. Furthermore, Fe/BFS17.5-60 exhibited moderate activity, even after three consecutive tests, and no change in the phase structure of the AAMS after the oxidation reaction was observed. Although prepared AAMs are interesting alternatives for catalytic water-phase applications, dissolution of Ca and Si, as well as small amounts of Al, was observed from AAMs during oxidation. In addition, the basic character of the material prevented higher removal of BPA. Therefore, additional attention should be focused on the stability and surface pH (e.g., pretreatment with acid) of AAMs in our future studies.

Author Contributions: Methodology, A.H. and J.P.; software, A.H; investigation, A.H. and T.H.; data curation, A.H.; writing—original draft preparation, A.H.; writing—review and editing, all authors; visualization, A.H.; project administration, U.L.; funding acquisition, U.L. and A.H. All authors have read and agreed to the published version of the manuscript.

Funding: This work was partly funded by the Renlund Foundation, within the project "New catalyst materials for wastewater treatment from industrial side streams".

Data Availability Statement: The data presented in this study are available within the article (tables and figures). The data presented in this study are available on request from the corresponding author.

Acknowledgments: Financial support from Renlund Foundation is gratefully acknowledged. Henrik Romar, Katariina Hautamäki, Tiina Leskelä, Ilkka Vesavaara, Jere Taipalus, Eemeli Koskela, Santeri Impiö, and Carlos Gonzales are acknowledged.

Conflicts of Interest: The authors declare no conflict of interest. The funders had no role in the design of the study; in the collection, analyses, or interpretation of data; in the writing of the manuscript, or in the decision to publish the results.

Abbreviations

AAM	Alkali-activated material
AOP	Advanced oxidation process
BET	Brunauer–Emmett–Teller
BFS	Blast furnace slag
BPA	Bisphenol A
CWPO	Catalytic wet peroxide oxidation
DFT	Density functional theory
DO	Dissolved oxygen
DRIFTS	Diffuse-reflectance infrared Fourier transform spectroscopy
Fe/BFS30-60	Iron-containing blast furnace slag-based catalyst
FESEM-EDS	Field emission scanning electron microscope with energy-dispersive X-ray spectroscopy
FTIR	Fourier transform infrared spectroscopy
HPLC	High-pressure liquid chromatography
ICDD	International Centre for Diffraction Data
ICP-OES	Inductively coupled plasma-optical emission spectroscopy
IR	Infrared
PDA	Photodiode array
PV	Pore volume
SSA	Specific surface areas
TFA	Trifluoracetic acid
TOC	Total organic carbon
UV	Ultraviolet
XRD	X-ray diffraction

References

1. Provis, J.L.; Van Deventer, J.S.J. 1—Introduction to geopolymers. *Geopolymers* **2009**, 1–11. [CrossRef]
2. Bernal, S.A.; Provis, J.L.; Fernández-Jiménez, A.; Krivenko, P.V.; Kavalerova, E.; Palacios, M.; Shi, C. Binder Chemistry—High Calcium Alkali-Activated Materials. In *Alkali Activated Materials: State-of-the-Art Report, RILEM TC 224-AAM*; Provis, J.L., van Deventer, J.S.J., Eds.; Springer: Dordrecht, The Netherlands, 2014; pp. 59–91.
3. Duxson, P.; Fernández-Jiménez, A.; Provis, J.L.; Lukey, G.C.; Palomo, A.; van Deventer, J.S.J. Geopolymer technology: The current state of the art. *J. Mater. Sci.* **2007**, *42*, 2917–2933. [CrossRef]
4. Davidovits, J. Geopolymers. *J. Therm. Anal.* **1991**, *37*, 1633–1656. [CrossRef]
5. Medpelli, D.; Seo, J.; Seo, D. Geopolymer with Hierarchically Meso-/Macroporous Structures from Reactive Emulsion Templating. *J. Am. Ceram. Soc.* **2014**, *97*, 70–73. [CrossRef]
6. Luukkonen, T.; Sarkkinen, M.; Kemppainen, K.; Rämö, J.; Lassi, U. Metakaolin geopolymer characterization and application for ammonium removal from model solutions and landfill leachate. *Appl. Clay. Sci.* **2016**, *119*, 266–276. [CrossRef]
7. Luukkonen, T.; Runtti, H.; Niskanen, M.; Tolonen, E.; Sarkkinen, M.; Kemppainen, K.; Rämö, J.; Lassi, U. Simultaneous removal of Ni(II), As(III), and Sb(III) from spiked mine effluent with metakaolin and blast-furnace-slag geopolymers. *J. Environ. Man.* **2016**, *166*, 579–588. [CrossRef]
8. Khan, M.I.; Min, T.K.; Azizli, K.; Sufian, S.; Ullah, H.; Man, Z. Effective removal of methylene blue from water using phosphoric acid based geopolymers: Synthesis, characterizations and adsorption studies. *RSC Adv.* **2015**, *5*, 61410–61420. [CrossRef]
9. Yousef, R.I.; El-Eswed, B.; Alshaaer, M.; Khalili, F.; Khoury, H. The influence of using Jordanian natural zeolite on the adsorption, physical, and mechanical properties of geopolymers products. *J. Hazard. Mater.* **2009**, *165*, 379–387. [CrossRef]
10. Sazama, P.; Bortnovsky, O.; Dědeček, J.; Tvarůžková, Z.; Sobalík, Z. Geopolymer based catalysts—New group of catalytic materials. *Catal. Today* **2011**, *164*, 92–99. [CrossRef]
11. Mejía de Gutiérrez, R.; Villaquirán-Caicedo, M.A.; Guzmán-Aponte, L.A. Alkali-activated metakaolin mortars using glass waste as fine aggregate: Mechanical and photocatalytic properties. *Constr. Build. Mater.* **2020**, *235*, 117510. [CrossRef]
12. Kang, L.; Zhang, Y.J.; Zhang, L.; Zhang, K. Preparation, characterization and photocatalytic activity of novel CeO2 loaded porous alkali-activated steel slag-based binding material. *Int. J. Hydrog. Energy* **2017**, *42*, 17341–17349. [CrossRef]
13. Sharma, S.; Medpelli, D.; Chen, S.; Seo, D. Calcium-modified hierarchically porous aluminosilicate geopolymer as a highly efficient regenerable catalyst for biodiesel production. *RSC Adv.* **2015**, *5*, 65454–65461. [CrossRef]
14. Guerra, P.; Kim, M.; Teslic, S.; Alaee, M.; Smyth, S.A. Bisphenol-A removal in various wastewater treatment processes: Operational conditions, mass balance, and optimization. *J. Environ. Manag.* **2015**, *152*, 192–200. [CrossRef] [PubMed]
15. Kitamura, S.; Suzuki, T.; Sanoh, S.; Kohta, R.; Jinno, N.; Sugihara, K.; Yoshihara, S.; Fujimoto, N.; Watanabe, H.; Ohta, S. Comparative Study of the Endocrine-Disrupting Activity of Bisphenol A and 19 Related Compounds. *Toxicol. Sci.* **2005**, *84*, 249–259. [CrossRef] [PubMed]

Melcer, H.; Klečka, G. Treatment of Wastewaters Containing Bisphenol A: State of the Science Review. *Water Environ. Res.* **2011**, *83*, 650–666. [CrossRef] [PubMed]

Chen, J.; Huang, X.; Lee, D. Bisphenol A removal by a membrane bioreactor. *Process. Biochem.* **2008**, *43*, 451–456. [CrossRef]

Shabtai, I.A.; Mishael, Y.G. Polycyclodextrin–Clay Composites: Regenerable Dual-Site Sorbents for Bisphenol A Removal from Treated Wastewater. *ACS Appl. Mater. Interfaces* **2018**, *10*, 27088–27097. [CrossRef] [PubMed]

Juhola, R.; Heponiemi, A.; Tuomikoski, S.; Hu, T.; Vielma, T.; Lassi, U. Preparation of Novel Fe Catalysts from Industrial By-Products: Catalytic Wet Peroxide Oxidation of Bisphenol A. *Top. Catal.* **2017**, *60*, 1387–1400. [CrossRef]

Salimi, M.; Esrafili, A.; Gholami, M.; Jonidi Jafari, A.; Rezaei Kalantary, R.; Farzadkia, M.; Kermani, M.; Sobhi, H.R. Contaminants of emerging concern: A review of new approach in AOP technologies. *Environ. Monit. Assess.* **2017**, *189*, 414. [CrossRef]

Kovačič, A.; Česen, M.; Laimou-Geraniou, M.; Lambropoulou, D.; Kosjek, T.; Heath, D.; Heath, E. Stability, biological treatment and UV photolysis of 18 bisphenols under laboratory conditions. *Environ. Res.* **2019**, *179*, 108738. [CrossRef]

Umar, M.; Roddick, F.; Fan, L.; Aziz, H.A. Application of ozone for the removal of bisphenol A from water and wastewater—A review. *Chemosphere* **2013**, *90*, 2197–2207. [CrossRef] [PubMed]

Sharma, J.; Mishra, I.M.; Kumar, V. Degradation and mineralization of Bisphenol A (BPA) in aqueous solution using advanced oxidation processes: UV/H2O2 and UV/S2O8− oxidation systems. *J. Environ. Manag.* **2015**, *156*, 266–275. [CrossRef] [PubMed]

Dudziak, M.; Burdzik, E. Oxidation of bisphenol A from simulated and real urban wastewater effluents by UV, O3 and UV/O3. *Null* **2016**, *57*, 1075–1083. [CrossRef]

Jia, C.; Wang, Y.; Zhang, C.; Qin, Q.; Kong, S.; Kouakou Yao, S. Photocatalytic Degradation of Bisphenol A in Aqueous Suspensions of Titanium Dioxide. *Environ. Eng. Sci.* **2012**, *29*, 630–637. [CrossRef]

Ahmadi, M.; Rahmani, H.; Takdastan, A.; Jaafarzadeh, N.; Mostoufi, A. A novel catalytic process for degradation of bisphenol A from aqueous solutions: A synergistic effect of nano-Fe3O4@Alg-Fe on O3/H2O2. *Process. Saf. Environ. Prot.* **2016**, *104*, 413–421. [CrossRef]

Perathoner, S.; Centi, G. Wet hydrogen peroxide catalytic oxidation (WHPCO) of organic waste in agro-food and industrial streams. *Top. Catal.* **2005**, *33*, 207–224. [CrossRef]

Tu, Y.; Tian, S.; Kong, L.; Xiong, Y. Co-catalytic effect of sewage sludge-derived char as the support of Fenton-like catalyst. *Chem. Eng. J.* **2012**, *185–186*, 44–51. [CrossRef]

Messele, S.A.; Soares, O.S.G.P.; Órfão, J.J.M.; Stüber, F.; Bengoa, C.; Fortuny, A.; Fabregat, A.; Font, J. Zero-valent iron supported on nitrogen-containing activated carbon for catalytic wet peroxide oxidation of phenol. *Appl. Catal. B Environ.* **2014**, *154—155*, 329–338. [CrossRef]

Dehkordi, A.M.; Ebrahimi, A.A. Catalytic Wet Peroxide Oxidation of Phenol in a New Two-Impinging-Jets Reactor. *Ind. Eng. Chem. Res.* **2009**, *48*, 10619–10626. [CrossRef]

Domínguez, C.M.; Ocón, P.; Quintanilla, A.; Casas, J.A.; Rodriguez, J.J. Graphite and carbon black materials as catalysts for wet peroxide oxidation. *Appl. Catal. B Environ.* **2014**, *144*, 599–606. [CrossRef]

Martin-Martinez, M.; Machado, B.F.; Serp, P.; Morales-Torres, S.; Silva, A.M.T.; Figueiredo, J.L.; Faria, J.L.; Gomes, H.T. Carbon nanotubes as catalysts for wet peroxide oxidation: The effect of surface chemistry. *Catal. Today* **2020**, *357*, 332–340. [CrossRef]

Juhola, R.; Heponiemi, A.; Tuomikoski, S.; Hu, T.; Prokkola, H.; Romar, H.; Lassi, U. Biomass-based composite catalysts for catalytic wet peroxide oxidation of bisphenol A: Preparation and characterization studies. *J. Environ. Chem. Eng.* **2019**, *7*, 103127. [CrossRef]

Yan, Y.; Jiang, S.; Zhang, H.; Zhang, X. Preparation of novel Fe-ZSM-5 zeolite membrane catalysts for catalytic wet peroxide oxidation of phenol in a membrane reactor. *Chem. Eng. J.* **2015**, *259*, 243–251. [CrossRef]

Valkaj, K.M.; Katović, A.; Zrnčević, S. Catalytic Properties of Cu/13X Zeolite Based Catalyst in Catalytic Wet Peroxide Oxidation of Phenol. *Ind. Eng. Chem. Res.* **2011**, *50*, 4390–4397. [CrossRef]

Galeano, L.A.; Gil, A.; Vicente, M.A. Effect of the atomic active metal ratio in Al/Fe-, Al/Cu- and Al/(Fe–Cu)-intercalating solutions on the physicochemical properties and catalytic activity of pillared clays in the CWPO of methyl orange. *Appl. Catal. B Environ.* **2010**, *100*, 271–281. [CrossRef]

Garrido-Ramirez, E.G.; Sivaiah, M.V.; Barrault, J.; Valange, S.; Theng, B.K.G.; Ureta-Zañartu, M.S.; Mora, M.d.l.L. Catalytic wet peroxide oxidation of phenol over iron or copper oxide-supported allophane clay materials: Influence of catalyst SiO2/Al2O3 ratio. *Microporous Mesoporous Mater.* **2012**, *162*, 189–198. [CrossRef]

Li, C.; He, Y.; Tang, Q.; Wang, K.; Cui, X. Study of the preparation of CdS on the surface of geopolymer spheres and photocatalyst performance. *Mater. Chem. Phys.* **2016**, *178*, 204–210. [CrossRef]

Fallah, M.; MacKenzie, K.J.D.; Hanna, J.V.; Page, S.J. Novel photoactive inorganic polymer composites of inorganic polymers with copper(I) oxide nanoparticles. *J. Mater. Sci.* **2015**, *50*, 7374–7383. [CrossRef]

Fallah, M.; MacKenzie, K.J.D.; Knibbe, R.; Page, S.J.; Hanna, J.V. New composites of nanoparticle Cu (I) oxide and titania in a novel inorganic polymer (geopolymer) matrix for destruction of dyes and hazardous organic pollutants. *J. Hazard. Mater.* **2016**, *318*, 772–782. [CrossRef]

Onisei, S.; Pontikes, Y.; Van Gerven, T.; Angelopoulos, G.N.; Velea, T.; Predica, V.; Moldovan, P. Synthesis of inorganic polymers using fly ash and primary lead slag. *J. Hazard. Mater.* **2012**, *205–206*, 101–110. [CrossRef] [PubMed]

42. Muñiz-Villarreal, M.S.; Manzano-Ramírez, A.; Sampieri-Bulbarela, S.; Gasca-Tirado, J.R.; Reyes-Araiza, J.L.; Rubio-Ávalos, J Pérez-Bueno, J.J.; Apatiga, L.M.; Zaldivar-Cadena, A.; Amigó-Borrás, V. The effect of temperature on the geopolymerizat process of a metakaolin-based geopolymer. *Mater. Lett.* **2011**, *65*, 995–998. [CrossRef]
43. Sindhunata; van Deventer, J.S.J.; Lukey, G.C.; Xu, H. Effect of Curing Temperature and Silicate Concentration on Fly-Ash-Bas Geopolymerization. *Ind. Eng. Chem. Res.* **2006**, *45*, 3559–3568. [CrossRef]
44. Roelofs, J.C.A.A.; Lensveld, D.J.; van Dillen, A.J.; de Jong, K.P. On the Structure of Activated Hydrotalcites as Solid Base Cataly for Liquid-Phase Aldol Condensation. *J. Catal.* **2001**, *203*, 184–191. [CrossRef]
45. Alzeer, M.I.M.; MacKenzie, K.J.D.; Keyzers, R.A. Porous aluminosilicate inorganic polymers (geopolymers): A new class environmentally benign heterogeneous solid acid catalysts. *Appl. Catal. A Gen.* **2016**, *524*, 173–181. [CrossRef]
46. Mozgawa, W.; Deja, J. Spectroscopic studies of alkaline activated slag geopolymers. *J. Mol. Struct.* **2009**, *924–926*, 434–4 [CrossRef]
47. Heinrich, F.; Schmidt, C.; Löffler, E.; Menzel, M.; Grünert, W. Fe–ZSM-5 Catalysts for the Selective Reduction of NO Isobutane—The Problem of the Active Sites. *J. Catal.* **2002**, *212*, 157–172. [CrossRef]
48. Waijarean, N.; MacKenzie, K.J.D.; Asavapisit, S.; Piyaphanuwat, R.; Jameson, G.N.L. Synthesis and properties of geopolym based on water treatment residue and their immobilization of some heavy metals. *J. Mater. Sci.* **2017**, *52*, 7345–7359. [CrossR
49. Gabrienko, A.A.; Danilova, I.G.; Arzumanov, S.S.; Toktarev, A.V.; Freude, D.; Stepanov, A.G. Strong acidity of silanol groups zeolite beta: Evidence from the studies by IR spectroscopy of adsorbed CO and 1H MAS NMR. *Microporous Mesoporous Mc* **2010**, *131*, 210–216. [CrossRef]
50. Sahu, P.K. A green approach to the synthesis of a nano catalyst and the role of basicity, calcination, catalytic activity and aging the green synthesis of 2-aryl bezimidazoles, benzothiazoles and benzoxazoles. *RSC Adv.* **2017**, *7*, 42000–42012. [CrossRef]
51. Mao, W.; Ma, H.; Wang, B. A clean method for solvent-free nitration of toluene over sulfated titania promoted by ceria cataly *J. Hazard. Mater.* **2009**, *167*, 707–712. [CrossRef]
52. Yunsheng, Z.; Wei, S.; Qianli, C.; Lin, C. Synthesis and heavy metal immobilization behaviors of slag based geopolymer. *J. Haz Mater.* **2007**, *143*, 206–213. [CrossRef] [PubMed]
53. Lee, W.K.W.; van Deventer, J.S.J. Use of Infrared Spectroscopy to Study Geopolymerization of Heterogeneous Amorphe Aluminosilicates. *Langmuir* **2003**, *19*, 8726–8734. [CrossRef]
54. Lei, Y.; Huo, J.; Liao, H. Fabrication and catalytic mechanism study of CeO2-Fe2O3-ZnO mixed oxides on double surfaces polyimide substrate using ion-exchange technique. *Mater. Sci. Semicond. Process.* **2018**, *74*, 154–164. [CrossRef]
55. Lei, Y.; Huo, J.; Liao, H. Microstructure and photocatalytic properties of polyimide/heterostructured NiO–Fe2O3–ZnO nanoco posite films via an ion-exchange technique. *RSC Adv.* **2017**, *7*, 40621–40631. [CrossRef]
56. Lu, L.; Li, J.; Ng, D.H.L.; Yang, P.; Song, P.; Zuo, M. Synthesis of novel hierarchically porous Fe3O4@MgAl–LDH magne microspheres and its superb adsorption properties of dye from water. *J. Ind. Eng. Chem.* **2017**, *46*, 315–323. [CrossRef]
57. Fajerwerg, K.; Debellefontaine, H. Wet oxidation of phenol by hydrogen peroxide using heterogeneous catalysis Fe-ZSM-5 promising catalyst. *Appl. Catal. B Environ.* **1996**, *10*, L229–L235. [CrossRef]
58. Centi, G.; Perathoner, S.; Torre, T.; Verduna, M.G. Catalytic wet oxidation with H2O2 of carboxylic acids on homogeneous a heterogeneous Fenton-type catalysts. *Catal. Today* **2000**, *55*, 61–69. [CrossRef]
59. Yan, Y.; Jiang, S.; Zhang, H. Catalytic wet oxidation of phenol with Fe–ZSM-5 catalysts. *RSC Adv.* **2016**, *6*, 3850–3859. [CrossR
60. Arnold, S.M.; Hickey, W.J.; Harris, R.F. Degradation of Atrazine by Fenton's Reagent: Condition Optimization and Prod Quantification. *Environ. Sci. Technol.* **1995**, *29*, 2083–2089. [CrossRef]
61. Hua, Z.; Ma, W.; Bai, X.; Feng, R.; Yu, L.; Zhang, X.; Dai, Z. Heterogeneous Fenton degradation of bisphenol A catalyzed efficient adsorptive Fe3O4/GO nanocomposites. *Environ. Sci. Pollut. Res.* **2014**, *21*, 7737–7745. [CrossRef]
62. Argyle, M.D.; Bartholomew, C.H. Heterogeneous Catalyst Deactivation and Regeneration: A Review. *Catalysts* **2015**, *5*, 145–2 [CrossRef]
63. Han, E.; Vijayarangamuthu, K.; Youn, J.; Park, Y.; Jung, S.; Jeon, K. Degussa P25 TiO2 modified with H2O2 under microwa treatment to enhance photocatalytic properties. *Catal. Today* **2018**, *303*, 305–312. [CrossRef]
64. Liu, J.; Xiong, Z.; Zhou, F.; Lu, W.; Jin, J.; Ding, S. Promotional effect of H2O2 modification on the cerium-tungsten-titaniu mixed oxide catalyst for selective catalytic reduction of NO with NH3. *J. Phys. Chem. Solids* **2018**, *121*, 360–366. [CrossRef]
65. Heponiemi, A.; Azalim, S.; Hu, T.; Vielma, T.; Lassi, U. Efficient removal of bisphenol A from wastewaters: Catalytic wet oxidation with Pt catalysts supported on Ce and Ce–Ti mixed oxides. *AIMS Mater. Sci.* **2019**, *6*, 25–44. [CrossRef]
66. Seaton, N.A.; Walton, J.P.R.B.; quirke, N. A new analysis method for the determination of the pore size distribution of por carbons from nitrogen adsorption measurements. *Carbon* **1989**, *27*, 853–861. [CrossRef]

Article

Eco-Friendly Cotton/Linen Fabric Treatment Using Aqueous Ozone and Ultraviolet Photolysis

Kengo Hamada [1,*], Tsuyoshi Ochiai [1,2], Yasuyuki Tsuchida [3], Kyohei Miyano [4], Yosuke Ishikawa [4], Toshinari Nagura [4] and Noritaka Kimura [3]

1. Kawasaki Technical Support Department, Local Independent Administrative Agency Kanagawa Institute of Industrial Science and Technology (KISTEC), Kanagawa 213-0012, Japan; pg-ochiai@newkast.or.jp
2. Photocatalysis International Research Center, Tokyo University of Science, 2641 Yamazaki, Noda, Chiba 278-8510, Japan
3. Department of Bioengineering, Nagaoka University of Technology, Niigata 940-2188, Japan; 2dedeenn6@gmail.com (Y.T.); nkimura@vos.nagaokaut.ac.jp (N.K.)
4. Business Strategy Department, Nisshinbo Textile Inc., Tokyo 103-8650, Japan; miyanokyohei@nisshinbo.co.jp (K.M.); ishikawa@nisshinbo.co.id (Y.I.); t.nagura@nisshinbo.co.jp (T.N.)
* Correspondence: k-hamada@kistec.jp; Tel.: +81-44-819-2105

Received: 30 September 2020; Accepted: 28 October 2020; Published: 2 November 2020

Abstract: Chemicals for the scouring and bleaching of fabrics have a high environmental load. In addition, in recent years, the high consumption of these products has become a problem in the manufacture of natural fabric products. Therefore, environmentally friendly, low-waste processes for fabric treatment are required. In this paper, we discuss the bleaching of fabrics using advanced oxidation processes (AOP). These processes use electrochemically generated aqueous ozone and ultraviolet (UV) irradiation to achieve bleaching. However, colour reversion often occurs. In this study, we suppressed unwanted colour reversion by treatment with rongalite. After treatment, changes in fabric colour were determined by measuring the colour difference and reflectance spectra. The best bleaching effect was obtained when ozone and UV irradiation treatments were combined, achieving results similar to those of a conventional bleaching method after 60 min of UV irradiation. In addition, the AOP treatment resulted in the simultaneous scouring of the fabric, as shown by the increased hydrophilicity of the fabric after AOP treatment. Thus, this AOP process represents a new fabric bleaching process that has an extremely low environmental impact.

Keywords: advanced oxidation processes; ozone; ultraviolet; bleaching; fabrics

1. Introduction

In the manufacture of natural fabric products, such as cotton and linen, alkaline chemicals and surfactants are used to remove contaminants (i.e., scouring). In addition, to bleach coloured components derived from natural products, an aqueous solution of sodium hypochlorite or heat treatment is used (i.e., bleaching) [1]. Therefore, the manufacture of natural fibre products is energy intensive, and the environmental load is high, especially because of the need for subsequent wastewater treatment [2–4]. An alternative to conventional bleaching with chemicals is advanced oxidation processes (AOP), which combine the treatment of hydrogen peroxide or aqueous ozone generated by electrolysis [5–9] with irradiation by ultraviolet (UV) light [10–12] and oxidation treatment using enzymes or microwave heating [13,14]. These methods do not use environmentally persistent agents and have low energy costs. In addition, oxidative bleaching with AOP allows the omission of the refining step used in conventional methods, again resulting in a more environmentally friendly method. However, after oxidative bleaching by AOP, the whiteness of the fabric deteriorates (colour reversion), especially during heating (ironing) after bleaching. Therefore, oxidative bleaching by AOP requires further optimisation.

In this study, we developed a bleaching technology with a low environmental load that combines aqueous electrochemically generated ozone and UV irradiation for the combined bleaching and scouring of cotton and linen. Changes in fabric colour were observed by measuring the colour difference and reflectance spectroscopy. The OH radicals generated by UV irradiation decompose the coloured components derived from natural products [15,16]; a chemical probe method [17,18] was used to estimate the amount of OH radicals generated from ozone by UV irradiation to understand the bleaching mechanism. Moreover, we proposed a reduction treatment with rongalite, a common reducing agent used in the textile industry, to suppress colour reversion after bleaching, which succeeded in minimising the colour reversion after ironing. Our method has excellent bleaching properties and extremely low environmental impact; thus, this method could replace conventional scouring and bleaching methods.

2. Results

2.1. Fabric Bleaching Using Aqueous Ozone under UV Irradiation

Figure 1a shows the relationship between processing time and the colour of cotton samples. For comparison, three types of treatment were performed: aqueous ozone + UV, aqueous ozone, and water + UV. The colour difference values decreased over time in all treatments. The treatment combining aqueous ozone and UV showed the greatest change, followed by those with aqueous ozone and water + UV. In addition, the change in colour difference was approximated as a first-order reaction, and the reaction rate constant was obtained (Figure 1b). The reaction rate constants for aqueous ozone + UV, aqueous ozone, and water + UV were 0.029, 0.015, and 0.009, respectively.

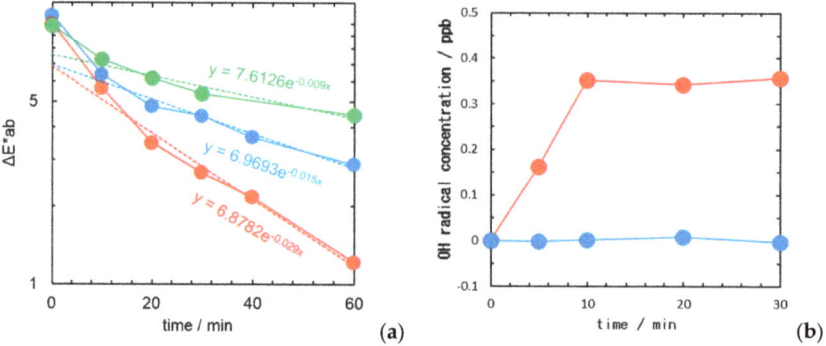

Figure 1. (a) Relationship between treatment time and colour difference (ΔE^*ab values are with respect to the colour of cotton bleached using a conventional chemical method) for each bleaching method (red: aqueous ozone + ultraviolet (UV), blue: aqueous ozone, green: water + UV). (b) Relationship between treatment time and OH radical concentration (red: aqueous ozone with UV, blue: aqueous ozone without UV).

The changes in the OH radical concentration were determined using a chemical probe fluorescence method. Figure 1b shows the relationship between the treatment time and OH radical concentration. No change in the OH radical concentration was observed in the aqueous ozone treatment without UV irradiation. In contrast, in the aqueous ozone treatment, the concentration of OH radicals increased with UV irradiation and became constant 10 min after the start of irradiation.

The changes in fabric colour were quantified using reflectance spectroscopy measurements. Figure 2 shows the reflectance spectra of cotton samples treated by AOP and conventional chemical bleaching for different periods. In the AOP-treated samples, the reflectance increased in the region below 550 nm with time, and a spectrum equivalent to that of the cotton fabrics bleached by the conventional method was achieved 60 min after the start of UV irradiation. Figure 3 shows photographs of the fabrics before bleaching, after AOP treatment for 60 min, and after conventional bleaching. The cotton fabrics before bleaching had a dull yellow colour. The cotton fabrics treated by AOP had a similar whiteness to that of the cotton bleached by the conventional method.

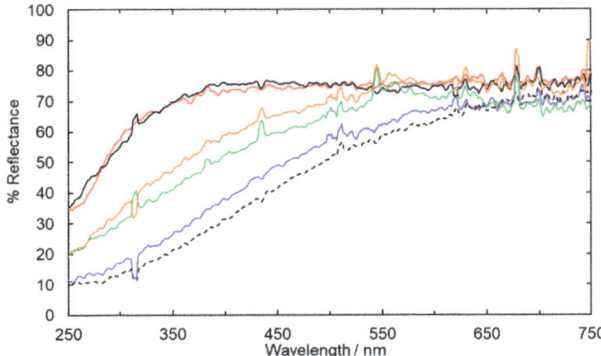

Figure 2. Reflectance spectra of cotton samples treated by advanced oxidation processes (AOP) for different periods (black dots, blue, green, yellow, and red indicate 0, 15, 30, 45, and 60 min, respectively) and conventional chemical bleaching (black).

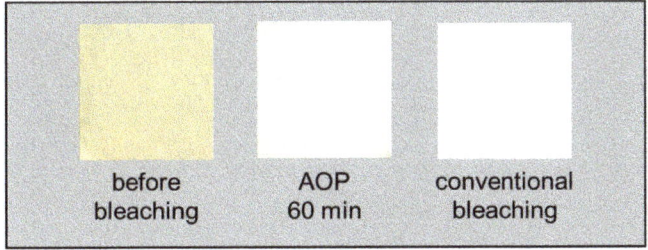

Figure 3. Photographs of the cotton samples before bleaching, after advanced oxidation processes (AOP) treatment for 60 min, and after conventional bleaching.

Figure 4 shows the reflectance spectra of linen samples treated using each bleaching treatment, as well as conventional chemical bleaching. The treatment with the combination of aqueous ozone and UV resulted in the largest increase in reflectance, followed by those with aqueous ozone and water + UV. Similar to that of the cotton samples, the percentage reflectance of the linen samples after AOP treatment increased. Notably, the samples treated by AOP for 120 min had improved reflectance over the conventional fabric.

Figure 5 shows photographs of the linen samples after bleaching treatment. The fabric before bleaching was grey. The bleaching process occurred in the centre, where UV irradiated aqueous ozone was in direct contact with the fabric. The linen fabrics treated with AOP for 120 min showed whiteness similar to that of the conventional method in the central spot.

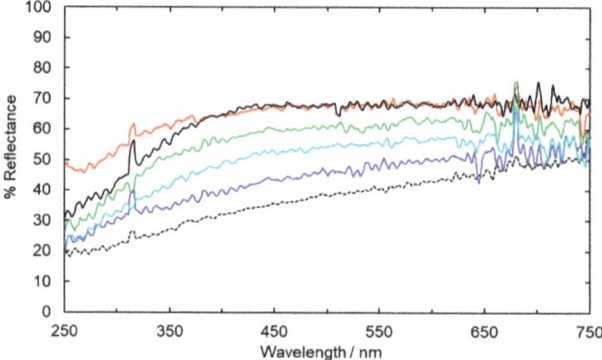

Figure 4. Reflectance spectra of linen samples treated by conventional chemical bleaching (black), no treatment (black dot), water + UV treatment for 60 min (blue), ozone water treatment for 60 min (light blue), AOP treatment for 60 min (green), and AOP treatment for 120 min (red).

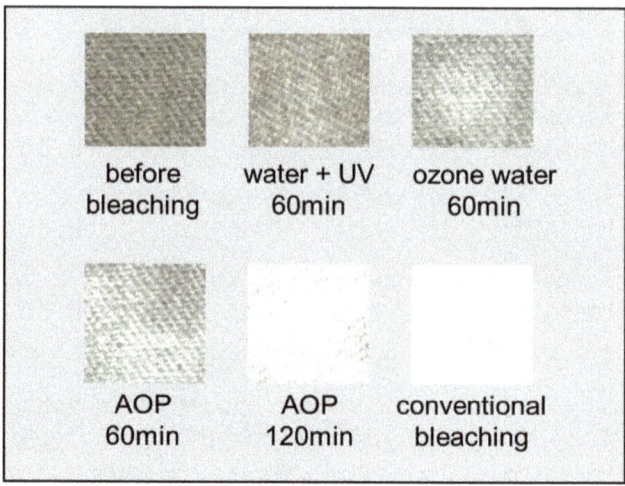

Figure 5. Photographs of the linen samples before bleaching, after water + ultraviolet (UV) treatment for 60 min, after aqueous ozone treatment for 60 min, after advanced oxidation processes (AOP) treatment for 60 min, after AOP treatment for 120 min, and after conventional bleaching.

Figure 6 shows the effect of AOP treatment on the fabric samples. The percentage reflectance increased in the region below 550 nm with and without scouring. The change in the reflectance spectrum of the cotton cloth was constant regardless of whether it was scoured. In addition, Figure 7 (left) shows a photograph of a drop of water on a cotton sample before scouring. Figure 7 (right) shows a photograph of a wetted fabric sample after bleaching by AOP treatment for 30 min. Before scouring and bleaching, the fabric was water repellent. After AOP treatment for 30 min, the cotton lost its water repellence, and the water drop penetrated the fabric.

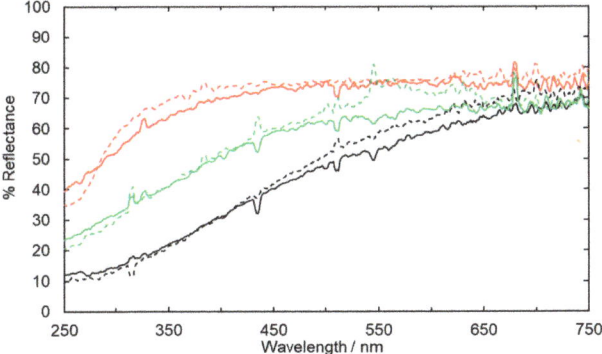

Figure 6. Reflectance spectra of the cotton samples after advanced oxidation processes (AOP) treatment for different periods (black, green, and red indicate 0, 30, and 60 min AOP treatment, respectively) with (dots) and without (solid line) scouring.

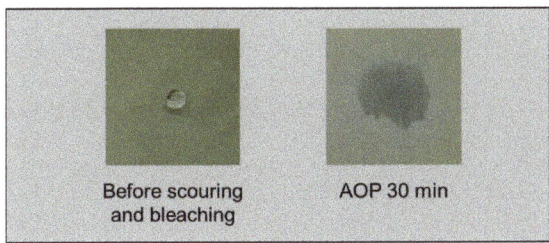

Figure 7. Photographs of cotton samples before scouring and bleaching and after advanced oxidation processes (AOP) treatment for 30 min treated with a drop of water.

Figure 8 shows the reflectance spectra of AOP-treated samples (0 and 60 min) before and after ironing at 200 °C. When the cotton was dried after AOP treatment, the reflectance decreased below 500 nm. When the ironed fabric was again subjected to AOP treatment, the spectrum was restored to the pre-ironing state. Ironing and AOP treatment were repeated five times, but only the results of the second ironing/AOP cycle are shown in Figure 8. The colour of the fabric changed from yellowish white to white and back after each of the five ironing/AOP cycles.

2.2. Suppression of Colour Reversion Using Rongalite Treatment

Figure 9 shows the reflectance spectra of cotton samples with and without rongalite treatment. The reflectance spectrum of the cotton fabric subjected to AOP treatment for 60 min increased with respect to that of the untreated sample. The reflectance spectrum of the cotton fabric ironed after AOP treatment decreased below 550 nm. On the other hand, the cotton fabric ironed after rongalite treatment did not show a decrease in the reflectance spectrum below 550 nm, and the reflectance spectra were maintained after AOP treatment. Figure 10 shows photographs of the cotton fabrics after AOP for 60 min, after ironing at 200 °C, and after ironing at 200 °C following rongalite treatment. The cotton sample after AOP was white. The cotton sample after ironing was not as white as that treated by AOP alone. The cotton sample after rongalite treatment and ironing was as white as that after AOP alone.

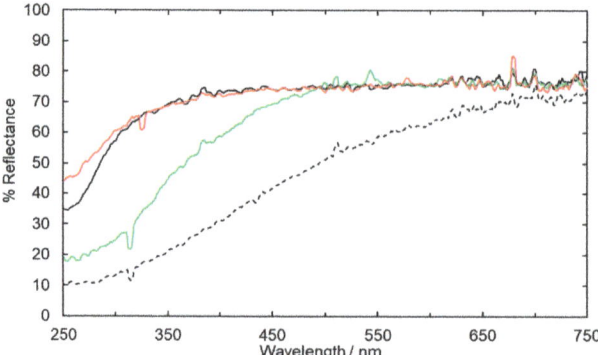

Figure 8. Reflectance spectra of cotton samples before (black dots) and after (black line) advanced oxidation processes (AOP) for 60 min, colour reversion after ironing at 200 °C (green), and after a second AOP treatment 15 min after ironing (red).

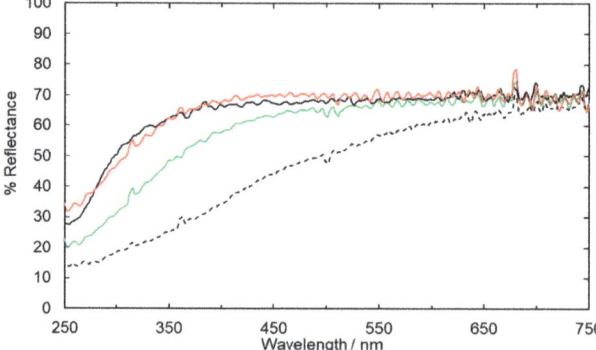

Figure 9. Reflectance spectra of cotton samples before (black dots) and after (black line) advanced oxidation processes (AOP) for 60 min and those of samples ironed at 200 °C with (red) and without (green) rongalite treatment.

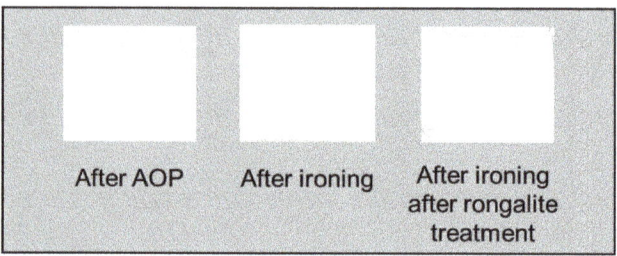

Figure 10. Photographs of cotton samples after advanced oxidation processes (AOP) for 60 min, after ironing at 200 °C, and after ironing at 200 °C following rongalite treatment.

Figure 11 shows the reflectance spectra of linen samples with and without rongalite treatment. The percentage reflectance of the linen fabrics subjected to AOP treatment for 120 min increased compared with that of the untreated sample. The reflectance spectrum of the linen fabric ironed after AOP treatment decreased at all wavelengths. In contrast, the cotton fabric ironed after rongalite treatment did not show a decrease, and the reflectance spectra were maintained after AOP treatment.

Figure 12 shows photographs of the linen fabrics after AOP for 120 min and after ironing at 200 °C, with and without rongalite treatment. The linen sample after AOP was white, but, after ironing, some whiteness was lost. After rongalite treatment and ironing, the sample remained as white as the sample after AOP alone.

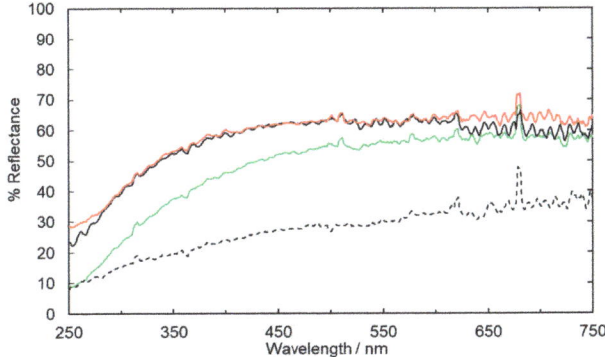

Figure 11. Reflectance spectra of linen samples without advanced oxidation processes (AOP; black dots) and after AOP for 120 min (black line) and of AOP-treated samples after ironing at 200 °C without (green) and with (red) rongalite treatment.

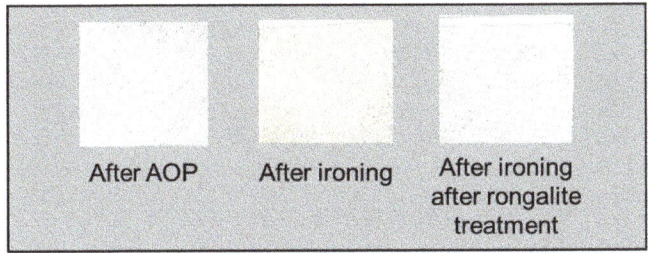

Figure 12. Photographs of the linen samples after advanced oxidation processes (AOP) for 120 min, after AOP and ironing at 200 °C, and after AOP and ironing at 200 °C following rongalite treatment.

3. Discussion

Figure 1a shows that the combination of aqueous ozone and UV light was most suitable for bleaching, although a bleaching effect was also obtained when using aqueous ozone alone and water + UV. Moreover, the reaction rate constant of the aqueous ozone + UV system was not a simple sum of those of the aqueous ozone and water + UV systems. Therefore, it appears that the cotton was bleached by the synergistic effect of ozone and UV irradiation. Figure 1b shows that OH radicals were generated by UV irradiation of the aqueous ozone. It is known that OH radicals decompose coloured components in fabrics during AOP [19,20], and the generation of OH radicals is likely responsible for the high reaction constant shown in Figure 1a. Specifically, excited singlet oxygen is produced by the UV irradiation of ozone at 310 nm, whereas excited triplet oxygen is produced by the visible light irradiation of ozone (> 460 nm). The bleaching OH radicals are generated from excited singlet oxygen [21]. Disodium terephthalate (NaTA) can be used as an indicator of the generated OH radicals, forming 2-terephthalic acid in the process [17]; approximately 0.35 ppb OH radicals were produced by UV irradiation. OH radicals are extremely reactive [22]; thus, their lifetime is very short. For these reasons, the OH radical concentration in Figure 1b is low, compared with the ozone concentration.

Figures 2 and 3 show that the AOP treatment that combined aqueous ozone and UV (60 min) had the same bleaching effect as the conventional method. As unbleached cotton contains coloured components, it had a low reflectance spectrum below 550 nm. Cotton contains approximately 90% cellulose and approximately 10% non-cellulosic matter, comprising proteins, waxes, pectin, and ash [1]. Therefore, on the basis of the results, these components were decomposed by AOP treatment to the same extent as in the conventional bleaching method. The linen before bleaching also contained coloured components, like cotton, and the low reflectance spectrum of the unbleached linen shown in Figure 4 is due to these components. As shown in Figures 4 and 5, in the case of linen, it took 120 min of AOP to bleach to the same extent as the conventional method. Thus, bleaching time is dependent on the amount of coloured components contained in the linen. Unlike cotton, linen contains only ~70% cellulose and a high proportion of non-cellulosic matter, which complicates the bleaching process. In addition, non-cellulosic matter contains a small amount of dark-coloured lignin [23], which suggests that linen will require a longer treatment time. As shown in Figure 6, AOP-treated cotton fabrics without scouring were bleached to the same degree as those with scouring and AOP treatment. In addition, Figure 7 shows that AOP treatment eliminated the water repellence of the cotton fabrics observed before scouring, resulting in hydrophilicity. As previously mentioned, before scouring, cotton contains wax [1], which yields water repellence. Therefore, the results in Figures 6 and 7 suggest that AOP treatment removes both coloured compounds and residual components (wax and oil) simultaneously. Thus, unlike conventional processes, our AOP process achieves scouring and bleaching at the same time.

In Figure 8, the colour return after the AOP process is shown by the reflectance spectra. In particular, the reflectance of the ironed sample was reduced, compared with that of the sample after AOP treatment. However, the reflectance recovered again after AOP treatment following ironing. However, repeated ironing resulted in the appearance of colour. The ironing and AOP treatment were repeated five times, but only the second cycle is shown in Figure 8. These results suggest that the colour reversion after ironing and AOP treatment is a reversible reaction. Cellulose, the main component of cotton, can be oxidised, and the oxidised form shows absorption in the near-UV to visible light region [24]. Even after AOP treatment, the cellulose in cotton may be oxidised, resulting in colour reversion. Therefore, in this study, rongalite, a reducing agent, was used to suppress colour reversion after AOP treatment and ironing. As shown in Figures 9 and 10, without rongalite treatment, reflectance decreased after ironing, whereas rongalite-treated fabrics did not show a decrease in reflectance in the wavelength region below 500 nm, even after ironing. In addition, even in the case of linen, rongalite-treated fabrics did not show a decrease in reflectance after ironing; the reflectance spectra after AOP treatment were similar to those before treatment (Figures 11 and 12).

Figure 13 shows the proposed mechanisms of colour reversion and suppression of colour reversion after rongalite treatment of fabrics. AOP treatment could damage the fabric [12,25], and previous studies on the cause of colour reversion after oxidative bleaching have reported that some of the OH groups of cellulose (particularly those at the 2nd and 3rd positions) are oxidised, resulting in the formation of C=O groups and some conjugated double bonds [24]. Thus, colour reversion could be suppressed by the reduction of the C=O groups generated in the cellulose fibres of OH groups after bleaching. Therefore, before ironing, the AOP-treated fabric was treated with a strongly reducing rongalite solution to reduce the C=O groups. Consequently, subsequent heat treatment (ironing) did not result in the formation of conjugated double bonds.

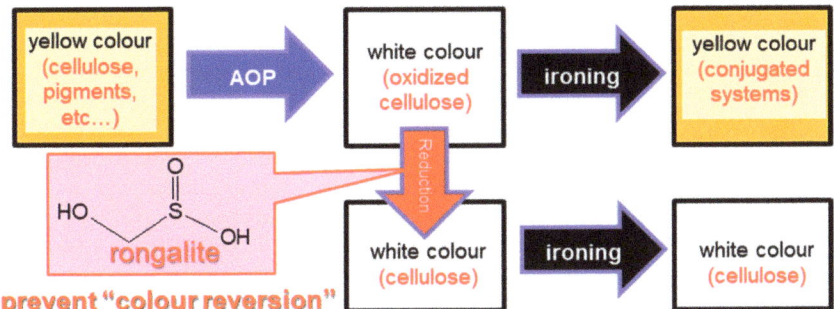

Figure 13. Proposed mechanisms of the suppression of colour reversion in fabrics and the effect of rongalite treatment.

4. Materials and Methods

Cotton and linen cloths were provided by Nisshinbo Textrail Co., Ltd. (Tokyo Japan). The colour differences of the fabrics were measured using a colour reader (CR-10, Konica Minolta Japan, Inc., Tokyo, Japan) and expressed as $L^*a^*b^*$ and ΔE^*ab values [26,27]. Here, L^* represents brightness, equalling zero for a black diffuser and 100 for a perfectly reflecting one; a^* represents colour on the red–green axis, being positive for red and negative for green; and b^* represents colour on the blue–yellow axis, being positive for yellow and negative for blue. The colour difference was calculated as $\Delta E^*ab = ((\Delta L^*)^2 + (\Delta a^*)^2 + (\Delta b^*)^2)^{1/2}$, where ΔL^* is the difference in brightness between two vivid surfaces and Δa^* and Δb^* are the differences in the colour coordinates a^* and b^*, respectively. Figure 14 shows the bleaching experiment in which aqueous ozone and UV irradiation were combined. UV light is emitted above a cloth sample placed on a glass plate. As a UV light source, an LA-310UV manufactured by Hayashi Clock Industry Co., Ltd. (Tokyo, Japan) was used. In addition, the electrolytically generated aqueous ozone was produced by an aqueous ozone generator (Quick O_3 Pico, AOD-TH2 manufactured by Aidenshi Co., Ltd., Nasushiobara, Japan) and dropped onto the fabric at the same position as the UV spot.

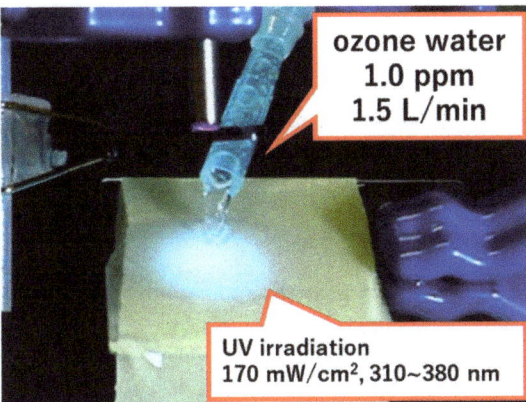

Figure 14. Photograph of a bleaching experiment. The fabric sample was set on a glass plate. Water containing electrochemically generated ozone was dropped onto the fabric, and the sample was irradiated with ultraviolet (UV) light.

The rongalite treatment was carried out by the dipping method. The sample after AOP treatment was immersed in a 0.5 wt% rongalite aqueous solution for 1 h, then drained and dried on a hot plate at

200 °C for 1 min. Figure 15 shows photographs and a schematic of the measurement of the amount of OH radicals in the aqueous ozone under UV irradiation. It is well known that OH radicals are generated by AOP treatment with aqueous ozone and UV irradiation [15,16,19,20]. In this study, the OH radical concentration in the ozone solution during AOP treatment was measured using a disodium terephthalate chemical probe [17,18]. As shown in Figure 15, aqueous ozone irradiated with UV was collected and then mixed with disodium terephthalate (NaTA, TCI Co., Ltd., Tokyo, Japan) until completely dissolved. Then, while irradiating the solution with UV light, the generated fluorescence was detected. An SEC-2000-UV/VIS Spectrometer manufactured by ALS Co., Ltd. (Tokyo, Japan) was used for UV light generation and fluorescence detection. Disodium terephthalate, used as an indicator, reacts with OH radicals in aqueous solution to produce 2-hydroxyterephthalic acid, which fluoresces at 425 nm under UV irradiation, allowing quantification of the generated OH radicals.

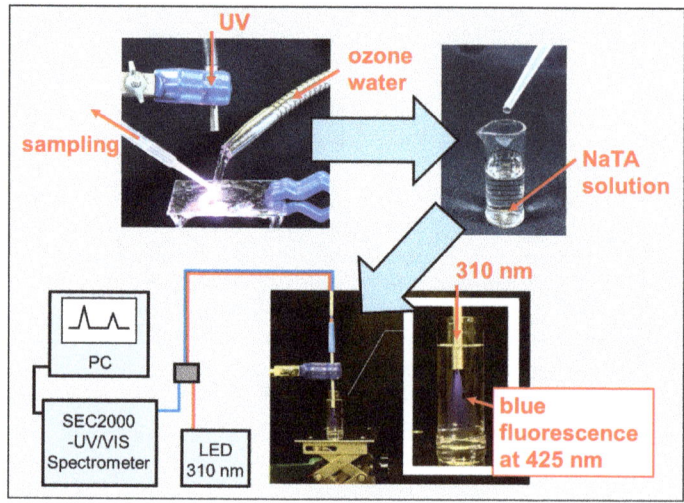

Figure 15. Photographs and schematic of OH radical quantification by fluorescence spectroscopy.

5. Conclusions

In this paper, we reported a two-step fabric treatment method based on AOP treatment using aqueous ozone and UV irradiation with rongalite treatment. The AOP process removed coloured components in fabric via decomposition with active oxygen species (ozone and the OH radicals generated in aqueous solution). The OH radical concentration in the aqueous solution was measured by a chemical probe, which can easily and quickly measure OH radicals on the spot, and it is applicable to industrial-scale process control. The reflectance spectra of the fabric samples revealed an increase in bleaching with increasing treatment time, and, after 60 min for cotton and 120 min for linen, bleaching comparable to that of the conventional method was achieved. In addition, we confirmed that AOP treatment is suitable for decomposing waxes and oils contained in the fabric before scouring. Therefore, this AOP process can replace conventional scouring and bleaching processes. Furthermore, we succeeded in preventing colour reversion by reducing cellulose oxidised by the AOP treatment with inexpensive rongalite. The AOP + rongalite treatment for fabrics is attractive because it produces little waste and consumes little energy compared with the conventional method. Thus, this method is a promising alternative to conventional bleaching in natural fibre processing.

Author Contributions: K.H., T.O., K.M., Y.I., T.N. and N.K. participated in the study design and conducted the study; K.H., T.O. and Y.T. collected and analysed the data. K.H. wrote the manuscript. All authors have read and agreed to the published version of the manuscript.

Funding: This research received no external funding.

Acknowledgments: We are grateful to Nisshinbo Textile Inc. for preparing the fabrics. We are also grateful to Ayato Koibuchi and Kosuke Miura (Nagaoka University of Technology) for help with the experiments and for insightful discussion. We would like to thank Editage (www.editage.com) for English language editing.

Conflicts of Interest: The authors declare no conflict of interest.

References

1. Abdel-Halim, E.S. An effective redox system for bleaching cotton cellulose. *Carbohydr. Polym.* **2012**, *90*, 316–321. [CrossRef] [PubMed]
2. Hebeish, A.; Hashem, M.; Shaker, N.; Ramadan, M.; El-Sadek, B.; Hady, M.A. New development for combined bioscouring and bleaching of cotton-based fabrics. *Carbohydr. Polym.* **2009**, *78*, 961–972. [CrossRef]
3. Lim, S.-H.; Gürsoy, N.Ç.; Hauser, P.; Hinks, D. Performance of a new cationic bleach activator on a hydrogen peroxide bleaching system. *Coloration Technol.* **2004**, *120*, 114–118. [CrossRef]
4. Shafie, A.E.; Fouda, M.M.G.; Hashem, M. One-step process for bio-scouring and peracetic acid bleaching of cotton fabric. *Carbohydr. Polym.* **2009**, *78*, 302–308. [CrossRef]
5. Yao, Y.; Kubota, Y.; Murakami, T.; Ochiai, T.; Ishiguro, H.; Nakata, K.; Fujishima, A. Electrochemical Inactivation Kinetics of Boron-Doped Diamond Electrode on Waterborne Pathogens. *J. Water Health* **2011**, *9*, 534–543. [CrossRef] [PubMed]
6. Ochiai, T.; Tago, S.; Hayashi, M.; Hirota, K.; Kondo, T.; Satomura, K.; Fujishima, A. Boron-doped diamond powder (BDDP)-based polymer composites for dental treatment using flexible pinpoint electrolysis unit. *Electrochem. Commun.* **2016**, *68*, 49–53. [CrossRef]
7. Ouchi, A. Photochemical Reactions and their Application to Textile Processing. *Oleoscience* **2015**, *15*, 455–460. [CrossRef]
8. Abdel-Halim, E.S.; Al-Deyab, S.S. One-step bleaching process for cotton fabrics using activated hydrogen peroxide. *Carbohydr. Polym.* **2013**, *92*, 1844–1849. [CrossRef]
9. Li, Q.; Ni, L.; Wang, J.; Quan, H.; Zhou, Y. Establishing an ultrasound-assisted activated peroxide system for efficient and sustainable scouring-bleaching of cotton/spandex fabric. *Ultrason. Sonochem.* **2020**, *68*, 105220. [CrossRef]
10. Perincek, S.; Bahtiyari, I.; Korlu, A.; Duran, K. New Techniques in Cotton Finishing. *Text. Res. J.* **2009**, *79*, 121–128. [CrossRef]
11. Perincek, S.D.; Duran, K.; Korlu, A.E.; Bahtiyari, İ.M. An Investigation in the Use of Ozone Gas in the Bleaching of Cotton Fabrics. *Ozone Sci. Eng.* **2007**, *29*, 325–333. [CrossRef]
12. Piccoli, H.H.; Ulson de Souza, A.A.; Ulson de Souza, S.M.A.G. Bleaching of Knitted Cotton Fabric Applying Ozone. *Ozone Sci. Eng.* **2015**, *37*, 170–177. [CrossRef]
13. Hashem, M.; Taleb, M.A.; El-Shall, F.N.; Haggag, K. New prospects in pretreatment of cotton fabrics using microwave heating. *Carbohydr. Polym.* **2014**, *103*, 385–391. [CrossRef] [PubMed]
14. Farooq, A.; Ali, S.; Abbas, N.; Fatima, G.A.; Ashraf, M.A. Comparative performance evaluation of conventional bleaching and enzymatic bleaching with glucose oxidase on knitted cotton fabric. *J. Clean. Prod.* **2013**, *42*, 167–171. [CrossRef]
15. He, Y.; Wang, X.; Xu, J.; Yan, J.; Ge, Q.; Gu, X.; Jian, L. Application of integrated ozone biological aerated filters and membrane filtration in water reuse of textile effluents. *Bioresour. Technol.* **2013**, *133*, 150–157. [CrossRef]
16. Buffle, M.-O.; Schumacher, J.; Meylan, S.; Jekel, M.; von Gunten, U. Ozonation and Advanced Oxidation of Wastewater: Effect of O_3 Dose, pH, DOM and HO•-Scavengers on Ozone Decomposition and HO• Generation. *Ozone Sci. Eng.* **2006**, *28*, 247–259. [CrossRef]
17. Hayashi, H.; Akamine, S.; Ichiki, R.; Kanazawa, S. Comparison of OH radical concentration generated by underwater discharge using two methods. *Int. J. Plasma Environ. Sci. Technol.* **2016**, *10*, 24–28.
18. Kanazawa, S.; Kawano, H.; Watanabe, S.; Furuki, T.; Akamine, S.; Ichiki, R.; Ohkubo, T.; Kocik, M.; Mizeraczyk, J. Observation of OH radicals produced by pulsed discharges on the surface of a liquid. *Plasma Sources Sci. Technol.* **2011**, *20*, 034010. [CrossRef]
19. Sharma, V.K.; Graham, N.J.D. Oxidation of Amino Acids, Peptides and Proteins by Ozone: A Review. *Ozone Sci. Eng.* **2010**, *32*, 81–90. [CrossRef]

20. Ikehata, K.; Jodeiri Naghashkar, N.; Gamal El-Din, M. Degradation of Aqueous Pharmaceuticals by Ozonation and Advanced Oxidation Processes: A Review. *Ozone: Sci. Eng.* **2006**, *28*, 353–414. [CrossRef]
21. Matsumi, Y.; Comes, F.J.; Hancock, G.; Hofzumahaus, A.; Hynes, A.J.; Kawasaki, M.; Ravishankara, A.R. Quantum yields for production of $O(^1D)$ in the ultraviolet photolysis of ozone: Recommendation based on evaluation of laboratory data. *J. Geophys. Res. Atmos.* **2002**, *107*, 1–12. [CrossRef]
22. Attri, P.; Kim, Y.H.; Park, D.H.; Park, J.H.; Hong, Y.J.; Uhm, H.S.; Kim, K.-N.; Fridman, A.; Choi, E.H. Generation mechanism of hydroxyl radical species and its lifetime prediction during the plasma-initiated ultraviolet (UV) photolysis. *Sci. Rep.* **2015**, *5*, 9332. [CrossRef] [PubMed]
23. Goswami, K.K. Bleaching of linen (*Linum usitatissimum*). *Indian J. Fibre Texlie Res.* **1993**, *18*, 82–86.
24. Yui, Y.; Tanaka, C.; Isogai, A. Functionalization of Cottton Fabrics by TEMPO-Mediated Oxidation. *SENI GAKKAISHI* **2013**, *69*, 222–228. [CrossRef]
25. Arooj, F.; Ahmad, N.; Shaikh, I.A.; Chaudhry, M.N. Application of ozone in cotton bleaching with multiple reuse of a water bath. *Text. Res. J.* **2013**, *84*, 527–538. [CrossRef]
26. Japanese Standards Association. *JIS Z 8722 Methods of Colour Measurement-Reflecting and Transmitting Objects*; Japanese Standards Association: Tokyo, Japan, 2009.
27. Ochiai, T.; Aoki, D.; Saito, H.; Akutsu, Y.; Nagata, M. Analysis of Adsorption and Decomposition of Odour and Tar Components in Tobacco Smoke on Non-Woven Fabric-Supported Photocatalysts. *Catalysts* **2020**, *10*, 304. [CrossRef]

Publisher's Note: MDPI stays neutral with regard to jurisdictional claims in published maps and institutional affiliations.

© 2020 by the authors. Licensee MDPI, Basel, Switzerland. This article is an open access article distributed under the terms and conditions of the Creative Commons Attribution (CC BY) license (http://creativecommons.org/licenses/by/4.0/).

Article

CeO$_2$ for Water Remediation: Comparison of Various Advanced Oxidation Processes

Roberto Fiorenza [1], Stefano Andrea Balsamo [1], Luisa D'Urso [1], Salvatore Sciré [1], Maria Violetta Brundo [2], Roberta Pecoraro [2], Elena Maria Scalisi [2], Vittorio Privitera [3] and Giuliana Impellizzeri [3,*]

[1] Department of Chemical Sciences, University of Catania, Viale A. Doria 6, 95125 Catania, Italy; rfiorenza@unict.it (R.F.); stefano.balsamo@phd.unict.it (S.A.B.); ldurso@unict.it (L.D.); sscire@unict.it (S.S.)
[2] Department of Biological, Geological and Environmental Science, University of Catania, Via Androne 81, 95124 Catania, Italy; mvbrundo@unict.it (M.V.B.); roberta.pecoraro@unict.it (R.P.); scalisimariaelena90@gmail.com (E.M.S.)
[3] CNR-IMM, Via S. Sofia 64, 95123 Catania, Italy; vittorio.privitera@cnr.it
* Correspondence: giuliana.impellizzeri@ct.infn.it; Tel.: + 39 095 3785345

Received: 30 March 2020; Accepted: 20 April 2020; Published: 21 April 2020

Abstract: Three different Advanced Oxidation Processes (AOPs) have been investigated for the degradation of the imidacloprid pesticide in water: photocatalysis, Fenton and photo-Fenton reactions. For these tests, we have compared the performance of two types of CeO$_2$, employed as a non-conventional photocatalyst/Fenton-like material. The first one has been prepared by chemical precipitation with KOH, while the second one has been obtained by exposing the as-synthetized CeO$_2$ to solar irradiation in H$_2$ stream. This latter treatment led to obtain a more defective CeO$_2$ (coded as "grey CeO$_2$") with the formation of Ce^{3+} sites on the surface of CeO$_2$, as determined by Raman and X-ray Photoelectron Spectroscopy (XPS) characterizations. This peculiar feature has been demonstrated as beneficial for the solar photo–Fenton reaction, with the best performance exhibited by the grey CeO$_2$. On the contrary, the bare CeO$_2$ showed a photocatalytic activity higher with respect to the grey CeO$_2$, due to the higher exposed surface area and the lower band-gap. The easy synthetic procedures of CeO$_2$ reported here, allows to tune and modify the physico-chemical properties of CeO$_2$, allowing a choice of different CeO$_2$ samples on the basis of the specific AOPs for water remediation. Furthermore, neither of the samples have shown any critical toxicity.

Keywords: ceria; pesticide; photocatalysis; photo-Fenton; AOPs

1. Introduction

Among the environmental questions of the present, water pollution by emergent contaminants, such as pharmaceuticals and pesticides, is a serious problem, making their removal a challenging task [1]. In particular, the use of pesticides has increased over the years to improve the production of agricultural goods and to satisfy the contextual growth of world population. Pesticides are a wide group of chemical compounds classified as persistent hazardous pollutants owing to a very high time of retention in water and giving rise to accumulation in sediment and in water effluents. They are also easily transferred over a long distance [2]. Their presence in the environment, especially in water, even at low concentrations, is a serious problem for both living organisms and human health.

Advanced Oxidation Processes (AOPs) are among the new, green and performing solutions for the removal of pesticides from water [3,4]. In these processes, the oxidation of the hazardous contaminants is obtained through the production of highly reactive radical species, such as $^{\bullet}$O$_2^-$, $^{\bullet}$O$_3^-$, or OH$^{\bullet}$. Different AOPs can be simultaneously utilized to avoid the generation of by-products in treated water [5]. In this context, the photocatalysis and the Fenton process are two of the most promising AOPs [6]. The degradation of pesticides in water by means of photocatalysis allows to efficiently remove these pollutants with a moderate formation of secondary products and the selectivity of the process can be enhanced if peculiar materials (such as molecularly imprinted photocatalysts) are employed [7–10]. The hydroxyl radicals are formed in this process after the irradiation of a semiconductor photocatalyst with UV or a solar/visible light source with the consequent formation of photoelectrons in the conduction band and photoholes in the valence band of the photocatalyst [11]. The Fenton process involves the reaction between Fe^{2+} and hydrogen peroxide to give the hydroxyl radicals (reaction 1):

$$Fe^{2+} + H_2O_2 \rightarrow Fe^{3+} + OH^\bullet + OH^- \tag{1}$$

The further reaction of the ferric ions with the excess of H_2O_2 re-generates the ferrous ions with the formation of the hydroperoxyl radicals ($HOO^\bullet$) (reaction 2):

$$Fe^{3+} + H_2O_2 \rightarrow Fe^{2+} + HOO^\bullet + H^+ \tag{2}$$

The regeneration of ferrous ions can be accelerated, enhancing also the efficiency of the overall degradation process, in the presence of visible or near ultraviolet irradiation (i.e., the photo-Fenton process, reactions 3-5), with the consequent formation of further hydroxyl radicals [12,13]. Furthermore, some Fe(III)–carboxylate complexes originated from the coordination of Fe^{3+}, and organic intermediates can adsorb in the UV–vis region, and other Fe^{2+} species can be formed through the ligand-to-metal charge transfer (LMCT) (reaction 4). Finally, also the zero-valent iron species can be considered a source of Fe^{2+} (reaction 5) [14].

$$Fe^{3+} + h\nu + H_2O \rightarrow Fe^{2+} + OH^\bullet + H^+ \tag{3}$$

$$[Fe^{3+}(RCO_2)]^{2+} + h\nu \rightarrow Fe^{2+} + CO_2 + R^\bullet \tag{4}$$

$$Fe^0 + h\nu \rightarrow Fe^{2+} + 2e^- \;(\lambda < 400 \text{ nm}) \tag{5}$$

The photo-Fenton process was successfully applied in the degradation of various pesticides and pharmaceuticals under solar light irradiation [15,16].

Among the various semiconductors used for the photocatalytic applications, recently, cerium oxide (CeO_2, commonly called *ceria*), a largely used catalyst in many thermo-catalytic reactions [17,18], was examined as an alternative to the most used metal oxide photocatalysts (such as TiO_2 and ZnO [19–22]). The most attractive properties of CeO_2 are: the lower band-gap (around 2.7–2.8 eV) compared to TiO_2 and ZnO, making the material sensitive to visible light; the presence of empty 4f energy levels that facilitate the electron transfers; the high stability in the reaction medium; the high oxygen mobility related to the reversible Ce^{4+}/Ce^{3+} transformation, and the ability to form nonstoichiometric oxygen-deficient CeO_{2-x} oxide [23]. The presence of defect centres in the CeO_2, together with the high oxygen mobility and the consequential redox properties can be exploited in the Fenton-like reactions, that in this case, are different to the radical-attacking mechanism of conventional iron-based Fenton process, are driven by the interaction between the hydrogen peroxide and the surface defective Ce^{3+} centres (reactions 6–8, [24,25]):

$$Ce^{3+} + H_2O_2 \rightarrow Ce^{4+} + OH^\bullet + OH^- \tag{6}$$

$$H_2O_2 + OH^\bullet \rightarrow H_2O + HOO^\bullet \tag{7}$$

$$Ce^{4+} + HOO^\bullet \rightarrow Ce^{3+} + H^+ + O_2 \tag{8}$$

The conventional iron species catalysts in the Fenton process require a strict operating pH range (between 3 and 4), thus increasing the overall process cost, whereas with CeO_2 it is possible to work at neutral pH [26].

On the basis of the above considerations, in this work we have studied the degradation of a largely used insecticide, i.e., the imidacloprid ($C_9H_{10}ClN_5O_2$), by the comparison between three different AOPs: solar photocatalysis, Fenton and solar photo-Fenton, taking advantage of the wide versatility of CeO_2 that can be used both as a photocatalyst and as a Fenton-like reagent.

The imidacloprid (hereafter called "IMI") is a neonicotinoid pesticide, which acts similarly to the natural insecticide nicotine [27]. Although IMI is not directly used in water, it is commonly transferred to water channels, and it presents a high leachability [28,29]. For its high toxicity, solubility, and stability, the presence of IMI in water even at low concentrations is a serious environmental concern.

The Fenton-like process through ceria is activated by the presence of non-stoichiometric Ce^{3+} centres on the surface of CeO_2 (reaction 6, [30,31]). One of the simpler methods to induce these defects is the exposure of CeO_2 to sunlight [32]. Indeed, the interaction of CeO_2 with the efficient UV solar photons (i.e., the photons with an energy higher than the band-gap of CeO_2) led to a release of the labile ceria surface oxygens with the formation of CeO_{2-x} defects. After the loss of oxygen, the cerium atoms adopted the most stable configuration, i.e., the Ce^{3+} oxidation state.

In this context, we have synthetized two different types of CeO_2: the first through one of the most employed preparation procedures for this oxide, as the precipitation with KOH from a cerium nitrate solution [23,33]; the second type using the same synthesis but irradiating the samples just after calcination with a solar lamp and in the presence of a H_2 flow, with the aim to further increase the surface defects on CeO_2. Interestingly, with this original modified strategy we have obtained a "grey CeO_2" instead of the typical yellow coloured ceria. The possibility to generate further defects on CeO_2 with solar exposure in a reducing atmosphere was especially exploited, due to a synergistic mechanism able to in situ provide the Ce^{3+} species, mainly in the solar photo-Fenton tests.

2. Results and Discussion

2.1. Characterizations of Bare CeO_2 and Grey (Modified) CeO_2

The first difference between the bare CeO_2 and the CeO_2 exposed to the solar irradiation (for 3 h) in H_2 flow at room temperature, is the change of the powder color. Figure 1 reports two photos of the synthetized materials: un-modified CeO_2 appears yellow in color, while modified CeO_2 is grey. This latter sample is coded as "grey CeO_2". Interestingly, we have noted that only the contemporaneous treatment with solar irradiation and H_2 flow could obtain the grey CeO_2, whereas each single treatment alone did not alter the structural and chemical properties of bare CeO_2.

Figure 1. Photo of the as-synthetized powders.

The differences in the physico-chemical properties of bare CeO_2 and grey CeO_2 were illustrated in Figure 2, where the XRD patterns (Figure 2a), the Raman (Figure 2b), and the FTIR (Figure 2c,d) spectra are reported.

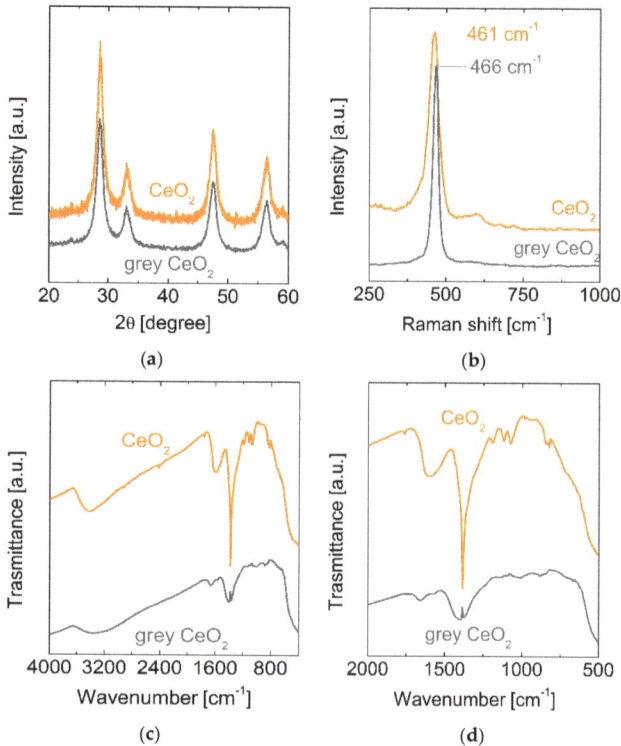

Figure 2. (a) XRD patterns, (b) Raman spectra, (c) FTIR spectra of the synthetized samples, and (d) FTIR zoom of the "carbonate" zone.

Both the samples exhibited the typical XRD pattern of ceria in the fluorite crystalline phase (Figure 2a), with the reflections at 2θ values of 28.6 (1 1 1), 33. 1 (2 0 0), 47.4 (2 2 0), and 56.4 (3 1 1) [34]. No substantial variation was detected in the grey CeO_2 compared to bare oxide, apart from a slight intensity decrease and a difference in the average crystallite size: 6.8 ± 0.8 nm for bare CeO_2 respect to 11.3 ± 1.1 nm for grey CeO_2, calculated using the Scherrer equation on the main diffraction peak of ceria 2θ = 28.6 (1 1 1). This size enhancement of grey CeO_2 was related to the occurrence of the formation of defects inside the crystalline structure of CeO_2 caused by the solar irradiation in the H_2 stream. Furthermore, in accordance with the literature data [26,35], the intensity diminution observed in the grey CeO_2 can be reasonable connected to a decrease in crystallinity due to the formation of CeO_{2-x} defects.

Interestingly, analyzing the Raman spectra of the samples (Figure 2b), the peak at 461 cm^{-1} of the CeO_2 was blue-shifted by 5 cm^{-1} in the grey CeO_2. The Raman peak at 461 cm^{-1} identifies the F_{2g} skeletal vibration of the cubic fluorite structure [36]. The position of this peak is influenced by the distortion of the Ce-O bonds [32]. Consequently, the treatment of grey CeO_2 led to a more defective structure with the modification of the cubic structure of CeO_2, resulting in the Raman shift. However, in the as-synthesized bare CeO_2, an imperfect crystalline stoichiometry was detected, being the small shoulder at about 600 cm^{-1} (more intense in the bare CeO_2), ascribed to Frenkel-type oxygen vacancies [37]. Other differences can be seen in the FTIR spectra (Figure 2c). The bands at about and 1620 cm^{-1} are attributed to the stretching and the bending of the O-H groups of residual water molecules respectively, whereas the group of bands in the range 1600–500 cm^{-1} are related to the presence of carbonates due to the interaction of the atmospheric carbon dioxide with ceria [38].

From the zoomed spectra illustrated in Figure 2d, it is possible to note the formation of different carbonate species. Specifically, for bare CeO_2, the high intense band at 1385 cm^{-1} is due to monodentate carbonates, whereas the bands at 1190 and 1120 cm^{-1} are related to the bridged carbonate species. Finally, the bands at 1071 and 839 cm^{-1} indicated the formation of hydrocarbonates [39–41]. In the grey CeO_2, the band assigned to the monodentate carbonate was broader and shifted at 1395 cm^{-1}, whereas the low intense bands at 1012 and 872 cm^{-1} can be also be assigned for this sample to the presence of hydrocarbonates, whereas there is no evidence of the formation of monodentate carbonates. It is clear that the surface interaction sites in the grey CeO_2 were changed compared to un-treated CeO_2. This can be reasonably related to the more defective surface of grey CeO_2.

The textural properties of the CeO_2 samples are displayed in the Figure 3. Both the materials displayed a N_2 adsorption–desorption isotherm of type III, with a H3 hysteresis loop (Figure 3a), indicating the presence of macro-meso slit-shaped pores [42]. The treatment with solar lamp in H_2 flow led to a decrease in the Brunauer–Emmett–Teller (BET) surface area. The grey CeO_2 exhibited a lower surface area (67 ± 1 m^2/g) than CeO_2 (81 ± 1 m^2/g). This decrease can be reasonably due to the agglomeration of CeO_2 particles caused by the irradiation treatment under solar lamp in H_2 flow, as further confirmed by the increase in mean crystalline size calculated by XRD. As a consequence, it was verified a shift towards large pores in the Barrett, Joyner and Halenda (BJH) pore size distribution curves (Figure 3b), with the mean pore size of the grey CeO_2 higher (58 ± 1 nm) with respect to bare CeO_2 (36 ± 1 nm). This size increase, verified by the grey CeO_2, is strictly correlated with the peculiar treatment of this latter sample, i.e., the simultaneous utilization of the simulated solar radiation and the H_2 stream. As stated before, according to the work of Aslam et al. [32], the solar light alone did not caused any change in the mean size of bare CeO_2; however, we used a more focused solar lamp than in ref. [32], which led to a slight heating of the sample (from room temperature to about 40 °C). On the contrary, the irradiation in a reductive atmosphere (H_2 stream) promoted the formation of numerous oxygen vacancies, a process characterized with an increase in the internal pressure inside the ceria crystalline structure with a consequent interatomic bond cleavage [43]. Reasonably, this process resulted in an agglomeration with a measurable size increase in the grey CeO_2 particle size. The same linear correlation between the increase in mean crystalline size, and the decrease in the BET surface area was already reported in the literature with other CeO_2-based samples [44,45]. The formation of defects did not alter the morphology of the CeO_2 materials, that, if prepared by chemical precipitation, are usually characterized by a random stacking of particles [23,32].

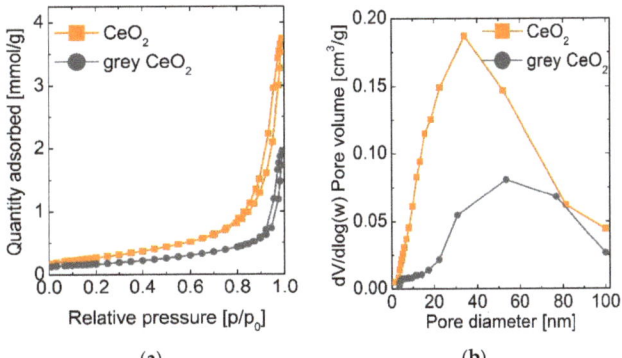

Figure 3. (a) N_2 adsorption–desorption isotherms of the CeO_2 samples; (b) pore size distribution curve of the analyzed samples evaluated by means of the Barrett, Joyner and Halenda (BJH) method.

The UV-vis Diffuse Reflectance spectra of the CeO_2 powders are displayed in Figure 4a where the reflectance function (Kubelka–Munk function) is plotted versus the wavelength. A slight variation in the absorption features was detected for grey CeO_2 with a shift at lower wavelengths that results in a slightly higher optical band-gap (3.1 ± 0.3 eV) compared to bare CeO_2 (2.7 ± 0.3 eV) estimated by graphing the modified Kubelka–Munk function versus the eV (Figure 4b) [46]. The lower band-gap of CeO_2 (activation wavelength ≤ 460 nm) is suitable to exploit, together with the UV portion, a part of visible component of the solar light, whereas the grey CeO_2 with a higher band-gap (activation wavelength ≤ 400 nm) will be preferentially activated by the solar UV photons.

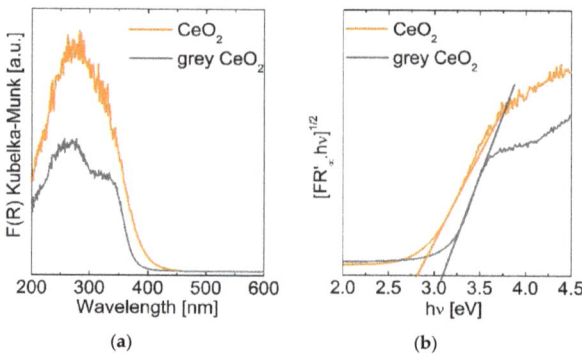

Figure 4. (a) UV-vis Diffuse Reflectance spectra of CeO_2 powders; (b) estimation of the optical band-gap of the samples by means of the modified Kubelka–Munk function.

For the photocatalytic degradation of the IMI and especially for the Fenton and photo-Fenton reactions, it is fundamental that the presence of Ce^{3+} defects on the surface of CeO_2. To establish the presence of these defect states, the XPS analysis was performed and the results are illustrated in Figure 5. In accordance with the literature data, the Ce $3d_{5/2}$ state involves the v, v', v" and v''' component, whereas the u, u', u" and u''' components are related to the Ce $3d_{3/2}$ state [47–50]. The v' and u' components indicate the presence of Ce^{3+}, whereas the peak at 916.4 eV (u''') for CeO_2 and at 916.8 eV for the grey CeO_2 are the typical fingerprint of Ce^{4+} [48–50]. It is clearly visible from Figure 5, as the component v' at 885.2 eV of Ce^{3+} is intense for grey CeO_2 whereas the same signal is absent in the bare CeO_2. The u' signal is covered to the u and u" components in both the samples. Furthermore, it is possible to note that, as the ratio between the v''' and u components is different and shifted of about 0.5 eV, as for the v component, compared to un-modified CeO_2. This is another indication of the modification of the ceria surface sites with the higher presence of Ce^{3+} states in the grey CeO_2 [50]. The irradiation with solar lamp in H_2 stream thus induced the formation of CeO_{2-x} defects on the surface of CeO_2, as also confirmed by the Raman spectroscopy, with a consequent modification to the surface chemical composition of ceria, as also indirectly corroborated by the FTIR with the formation of different carbonate species in the two CeO_2 samples.

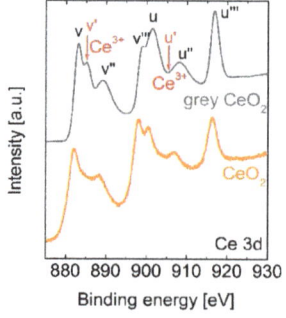

Figure 5. XPS spectra of the CeO_2 samples.

2.2. (Photo)catalytic Activity

We have compared the (photo)catalytic activity of the synthetized sample in the degradation of the IMI pesticide. Three different AOPs were investigated: a) the photocatalytic oxidation (Figure 6a) utilized as an irradiation source a solar lamp; b) the Fenton reaction (Figure 6b), adding 5 mL of H_2O_2 (3%, 0.9M) in the reaction mixture, c) the photo-Fenton reaction (Figure 6c) utilizing both the solar lamp and the hydrogen peroxide.

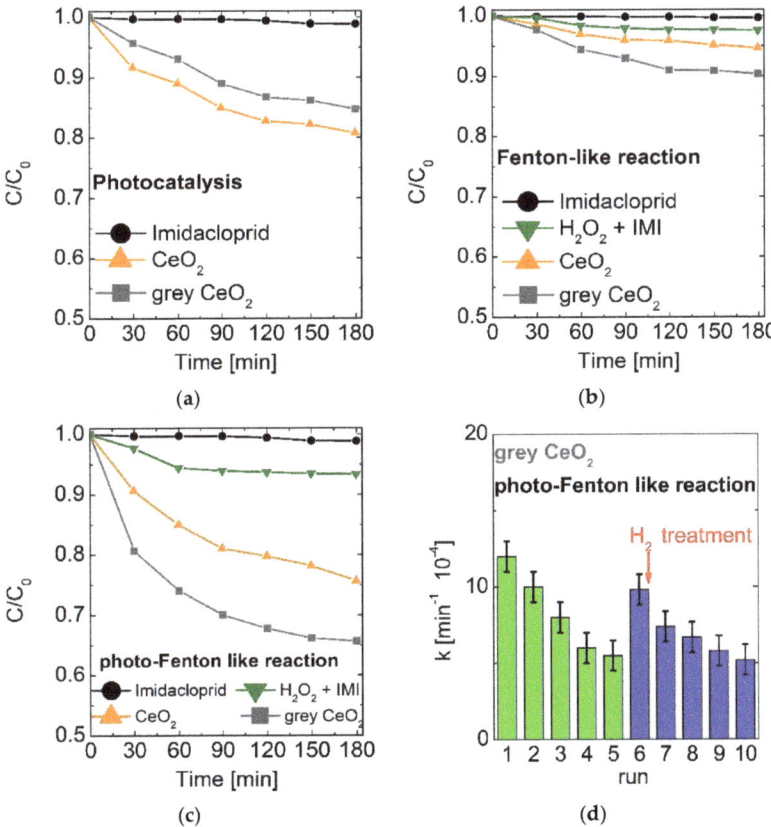

Figure 6. (a) Photocatalytic degradation of imidacloprid (IMI) under solar light irradiation, (b) Fenton-like reaction, (c) photo-Fenton like reaction on the CeO_2-based samples, (d) photo-Fenton like reaction utilizing grey CeO_2 in different runs.

The solar photodegradation of pesticides required harder conditions in comparison to the degradation of other pollutants (for example, dyes) [51,52]. As a result, even utilizing TiO_2 (the most investigated photocatalyst), the degradation efficiency is not so high [7,53]. Furthermore, in accordance with our preceding work [7], and the literature data [54,55], as confirmed for all the AOPs investigated, the IMI degradation is characterized by the formation of various by-products as amine and chloro-pyridine species. The reaction mechanism involves the breaking of C–N and the N–N bonds followed by the formation of small molecules, such as chlorine dioxide, nitrogen oxides species, water and carbon dioxide [7,54,55]. The reported degradation percentage of IMI (i.e., the variation of the IMI concentration respect to the initial IMI concentration) was low even through photocatalysis [54–56], solar photo-Fenton [57], or UV-A photolysis [28]. In particular, with UV irradiation it is possible to obtain a complete photolysis of IMI after a long time of irradiation (about 10 h) [28], whereas in our precedent work [7] with molecularly imprinted TiO$_2$ samples, it was possible to selectively photodegrade IMI even in a pesticide mixture, although the degradation efficiency did not exceed 40% with a partial mineralization of ~35% (evaluated by the Total Organic Carbon, TOC, analysis) after 3 h of UV irradiation. Kitsiou et al. [57] found that the reaction efficiency can be improved utilizing a combination of photo-Fenton under UV-A irradiation and TiO$_2$, due to the synergism between the homogenous iron catalyst and the heterogeneous TiO$_2$ photocatalyst (~80% of degradation after 2 h of UV-A irradiation and ~60% of TOC mineralization), whereas only the solar homogenous photo-Fenton with iron reached ~50% for both degradation and removal of organic carbon after 3 h of UV-A irradiation. The most promising result was obtained by Sharma et al. [54] with a particular TiO$_2$ supported on mesoporous silica SBA-15, that allowed to achieve ~90% IMI degradation after 3 h of solar irradiation. In this contest, the obtained (photo)catalytic performances of CeO$_2$ for the degradation of IMI described in this work are in line with the results obtained with the TiO$_2$-based materials.

Figure 6a reports the photocatalytic degradation of the synthetized powders. In the test without catalysts, (black line in Figure 6a) no substantial variations in the initial concentration of IMI was measured, as expected. On the other hand, after 3 h of solar light irradiation the bare CeO$_2$ was able to degrade around the 20% of the initial concentration of IMI, whereas the grey CeO$_2$ showed a slightly lower performance (~16%). This can be reasonably explained considering the lower surface area and/or the slightly higher band-gap of grey CeO$_2$ with respect to the un-modified CeO$_2$.

The catalytic activity through the Fenton reaction (Figure 6b) is significantly lower compared to the photocatalytic tests (the test was carried out without irradiation). In these tests, no substantial degradation of IMI was measured in the run carried out without catalysts, but with H$_2$O$_2$ (Figure 6b, olive line).

As explained in the Introduction (see reactions 6-8), the Fenton process requires the presence and the fast regeneration of Ce^{3+} defect sites. For this reason, differently to the photocatalytic tests, the grey CeO$_2$ is more active than the bare CeO$_2$. As detected by Raman and XPS measurements, the un-modified CeO$_2$ exhibited a much lower presence of defect centres with respect to grey CeO$_2$. Conversely, despite the major presence of surface defects in the grey CeO$_2$, the degradation percentage measured on grey CeO$_2$ after 3 h of reaction in the Fenton-like test (~10%) was lower compared to the degradation efficiency of the photocatalytic test (~16%) obtained with the same sample, which pointed to the slow regeneration of the Ce^{3+} sites.

As reported, the Fenton-like reaction with CeO$_2$ involves the formation of peroxide species on the surface of ceria due to the complexation of H$_2$O$_2$ with Ce^{3+} sites [25,30,58]. These peroxide species are chemically stable and can saturate the surface of CeO$_2$, hindering the adsorption and subsequent oxidation of the organic target contaminant [25,30,58]. Indeed, a higher concentration of H$_2$O$_2$ (superior to 3%, i.e., 0,9 M) both in the Fenton and in the photo-Fenton like reactions led to a considerable decrease in the catalytic activity of the grey CeO$_2$ (the highest performing sample for these tests, Figure S1). The regeneration and the further formation of Ce^{3+} defect centres are, in this way, crucial steps.

Interestingly, in the photo-Fenton like reaction (Figure 6c), the grey CeO$_2$ displayed the best performance (~35% of degradation), comparing all the investigated AOPs with the two CeO$_2$ samples. The solar irradiation could boost the further formation of Ce^{3+} sites. An indirect confirmation is derived from the slight enhancement of the catalytic activity of bare CeO$_2$ (25%) compared to the solar photocatalytic test (20%) that can be attributed to the formation of in situ oxygen vacancies in the surface of bare CeO$_2$ which can react with the hydrogen peroxide.

As reported in the literature [25,32], the irradiation of ceria with photons which possess energy higher than the CeO$_2$ band-gap can exploit the following reaction:

$$CeO_2 + h\nu \ (E \geq E_g) \rightarrow Ce^{+3,+4}O_{2-x} + x/2 \ O_2 \quad (9)$$

The interaction of the highly energetic photons with the surface of CeO$_2$ leads to the loss of surface oxygen, thus allowing the formation of the Ce^{3+} states. The same reaction was exploited during the preparation of grey CeO$_2$, where the formation of Ce^{3+} was further increased due to the reducing atmosphere. Therefore, with the grey CeO$_2$, owing to a higher number of defective centres compared to bare ceria (as shown by XPS and Raman analyses), it is possible to reach the best performance in the degradation of IMI by the photo-Fenton-like reaction. Furthermore, the contemporaneous presence of the hydrogen peroxide and the solar irradiation enhances the formation of hydroxyl radicals through the photolytic decomposition of H$_2$O$_2$, as confirmed by the experiment carried out without a catalyst (H$_2$O$_2$ + IMI) that led to a slight variation in the initial concentration of IMI (Figure 6c, olive line).

It is important to highlight that, in the photocatalytic tests, the occurrence of the photon interaction (reaction 9, reported above) can be exploited, but it has a minor role in determining the final performance. The presence of Ce^{3+} was usually connected in the literature [59–61] to an improvement in the photocatalytic performance, especially under visible light irradiation, due to the presence of as-formed oxygen vacancies that shift the absorption of CeO$_2$ towards the visible-light region, improving the separation of the photogenerated charge carriers. However, the mean crystalline size and consequently the active surface area of the photocatalyst contribute considerably to the overall photocatalytic activity [34,62], as in our case, where the influence of the surface area is more preponderant than the effect of defects. For this reason, in the solar photocatalytic test, the bare CeO$_2$ (BET surface area of 81 m^2/g) was more active than grey CeO$_2$ (BET surface area of 67 m^2/g).

Finally, to test the reusability of grey CeO$_2$, different photo-Fenton like reaction runs were performed on the same sample. In Figure 6d, the variation in the kinetic constant (referring to a first order kinetic [7] is reported with respect to the various runs. After five runs, the kinetic constant decreases from 12 ± 1·10^{-4} min^{-1} to 5.5 ± 0.6·10^{-4} min^{-1}, highlighting that the continuous redox Ce^{3+}→Ce^{4+} process on the surface of grey ceria led to a progressive deactivation of the catalyst, reasonably for the saturation of the surface sites with the products of IMI degradation. Nevertheless, when the same sample were pre-treated before the tests in the H$_2$ flow at room temperature for 1 h, it was possible to exploit an almost total reversibility of grey CeO$_2$ in the photo-Fenton-like reaction. In fact, the kinetic constant raised to 10 ± 1·10^{-4} min^{-1} in run 6 and went back to 5.0 ± 0.5·10^{-4} min^{-1} after the subsequent other four runs, pointing to the crucial role of H$_2$ in the restoring the Ce^{3+} sites on grey CeO$_2$.

The comparison with the commercial TiO$_2$ P25 Degussa (Figure 7) showed, as in the photocatalytic test, that bare CeO$_2$ had a comparable catalytic behaviour with respect to TiO$_2$ (Figure 7a), whereas this latter sample exhibited no substantial activity in the Fenton-like test (Figure 7b), and a lower activity compared to CeO$_2$ and grey CeO$_2$ in the photo-Fenton test (Figure 7c). This pointed out that, both for Fenton and photo-Fenton like reactions, the CeO$_2$-based materials are better-performing than the commercial TiO$_2$. As reported [63,64], the bare TiO$_2$ without structural (i.e., incorporation of surface defects) or chemical (as the formation of composites with iron oxides) modifications is not able to promote Fenton-like reactions.

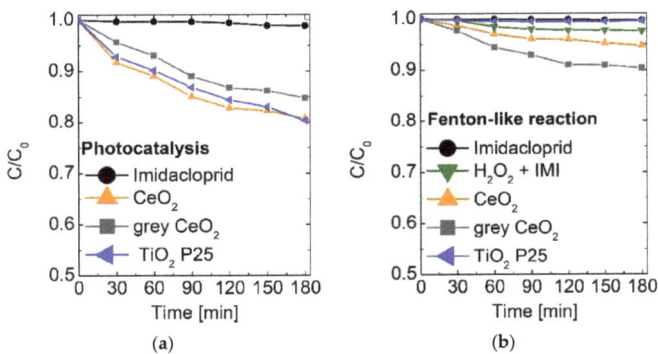

Figure 7. Cont.

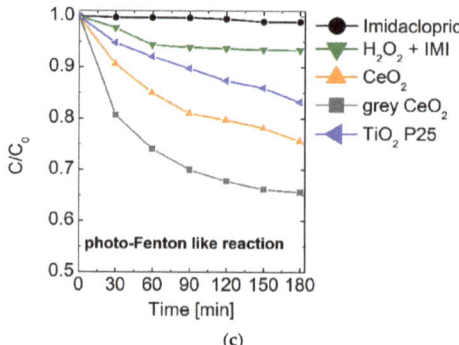

Figure 7. (a) Photocatalytic degradation of IMI under solar light irradiation, (b) Fenton-like reaction, (c) photo-Fenton like reaction on the analyzed samples.

These data demonstrate the possibility of modifying and tuning the physico-chemical properties of CeO_2 with simple treatments, such as the solar light irradiation in a H_2 stream, so to maximize the catalytic performance. It is important to highlight, finally, that the CeO_2 sample simply treated with H_2 or irradiated with a solar lamp without an H_2 stream did not show substantial changes compared to bare CeO_2 in the degradation performance of IMI in all the AOPs investigated.

2.3. Tocixity Tests

Artemia salina dehydrated cysts were employed for the acute toxicity test.

Artemia salina nauplii can readily ingest fine particles smaller than 50 µm [65], and it is a non-selective filter-feeder organism. For these reasons, it is currently considered as a good model organism to assess in vivo nanoparticles toxicity, as previously demonstrated [66].

Low mortality percentages were evidenced after 24 and 48 h of exposure (Table 1), at different concentrations of both powders (bare CeO_2 and grey CeO_2). Statistical analysis, carried out by one-way ANOVA test, gave no significant values for all the immobilization percentages nor treated groups after 24 and 48 h of exposure nor between treated and control (Ctrl, i.e., without metal oxide particles) groups ($p > 0.05$). The percentages of immobilized nauplii are reported in the Table 1. These data pointed to the low critical toxicity of the examined powders. Furthermore, it is possible to note that the modification of CeO_2 led to have a lower mortality with respect to the bare CeO_2.

Table 1. Percentages of immobilized nauplii after exposition to CeO_2 (bare CeO_2 and grey CeO_2) at three different concentrations for 24 and 48 h.

Sample	Ctrl	10^{-1}	10^{-2}	10^{-3}
Bare CeO_2	1.6% (24 h)	11.6% (24 h)	6.6% (24 h)	3.3% (24 h)
	6.6% (48 h)	28.3% (48 h)	23.3% (48 h)	21.6% (48 h)
Grey CeO_2	3.3% (24 h)	8.3 % (24 h)	6.6% (24 h)	4.0% (24 h)
	8.3% (48 h)	23.3% (48 h)	18.3% (48 h)	15.0% (48 h)

Figure 8 shows *Artemia salina* nauplii treated with bare CeO_2 for 24 h, 48 h and untreated nauplii (i.e., the controls).

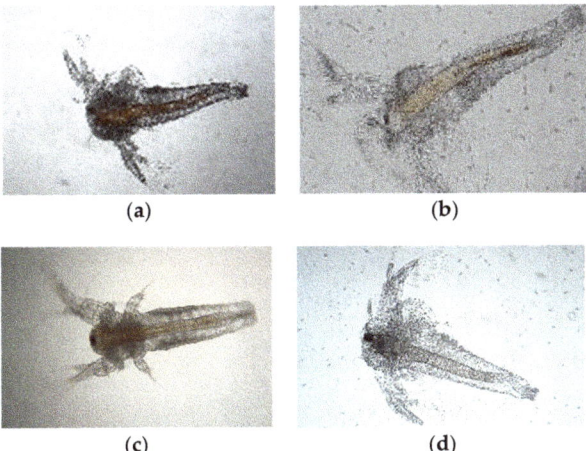

Figure 8. *Artemia salina* nauplii: nauplii exposed to bare CeO$_2$ for 24 h (**a**), nauplii exposed to bare CeO$_2$ for 48 h (**b**), control at 24 h (**c**), control at 48 h (**d**).

3. Materials and Methods

3.1. Samples Preparation

The bare CeO$_2$ was prepared via chemical precipitation from Ce(NO$_3$)$_3$·6H$_2$O (Fluka, Buchs, Switzerland) at pH > 8 utilizing a solution of KOH (1M, Fluka, Buchs, Switzerland). The obtained slurry was maintained under stirring at 80 °C for 3 h. After digestion for 24 h, it was filtered, washed with deionized water several times, and dried at 100 °C for 12 h. Finally, the powders were calcined in air at 450 °C for 3 h. The modified CeO$_2$ (grey CeO$_2$) was obtained with the same synthetic procedure reported above, but exposing the powders after calcination to the light of a solar lamp for 3 h (OSRAM Vitalux 300 W, 300–2000 nm; OSRAM Opto Semiconductors GmbH, Leibniz, Regensburg Germany) in a hydrogen stream (20 cc/min) at room temperature.

3.2. Samples Characterization

X-ray powder diffraction (XRD) measures were performed with a PANalytical X'pertPro X-ray diffractometer (Malvern PANalytical, Enigma Business Park, Grovewood Road, Malvern United Kingdom), employing a Cu Kα radiation. The identification of the crystalline phases was made comparing the diffractions with those of standard materials reported in the JCPDS Data File. Raman spectra were carried out with a WITec alpha 300 confocal Raman system (WITec Wissenschaftliche Instrumente und Technologie GmbH Ulm, Germany) with an excitation source at 532 nm under the same experimental condition reported in the ref. [67]. Fourier Transform Infrared Spectroscopy (FTIR) spectra were obtained in the range 4000–400 cm^{-1} using a Perkin Elmer FT-IR System 2000 (Perkin-Elmer, Waltham, MA, USA). The background spectrum was carried out with KBr. The textural properties of the samples were measured by N$_2$ adsorption–desorption at −196 °C with a Micromeritics Tristar II Plus 3020 (Micromeritics Instrument Corp. Norcross, USA), out-gassing the analysed materials at 100 °C overnight. UV-Vis-Diffuse Reflectance (UV-Vis DRS) spectra were obtained in the range of 200–800 nm using a Cary 60 spectrometer (Agilent Stevens Creek Blvd. Santa Clara, United States). X-ray photoelectron spectroscopy (XPS) measurements were recorded using a K-Alpha X-ray photoelectron spectrometer (Thermo Fisher Scientific Waltham, MA USA), utilizing the C 1 peak at 284.9 eV (ascribed to adventitious carbon) as a reference.

3.3. (Photo)catalytic Experiments

The photocatalytic tests were performed utilizing a solar lamp (OSRAM Vitalux 300 W, 300–2000 nm, OSRAM Opto Semiconductors GmbH, Leibniz, Regensburg Germany) irradiating a jacketed Pyrex batch reactor, kept at 25 °C. A total of 50 mg of powder was suspended in 50 mL of the reactant solution containing 5×10^{-5} M of IMI (Sigma-Aldrich, Buchs, Switzerland). The reaction mixture was stirred for 120 min in the dark so as to achieve the adsorption/desorption equilibrium. During the tests, aliquots of the suspension were withdrawn at a given time interval to measure the IMI concentration by means of Cary 60 UV–vis spectrophotometer (Agilent Stevens Creek Blvd. Santa Clara, United States). The IMI degradation was evaluated by following the absorbance peaks at 270 nm in the Lambert–Beer regime, reporting the C/C_0 ratio as a function of time t, where C is the concentration of the contaminant at the time t, while C_0 is the starting concentration of the pollutant. The Fenton-like reaction was carried out with the same apparatus described above, adding 5 mL of hydrogen peroxide (3%, 0.9 M Fluka, Buchs, Switzerland) in the reactor without irradiation; in the photo-Fenton-like tests the solar light irradiation was employed, too. In all the catalytic tests the experimental error was 1%, i.e., within the symbol size.

3.4. Toxicity Tests

Artemia salina dehydrated cysts were used for the acute toxicity test. Cysts (Hobby, Germany) were hydrated in ASPM seawater solution (ASPM is an artificial seawater made of: NaCl = 26.4 g, KCl = 0.84 g, $CaCl_2 \cdot H_2O$ = 1.67 g, $MgCl \cdot H_2O$ = 4.6 g, $MgSO_4 \cdot 7H_2O$ = 5.58 g, $NaHCO_3$ = 0.17 g, and H_3BO_3 = 0.03 g) maintaining standard laboratory conditions (1.500 lux daylight; 26 ± 1 °C; continuous aeration), then nauplii hatched within 24 h.

Two stock solutions of CeO_2 (bare CeO_2 and grey CeO_2) were prepared after dilution in ASPM solution. Then, fresh suspensions with different concentrations of powders (10^{-1}, 10^{-2}, 10^{-3} mg/mL) were made starting from the stock suspensions (1 mg/10 mL). These solutions were vortexed for 30 s. One nauplius per well in 96-well microplates, was added with 200 µL of each different concentration of powder solutions. They were incubated at 26 °C for 24/48 h. The number of surviving nauplii in each well was counted under a stereomicroscope after 24/48 h. A control group was also set up with ASPM seawater solution only. Larvae were not fed during the bioassays.

At the end of the test, the endpoints (immobility, i.e., death) were evaluated with a stereomicroscope (Leica EZ4, Leica Microsystems Srl, Buccinasco (MI), Italy): a nauplium was considered to be immobile or dead if it could not move its antennae after slight agitation of the water for 10 seconds. Larvae that were completely motionless were counted as dead, and the percentages of mortality compared to the control were calculated. The death % of the crustacean for each concentration was calculated as follows: (n. dead nauplii/n. total animal treated 100). Data were analyzed for differences between the control and treatments using one-way ANOVA followed by Tukey's test, where $p < 0.05$ is considered significant and $p < 0.01$ extremely significant.

4. Conclusions

Different AOPs (photocatalysis, Fenton and photo-Fenton reactions) were investigated for the degradation of the IMI insecticide. Two CeO_2 samples were tested, one prepared with the conventional precipitation route and another one modifying the as-prepared CeO_2 by exposing it to solar light irradiation in a hydrogen stream. The latter treatment allowed to obtain a more defective ceria with increased performance in the solar photo-Fenton reaction. This hybrid AOP obtained the best degradation rate of the IMI. The photon interaction with the surface of CeO_2 led to a loss of oxygen with the formation of Ce^{3+} centers, which is essential to boost the degradation rate of the pesticide degradation through the photo-Fenton process. The solar irradiation in a reducing atmosphere could obtain a defective ceria that can be exploited to explore new synergisms between different AOPs for wastewater purification over non-conventional photocatalysts/Fenton materials. Furthermore, the investigated materials did not exhibit critical toxicity.

Supplementary Materials: The following are available online at http://www.mdpi.com/2073-4344/10/4/446/s1, Figure S1: Influence of the H_2O_2 concentration.

Author Contributions: Conceptualization R.F. and G.I.; Catalytic tests R.F. and S.A.B.; Investigations and writing R.F.; Raman measurements L.D.; Toxicity tests M.V.B.; R.P. and E.M.S.; Review and editing G.I., S.S. and V.P. All authors have read and agreed to the current version of the manuscript.

Funding: This work was partially founded by the Micro WatTS project, Interreg V. A. Italia-Malta (CUP: B61G18000070009). R.F. thanks the PON project "AIM" founded by the European Social Found (ESF) CUP: E66C18001220005 for the financial support. R.P. thanks the PON project "AIM" founded by the European Social Found (ESF) CUP: E66C18001300007 for the financial support.

Conflicts of Interest: The authors declare no conflict of interest.

References

1. Salimi, M.; Esrafili, A.; Gholami, M.; Jonidi Jafari, A.; Rezaei Kalantary, R.; Farzadkia, M.; Kermani, M.; Sobhi, H.R. Contaminants of emerging concern: A review of new approach in AOP technologies. *Environ. Monit. Assess.* **2017**, *189*, 414. [CrossRef] [PubMed]
2. Katsikantami, I.; Colosio, C.; Alegakis, A.; Tzatzarakis, M.N.; Vakonaki, E.; Rizos, A.K.; Sarigiannis, D.A.; Tsatsakis, A.M. Estimation of daily intake and risk assessment of organophosphorus pesticides based on biomonitoring data—The internal exposure approach. *Food Chem. Toxicol.* **2019**, *123*, 57–71. [CrossRef] [PubMed]
3. Malakootian, M.; Shahesmaeili, A.; Faraji, M.; Amiri, H.; Silva Martinez, S. Advanced oxidation processes for the removal of organophosphorus pesticides in aqueous matrices: A systematic review and meta-analysis. *Process Saf. Environ. Prot.* **2020**, *134*, 292–307. [CrossRef]
4. Mazivila, S.J.; Ricardo, I.A.; Leitão, J.M.M.; Esteves da Silva, J.C.G. A review on advanced oxidation processes: From classical to new perspectives coupled to two- and multi-way calibration strategies to monitor degradation of contaminants in environmental samples. *Trends Environ. Anal. Chem.* **2019**, *24*, e00072. [CrossRef]
5. Dewil, R.; Mantzavinos, D.; Poulios, I.; Rodrigo, M.A. New perspectives for Advanced Oxidation Processes. *J. Environ. Manag.* **2017**, *195*, 93–99. [CrossRef] [PubMed]
6. Giménez, J.; Bayarri, B.; González, Ó.; Malato, S.; Peral, J.; Esplugas, S. Advanced Oxidation Processes at Laboratory Scale: Environmental and Economic Impacts. *ACS Sustain. Chem. Eng.* **2015**, *3*, 3188–3196. [CrossRef]
7. Fiorenza, R.; Di Mauro, A.; Cantarella, M.; Iaria, C.; Scalisi, E.M.; Brundo, M.V.; Gulino, A.; Spitaleri, L.; Nicotra, G.; Dattilo, S.; et al. Preferential removal of pesticides from water by molecular imprinting on TiO_2 photocatalysts. *Chem. Eng. J.* **2020**, *379*, 122309. [CrossRef]
8. Fiorenza, R.; Di Mauro, A.; Cantarella, M.; Privitera, V.; Impellizzeri, G. Selective photodegradation of 2,4-D pesticide from water by molecularly imprinted TiO_2. *J. Photochem. Photobiol. A Chem.* **2019**, *380*. [CrossRef]
9. Fiorenza, R.; Di Mauro, A.; Cantarella, M.; Gulino, A.; Spitaleri, L.; Privitera, V.; Impellizzeri, G. Molecularly imprinted N-doped TiO_2 photocatalysts for the selective degradation of o-phenylphenol fungicide from water. *Mater. Sci. Semicond. Process.* **2020**, *112*, 105019. [CrossRef]
10. Cantarella, M.; Di Mauro, A.; Gulino, A.; Spitaleri, L.; Nicotra, G.; Privitera, V.; Impellizzeri, G. Selective photodegradation of paracetamol by molecularly imprinted ZnO nanonuts. *Appl. Catal. B Environ.* **2018**, *238*, 509–517. [CrossRef]
11. Iervolino, G.; Zammit, I.; Vaiano, V.; Rizzo, L. Limitations and Prospects for Wastewater Treatment by UV and Visible-Light-Active Heterogeneous Photocatalysis: A Critical Review. *Top. Curr. Chem.* **2020**, *378*. [CrossRef] [PubMed]
12. Herney-Ramirez, J.; Vicente, M.A.; Madeira, L.M. Heterogeneous photo-Fenton oxidation with pillared clay-based catalysts for wastewater treatment: A review. *Appl. Catal. B Environ.* **2010**, *98*, 10–26. [CrossRef]
13. Babuponnusami, A.; Muthukumar, K. A review on Fenton and improvements to the Fenton process for wastewater treatment. *J. Environ. Chem. Eng.* **2014**, *2*, 557–572. [CrossRef]
14. Liu, X.; Zhou, Y.; Zhang, J.; Luo, L.; Yang, Y.; Huang, H.; Peng, H.; Tang, L.; Mu, Y. Insight into electro-Fenton and photo-Fenton for the degradation of antibiotics: Mechanism study and research gaps. *Chem. Eng. J.* **2018**, *347*, 379–397. [CrossRef]
15. Dutta, A.; Das, N.; Sarkar, D.; Chakrabarti, S. Development and characterization of a continuous solar-collector-reactor for wastewater treatment by photo-Fenton process. *Sol. Energy* **2019**, *177*, 364–373. [CrossRef]

16. De la Obra, I.; Ponce-Robles, L.; Miralles-Cuevas, S.; Oller, I.; Malato, S.; Sánchez Pérez, J.A. Microcontaminant removal in secondary effluents by solar photo-Fenton at circumneutral pH in raceway pond reactors. *Catal. Today* **2017**, *287*, 10–14. [CrossRef]
17. Sciré, S.; Fiorenza, R.; Gulino, A.; Cristaldi, A.; Riccobene, P.M. Selective oxidation of CO in H_2-rich stream over ZSM5 zeolites supported Ru catalysts: An investigation on the role of the support and the Ru particle size. *Appl. Catal. A Gen.* **2016**, *520*, 82–91. [CrossRef]
18. Fiorenza, R.; Spitaleri, L.; Gulino, A.; Sciré, S. High-Performing Au-Ag Bimetallic Catalysts Supported on Macro-Mesoporous CeO_2 for Preferential Oxidation of CO in H_2-Rich Gases. *Catalysts* **2020**, *10*, 49. [CrossRef]
19. Fiorenza, R.; Condorelli, M.; D'Urso, L.; Compagnini, G.; Bellardita, M.; Palmisano, L.; Sciré, S. Catalytic and Photothermo-catalytic Applications of TiO_2-CoO_x Composites. *J. Photocatal.* **2020**, *1*. [CrossRef]
20. Scuderi, V.; Amiard, G.; Sanz, R.; Boninelli, S.; Impellizzeri, G.; Privitera, V. TiO_2 coated CuO nanowire array: Ultrathin p–n heterojunction to modulate cationic/anionic dye photo-degradation in water. *Appl. Surf. Sci.* **2017**, *416*, 885–890. [CrossRef]
21. Di Mauro, A.; Fragalà, M.E.; Privitera, V.; Impellizzeri, G. ZnO for application in photocatalysis: From thin films to nanostructures. *Mater. Sci. Semicond. Process.* **2017**, *69*, 44–51. [CrossRef]
22. Vela, N.; Calín, M.; Yáñez-Gascón, M.J.; el Aatik, A.; Garrido, I.; Pérez-Lucas, G.; Fenoll, J.; Navarro, S. Removal of Pesticides with Endocrine Disruptor Activity in Wastewater Effluent by Solar Heterogeneous Photocatalysis Using ZnO/Na2S2O8. *Water Air Soil Pollut.* **2019**, *230*, 134. [CrossRef]
23. Bellardita, M.; Fiorenza, R.; Palmisano, L.; Sciré, S. Photocatalytic and photo-thermo-catalytic applications of cerium oxide based material. In *Cerium Oxide (CeO_2): Synthesis, Properties and Applications*; a Volume in Metal Oxides Series; Elsevier: Amsterdam, The Netherlands, 2020; ISBN 978-0-12-815661-2. [CrossRef]
24. Heckert, E.G.; Seal, S.; Self, W.T. Fenton-like reaction catalyzed by the rare earth inner transition metal cerium. *Environ. Sci. Technol.* **2008**, *42*, 5014–5019. [CrossRef] [PubMed]
25. Issa Hamoud, H.; Azambre, B.; Finqueneisel, G. Reactivity of ceria–zirconia catalysts for the catalytic wet peroxidative oxidation of azo dyes: Reactivity and quantification of surface Ce(IV)-peroxo species. *J. Chem. Technol. Biotechnol.* **2016**, *91*, 2462–2473. [CrossRef]
26. Zhang, N.; Tsang, E.P.; Chen, J.; Fang, Z.; Zhao, D. Critical role of oxygen vacancies in heterogeneous Fenton oxidation over ceria-based catalysts. *J. Colloid Interface Sci.* **2020**, *558*, 163–172. [CrossRef] [PubMed]
27. Hayat, W.; Zhang, Y.; Hussain, I.; Huang, S.; Du, X. Comparison of radical and non-radical activated persulfate systems for the degradation of imidacloprid in water. *Ecotoxicol. Environ. Saf.* **2020**, *188*, 109891. [CrossRef] [PubMed]
28. Wamhoff, H.; Schneider, V. Photodegradation of imidacloprid. *J. Agric. Food Chem.* **1999**, *47*, 1730–1734. [CrossRef] [PubMed]
29. Tišler, T.; Jemec, A.; Mozetič, B.; Trebše, P. Hazard identification of imidacloprid to aquatic environment. *Chemosphere* **2009**, *76*, 907–914. [CrossRef]
30. Zang, C.; Yu, K.; Hu, S.; Chen, F. Adsorption-depended Fenton-like reaction kinetics in CeO_2-H_2O_2 system for salicylic acid degradation. *Colloids Surf. A Physicochem. Eng. Asp.* **2018**, *553*, 456–463. [CrossRef]
31. Xu, L.; Wang, J. Magnetic nanoscaled Fe_3O_4/CeO_2 composite as an efficient fenton-like heterogeneous catalyst for degradation of 4-chlorophenol. *Environ. Sci. Technol.* **2012**, *46*, 10145–10153. [CrossRef]
32. Aslam, M.; Qamar, M.T.; Soomro, M.T.; Ismail, I.M.I.; Salah, N.; Almeelbi, T.; Gondal, M.A.; Hameed, A. The effect of sunlight induced surface defects on the photocatalytic activity of nanosized CeO_2 for the degradation of phenol and its derivatives. *Appl. Catal. B Environ.* **2016**, *180*, 391–402. [CrossRef]
33. Fiorenza, R.; Spitaleri, L.; Gulino, A.; Sciré, S. Ru–Pd bimetallic catalysts supported on CeO_2-MnO_x oxides as efficient systems for H_2 purification through CO preferential Oxidation. *Catalysts* **2018**, *8*, 203. [CrossRef]
34. Fiorenza, R.; Bellardita, M.; Barakat, T.; Sciré, S.; Palmisano, L. Visible light photocatalytic activity of macro-mesoporous TiO_2-CeO_2 inverse opals. *J. Photochem. Photobiol. A Chem.* **2018**, *352*, 25–34. [CrossRef]
35. Choudhury, B.; Choudhury, A. Ce^{3+} and oxygen vacancy mediated tuning of structural and optical properties of CeO_2 nanoparticles. *Mater. Chem. Phys.* **2012**, *131*, 666–671. [CrossRef]
36. Ma, L.; Wang, D.; Li, J.; Bai, B.; Fu, L.; Li, Y. Ag/CeO_2 nanospheres: Efficient catalysts for formaldehyde oxidation. *Appl. Catal. B Environ.* **2014**, *148–149*, 36–43. [CrossRef]
37. Popovic, Z.V.; Dohcevic-Mitrovic, Z.; Konstantinovic, M.J.; Scepanovic, M. Raman scattering characterization of nanopowders and nanowires (rods). *J. Raman Spectrosc.* **2007**, *38*, 750–755. [CrossRef]

38. Natile, M.M.; Boccaletti, G.; Glisenti, A. Properties and reactivity of nanostructured CeO_2 powders: Comparison among two synthesis procedures. *Chem. Mater.* **2005**, *17*, 6272–6286. [CrossRef]
39. Davydov, A. *Molecular Spectroscopy of Oxide Catalyst Surfaces*; Sheppard, N.T., Ed.; JohnWiley & Sons Ltd.: Chichester, UK, 2003.
40. Liu, B.; Li, C.; Zhang, Y.; Liu, Y.; Hu, W.; Wang, Q.; Han, L.; Zhang, J. Investigation of catalytic mechanism of formaldehyde oxidation over three-dimensionally ordered macroporous Au/CeO_2 catalyst. *Appl. Catal. B Environ.* **2012**, *111*, 467–475. [CrossRef]
41. Finos, G.; Collins, S.; Blanco, G.; Del Rio, E.; Cíes, J.M.; Bernal, S.; Bonivardi, A. Infrared spectroscopic study of carbon dioxide adsorption on the surface of cerium-gallium mixed oxides. *Catal. Today* **2012**, *180*, 9–18. [CrossRef]
42. Sing, K.S.W.; Everet, D.H.; Haul, R.A.W. Reporting Physisorption Data for gas/solid system with Special Reference to the Determination of Surface Area and Porosity. *Pure Appl. Chem.* **1985**, *57*, 603–619. [CrossRef]
43. Verma, R.; Samdarsh, S.K.; Bojja, S.; Paul, S.; Choudhury, B. A novel thermophotocatalyst of mixed-phase cerium oxide (CeO_2/Ce_2O_3) homocomposite nanostructure: Role of interface and oxygen vacancies. *Sol. Energy Mater. Sol. Cells* **2015**, *141*, 414–422. [CrossRef]
44. Jiang, D.; Wang, W.; Zhang, L.; Zheng, Y.; Wang, Z. Insights into the Surface-Defect Dependence of Photoreactivity over CeO_2 Nanocrystals with Well-Defined Crystal Facets. *ACS Catal.* **2015**, *5*, 4851–4858. [CrossRef]
45. Mena, E.; Rey, A.; Rodríguez, E.M.; Beltrán, F.J. Nanostructured CeO_2 as catalysts for different AOPs based in theapplication of ozone and simulated solar radiation. *Catal. Today* **2017**, *280*, 74–79. [CrossRef]
46. Kim, Y.I.; Atherton, S.J.; Brigham, E.S.; Mallouk, T.E. Sensitized layered metal oxide semiconductor particles for photochemical hydrogen evolution from nonsacrificial electron donors. *J. Phys. Chem.* **1993**, *97*, 11802–11810. [CrossRef]
47. Gao, X.; Jiang, Y.; Zhong, Y.; Luo, Z.; Cen, K. The activity and characterization of CeO_2-TiO_2 catalysts prepared by the sol—Gel method for selective catalytic reduction of NO with NH_3. *J. Hazard. Mater.* **2010**, *174*, 734–739. [CrossRef] [PubMed]
48. Gamboa-Rosales, N.K.; Ayastuy, J.L.; González-Marcos, M.P.; Gutiérrez-Ortiz, M.A. Oxygen-enhanced water gas shift over ceria-supported Au-Cu bimetallic catalysts prepared by wet impregnation and deposition-precipitation. *Int. J. Hydrogen Energy* **2012**, *37*, 7005–7016. [CrossRef]
49. Fiorenza, R.; Crisafulli, C.; Condorelli, G.G.; Lupo, F.; Scirè, S. Au-Ag/CeO_2 and Au-Cu/CeO_2 Catalysts for Volatile Organic Compounds Oxidation and CO Preferential Oxidation. *Catal. Lett.* **2015**, *145*, 1691–1702. [CrossRef]
50. Bêche, E.; Charvin, P.; Perarnau, D.; Abanades, S.; Flamant, G. Ce 3d XPS investigation of cerium oxides and mixed cerium oxide ($Ce_xTi_yO_z$). *Surf. Interface Anal.* **2008**, *40*, 264–267. [CrossRef]
51. Ahmed, S.; Rasul, M.G.; Brown, R.; Hashib, M.A. Influence of parameters on the heterogeneous photocatalytic degradation of pesticides and phenolic contaminants in wastewater: A short review. *J. Environ. Manag.* **2011**, *92*, 311–330. [CrossRef]
52. Farré, M.J.; Franch, M.I.; Malato, S.; Ayllón, J.A.; Peral, J.; Doménech, X. Degradation of some biorecalcitrant pesticides by homogeneous and heterogeneous photocatalytic ozonation. *Chemosphere* **2005**, *58*, 1127–1133. [CrossRef]
53. Verma, A.; Toor, A.P.; Prakash, N.T.; Bansal, P.; Sangal, V.K. Stability and durability studies of TiO_2 coated immobilized system for the degradation of imidacloprid. *New J. Chem.* **2017**, *41*, 6296–6304. [CrossRef]
54. Sharma, M.V.P.; Kumari, D.V.; Subrahmanyam, M. TiO_2 supported over SBA-15: An efficient photocatalyst for the pesticide degradation using solar light. *Chemosphere* **2008**, *73*, 1562–1569. [CrossRef] [PubMed]
55. Agüera, A.; Almansa, E.; Malato, S.; Maldonado, M.I.; Fernández-Alba, A.R. Evaluation of photocatalytic degradation of Imidacloprid in industrial water by GC-MS and LC-MS. *Analusis* **1998**, *26*, 245–251. [CrossRef]
56. Redlich, D.; Shahin, N.; Ekici, P.; Friess, A.; Parlar, H. Kinetic study of the photoinduced degradation of imidacloprid in aquatic media. *Clean Soil Air Water* **2007**, *35*, 452–458. [CrossRef]
57. Kitsiou, V.; Filippidis, N.; Mantzavinos, D.; Poulios, I. Heterogeneous and homogeneous photocatalytic degradation of the insecticide imidacloprid in aqueous solutions. *Appl. Catal. B Environ.* **2009**, *86*, 27–35. [CrossRef]
58. Zang, C.; Zhang, X.; Hu, S.; Chen, F. The role of exposed facets in the Fenton-like reactivity of CeO_2 nanocrystal to the Orange II. *Appl. Catal. B Environ.* **2017**, *216*, 106–113. [CrossRef]
59. Yuan, S.; Xu, B.; Zhang, Q.; Liu, S.; Xie, J.; Zhang, M.; Ohno, T. Development of the Visible-Light Response of CeO_{2-x} with a high Ce^{3+} Content and Its Photocatalytic Properties. *ChemCatChem* **2018**, *10*, 1267–1271. [CrossRef]

60. Xiu, Z.; Xing, Z.; Li, Z.; Wu, X.; Yan, X.; Hu, M.; Cao, Y.; Yang, S.; Zhou, W. Ti^{3+}-TiO_2/Ce^{3+}-CeO_2 Nanosheet heterojunctions as efficient visible-light driven photocatalysts. *Mater. Res.* **2018**, *100*, 191–197. [CrossRef]
61. Saravanan, R.; Agarwal, S.; Gupta, V.K.; Khan, M.M.; Gracia, F.; Mosquera, E.; Narayanan, V.; Stephen, A. Line defect Ce^{3+} induced Ag/CeO_2/ZnO nanostructure for visible-light photocatalytic activity. *J. Photochem. Photobiol. A Chem.* **2018**, *353*, 499–506. [CrossRef]
62. Černigoj, U.; Štangar, U.L.; Jirkovský, J. Effect of dissolved ozone or ferric ions on photodegradation of thiacloprid in presence of different TiO_2 catalysts. *J. Hazard. Mater.* **2010**, *177*, 399–406. [CrossRef] [PubMed]
63. Zhang, A.Y.; Lin, T.; He, Y.Y.; Mou, Y.X. Heterogeneous activation of H_2O_2 by defect-engineered TiO_{2-x} single crystals for refractory pollutants degradation: A Fenton-like mechanism. *J. Hazard. Mater.* **2016**, *311*, 81–90. [CrossRef] [PubMed]
64. Du, D.; Shi, W.; Wang, L.; Zhan, J. Yolk-shell structured Fe_3O_4@void@TiO_2 as a photo-Fenton-like catalyst for the extremely efficient elimination of tetracycline. *Appl. Catal. B Environ.* **2017**, *200*, 484–492. [CrossRef]
65. Zhu, X.; Chang, Y.; Chen, Y. Toxicity and bioaccumulation of TiO_2 nanoparticle aggregates in *Daphnia magna*. *Chemosphere* **2010**, *78*, 209–215. [CrossRef] [PubMed]
66. Cantarella, M.; Gorrasi, G.; Di Mauro, A.; Scuderi, M.; Nicotra, G.; Fiorenza, R.; Scirè, S.; Scalisi, M.E.; Brundo, M.V.; Privitera, V.; et al. Mechanical milling: A sustainable route to induce structural transformations in MoS_2 for applications in the treatment of contaminated water. *Sci. Rep.* **2019**, *9*, 974. [CrossRef] [PubMed]
67. Fiorenza, R.; Bellardita, M.; D'Urso, L.; Compagnini, G.; Palmisano, L.; Scirè, S. Au/TiO_2-CeO_2 catalysts for photocatalytic water splitting and VOCs oxidation reactions. *Catalysts* **2016**, *6*, 121. [CrossRef]

© 2020 by the authors. Licensee MDPI, Basel, Switzerland. This article is an open access article distributed under the terms and conditions of the Creative Commons Attribution (CC BY) license (http://creativecommons.org/licenses/by/4.0/).

Hydrothermal and Co-Precipitated Synthesis of Chalcopyrite for Fenton-like Degradation toward Rhodamine B

Po-Yu Wen [1], Ting-Yu Lai [1], Tsunghsueh Wu [2,*] and Yang-Wei Lin [1,*]

1. Department of Chemistry, National Changhua University of Education, 1 Jin-De Road, Changhua City 50007, Taiwan; bba02123@gmail.com (P.-Y.W.); superhuman860731@gmail.com (T.-Y.L.)
2. Department of Chemistry, University of Wisconsin-Platteville, 1 University Plaza, Platteville, WI 53818-3099, USA
* Correspondence: wut@uwplatt.edu (T.W.); linywjerry@cc.ncue.edu.tw (Y.-W.L.); Tel.: +1-608-342-6018 (T.W.); +886-4-7232105-3553 (Y.-W.L.)

Abstract: In this study, Chalcopyrite ($CuFeS_2$) was prepared by a hydrothermal and co-precipitation method, being represented as H-$CuFeS_2$ and C-$CuFeS_2$, respectively. The prepared $CuFeS_2$ samples were characterized by scanning electron microscope (SEM), transmission electron microscope (TEM), energy dispersive X-ray spectroscopy mapping (EDS-mapping), powder X-ray diffractometer (XRD), X-ray photoelectron spectrometry (XPS), and Raman microscope. Rhodamine B (RhB, 20 ppm) was used as the target pollutant to evaluate the degradation performance by the prepared $CuFeS_2$ samples. The H-$CuFeS_2$ samples (20 mg) in the presence of $Na_2S_2O_8$ (4 mM) exhibited excellent degradation efficiency (98.8% within 10 min). Through free radical trapping experiment, the major active species were $\bullet SO_4^-$ radicals and $\bullet OH$ radicals involved the RhB degradation. Furthermore, $\bullet SO_4^-$ radicals produced from the prepared samples were evaluated by iodometric titration. In addition, one possible degradation mechanism was proposed. Finally, the prepared H-$CuFeS_2$ samples were used to degrade different dyestuff (rhodamine 6G, methylene blue, and methyl orange) and organic pollutant (bisphenol A) in the different environmental water samples (pond water and seawater) with 10.1% mineral efficiency improvement comparing to traditional Fenton reaction.

Keywords: hydrothermal preparation; co-precipitation; $CuFeS_2$; Fenton-like reaction; degradation; environmental water samples

1. Introduction

Since the industrial revolution, the development of various industries has made life in human society more convenient, but has also caused many environmental problems. Wastewater, such as cooling water and clean water for equipment, is discharged from various industrial processes. The constituents in any wastewater are diverse and complex consisting of raw materials, intermediate products, by-products, and end products. Charging these compounds directly to the environment can have detrimental consequences. For example, oxygen-containing organic compounds such as aldehydes, ketones, and ethers are reductive, meaning that they are capable of consuming dissolved oxygen in the water to a low-level endangering aquatic organisms. Wastewater can also contain a large amount of nitrogen, phosphorus, and potassium which can promote the growth of algae and triggering eutrophication pollution in water bodies [1]. Released toxic substances from wastewater can bio-accumulate in fish and eventually pass to people who consumed it. Thus, wastewater treatment is an important process to avoid these consequences. Within the treatment options, physical, biological, and chemical methods are mainly used to treat wastewater by removing pollutants in the water and reducing organic pollutants and eutrophic substances in the water [2–8].

Among many treatment processes, the in-site chemical oxidation method is to inject and mix oxidants into the underground environment aiming to degrade pollutants in groundwater and soil [9–11]. Under ideal conditions, this chemical treatment c convert organic pollutants into less toxic molecules, such as carbon dioxide, water, a inorganic salts. The commonly used oxidants are permanganate (MnO_4^-), Fenton reage (Fe^{2+}/H_2O_2), and ozone (O_3) [12–15]. The mechanism of Fenton reagent (Fe^{2+}/H_2O_2) reactions has been known to produce •OH radicals, which can cleave C–H bonds of orga compounds, turning them into environmentally benign final products [16,17]. Howev this Fenton method has some drawbacks including the specific pH working range a Fe sludge precipitation at the end of the Fenton reaction [18,19]. Another alternative w to generate radicals with a wider pH working range is by using persulfate salts [20–2 Persulfate salts can come from two types: peroxymonosulfate (HSO_5^-) and peroxydisulfa ($S_2O_8^{2-}$), both of which contain an O–O bond (peroxide group) capable of generati •SO_4^- radicals and •OH radicals in Fenton-like reaction for degradation of organic co pounds [27,28]. Persulfate salts are strong oxidant ($E^0 = 2.1$ V), yet they are very stable transportation and prolong storage making them very attractive oxidants for undergrou water treatment [29].

There are a few ways of activating persulfate to generate radicals, such as therm decomposition, alkaline activation, transition metal ions activation, and heterogeneo catalysis [30]. Among them all, transition metal ions activation is considered the simple and most benign method with no external energy requirement and recyclability of transiti metal ions [31–34]. Cobalt ions are commonly used in activating persulfate in resear but their hazardous nature makes them unsuitable for water treatment [35,36]. Thus, it necessary to find an alternative transition metal catalyst that can be used in water treatme

Recently, Cu/Fe-bearing solids such as chalcopyrite ($CuFeS_2$) have been widely us as catalysts in advanced oxidation processes (AOPs) for wastewater treatment [37–41]. F instance, Dotto et al. demonstrated the ability of their prepared citrate-$CuFeS_2$ materials degrade 90% of bisphenol A (BPA) in a 15-min Fenton process [42]. Their novel CuFe samples were prepared with a microwave reactor (1400 W, 200 °C, 7 min). Pastra Martinez et al. used the mineral of $CuFeS_2$ mined from Jendouba, Tunisia, to cataly tyrosol degradation (85.0% degradation within 60 min) by using a UV light-assisted Fent reaction [43]. However, for ground water treatment, this method requires UV light external energy requirement. Chang et al. proposed that the microwave-assisted synthe of $CuFeS_2/Ag_3PO_4$ with enhanced rhodamine B (RhB) degradation (96% degradati within 1 min) under visible light-Fenton process [44]. However, these methods also ne light irradiation to improve the degradation performance of $CuFeS_2$.

Herein, we synthesized $CuFeS_2$ samples through hydrothermal and co-precipitat method to realize the advantages in material preparation, stability of materials, and deg dation performance in new water treatment option. We expected higher temperatu and pressure treatment (hydrothermal process) to make the prepared particles with sm size and high special surface area compared to the co-precipitated process, resulting higher catalytic activity [45–47]. In order to prove this, the prepared $CuFeS_2$ samples in t presence of $Na_2S_2O_8$ were used to evaluate the degradation efficiency of various dyest (RhB, rhodamine 6G [R6G], methylene blue [MB], methyl orange [MO], and BPA). T degradation mechanism of $CuFeS_2$ was elucidated and the reactive species were identifi Finally, the practical applications of $CuFeS_2$ samples in the treatment of environmen samples were demonstrated.

2. Results and Discussion

2.1. Characterization of the $CuFeS_2$ Samples

The morphology and composition of the prepared H-$CuFeS_2$ and C-$CuFeS_2$ samp were analyzed through SEM and EDS-mapping (Figure 1). As shown in Figure 1, H-CuFe and C-$CuFeS_2$ samples appear as sphere-like structures, with the average diameter rangi 25–40 nm and 95–125 nm, respectively. The smaller particle size of the H-$CuFeS_2$ sa ples can be the result of hydrothermal treatment which hindered the particle growth. C the other hand, high concentration of $N_2H_4 \cdot H_2O$ was used as reducing agent to prepa

C-CuFeS$_2$ samples, resulting in particle agglomeration. In addition, Ostwald ripening may occur during heating procedure. Therefore, small C-CuFeS$_2$ samples dissolved and redeposited onto larger C-CuFeS$_2$ samples. From the results of energy dispersive spectrometer (EDS)-mapping (green color, S elements; blue color, Fe elements; and red color, Cu elements), the presence of Cu, Fe, and S elements in both CuFeS$_2$ samples were confirmed and dispersed well in their crystals.

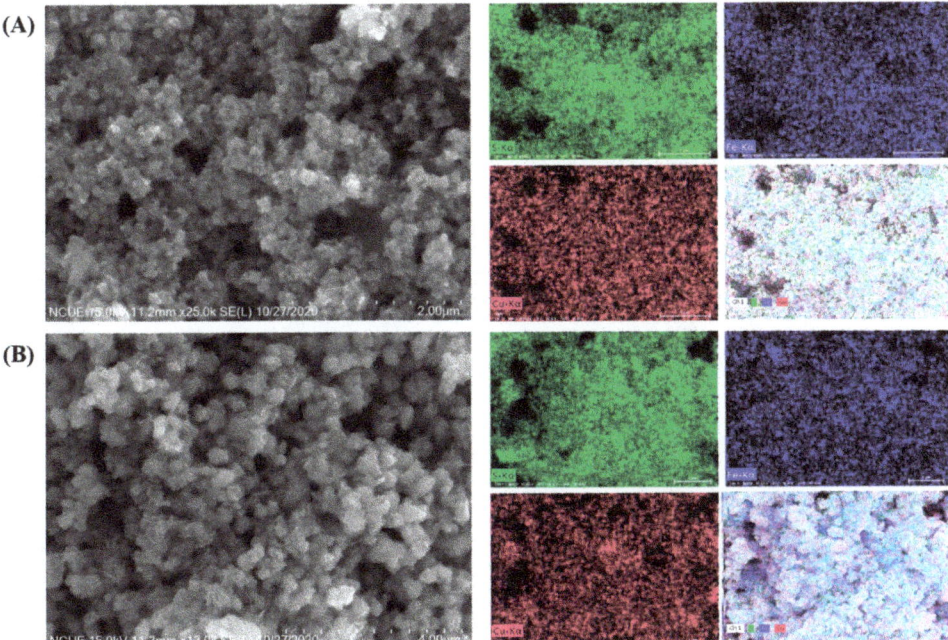

Figure 1. SEM images and EDS-mapping of (**A**) H-CuFeS$_2$ and (**B**) C-CuFeS$_2$ samples.

Figures 2 and 3 showed the TEM images and EDS spectra of the CuFeS$_2$ samples. The diameters of both CuFeS$_2$ samples from TEM images were consistent with the SEM results. We also found both CuFeS$_2$ samples possessed 0.31 nm and 0.23 nm of lattice lines, corresponding to the crystal planes of (112) and (204). The EDS spectra of the prepared CuFeS$_2$ samples confirm the presence of Cu, Fe, and S elements in their crystals, accordingly. The atomic ratios (Cu:Fe:S) for the H-CuFeS$_2$ and C-CuFeS$_2$ samples were determined to be 1.1:1:1.8 and 1.4:1:2.0, respectively. High content of Cu elements in the C-CuFeS$_2$ in the sample is consistent with lower solubility predicted from smaller Ksp value of Cu$_2$S when comparing with Fe$_2$S$_3$ (Ksp of Fe$_2$S$_3$: 3.7×10^{-19}, Ksp of Cu$_2$S: 2.0×10^{-47}).

XRD was used to investigate the crystal structure of the prepared CuFeS$_2$ samples. The XRD patterns of the prepared CuFeS$_2$ samples are shown in Figure 4A. The diffraction peaks at 29.5°, 49.1°, and 58.6° were identified and assigned to the (112), (204), (312), (204), and (312) faces of the tetragonal chalcopyrite CuFeS$_2$, respectively (PDF 83-0983). Through the Scherrer equation, the average crystal size of H-CuFeS$_2$ and C-CuFeS$_2$ was 20.36 and 11.2 nm, respectively. The Raman spectra of the prepared CuFeS$_2$ samples are shown in Figure 4B. The Raman shifts at 212 cm^{-1}, 276 cm^{-1}, and 379 cm^{-1} correspond to the S element, Cu(I)-S, and Fe(III)-S stretching vibration, respectively.

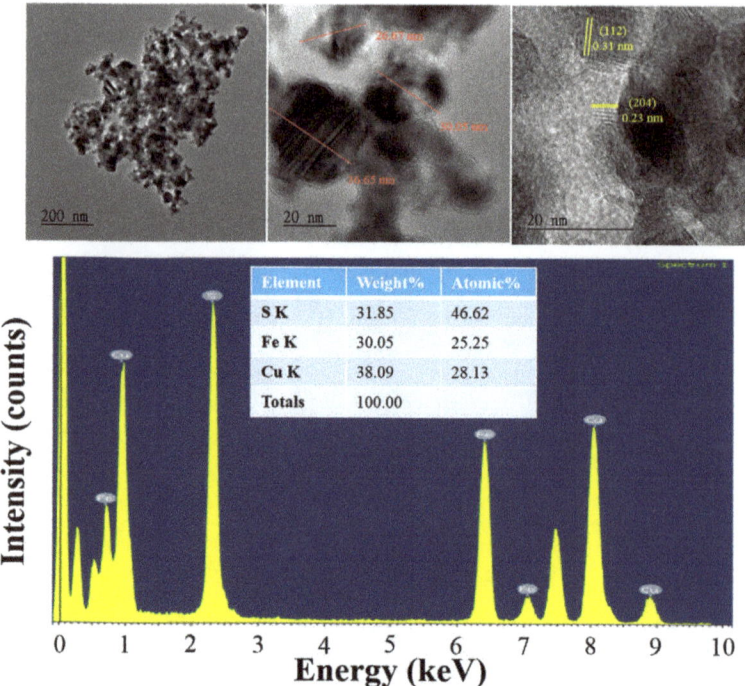

Figure 2. TEM images and EDS spectra of H-CuFeS$_2$ samples.

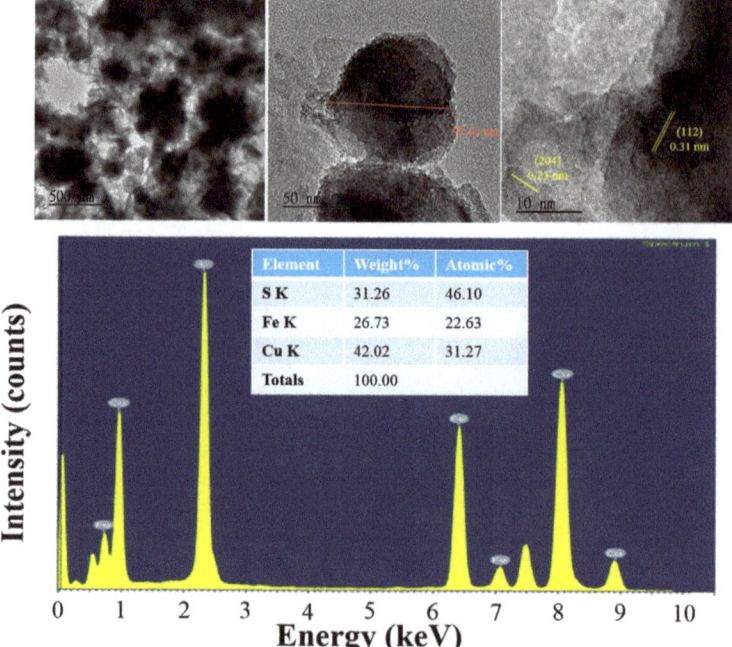

Figure 3. TEM images and EDS spectra of C-CuFeS$_2$ samples.

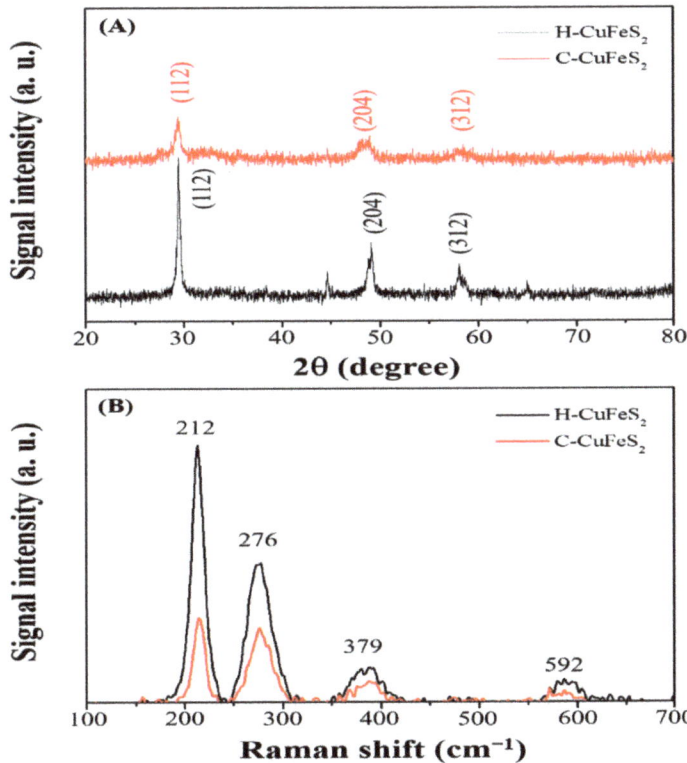

Figure 4. (**A**) XRD and (**B**) Raman spectra of H-CuFeS$_2$ (black), and C-CuFeS$_2$ (red) samples.

As another quality assurance method, XPS analysis of the prepared CuFeS$_2$ samples (Figures 5 and 6) revealed that it contains three elements: Cu, Fe, and S [48,49]. High-resolution XPS revealed Cu2p, Fe2p, and S2p in the H-CuFeS$_2$ samples as shown in Figure 5B–D, respectively. In Figure 5B, the peaks at 931.9 and 951.7 eV correspond to Cu$^+$ 2p$_{3/2}$ and Cu$^+$ 2p$_{1/2}$, respectively, whereas those at 933.2 and 953.0 eV correspond to Cu^{2+} 2p$_{3/2}$ and Cu^{2+} 2p$_{1/2}$, respectively. The peaks at 711.7 and 724.9 eV correspond to Fe^{2+} 2p$_{3/2}$ and Fe^{2+} 2p$_{1/2}$, respectively, whereas those at 715.2 and 734.3 eV correspond to Fe^{3+} 2p$_{3/2}$ and Fe^{3+} 2p$_{1/2}$, respectively (Figure 5C). The peaks at 162.5 and 167.8 eV correspond to S^{2-} 2p and S^{6+} 2p, respectively (Figure 5D). For C-CuFeS$_2$ samples, the peaks at 931.9 and 951.6 eV correspond to Cu$^+$ 2p$_{3/2}$ and Cu$^+$ 2p$_{1/2}$, respectively, whereas those at 934.1 and 952.7 eV correspond to Cu^{2+} 2p$_{3/2}$ and Cu^{2+} 2p$_{1/2}$, respectively (Figure 6B). The peaks at 711.4 and 724.8 eV correspond to Fe^{2+} 2p$_{3/2}$ and Fe^{2+} 2p$_{1/2}$, respectively, whereas those at 714.6 and 734.2 eV correspond to Fe^{3+} 2p$_{3/2}$ and Fe^{3+} 2p$_{1/2}$, respectively (Figure 6C). The peaks at 162.5 and 168.9 eV correspond to S^{2-} 2p and S^{6+} 2p, respectively (Figure 6D).

According to its peak area, the percentage of different oxidation states of each element in the prepared CuFeS$_2$ samples can be estimated. In H-CuFeS$_2$ samples, elemental compositions were found 82.3% Cu$^+$ and 17.6% Cu^{2+} from Cu analysis, 66.9% Fe^{2+} and 33.1% Fe^{3+} from Fe analysis, and 74.2% S^{2-} and 25.7% S^{6+} from sulfur analysis. In C-CuFeS$_2$ samples, elemental composition was found to be 90.6% Cu$^+$ vs. 9.3% Cu^{2+} for Cu, 60.8% Fe^{2+} vs. 39.2% Fe^{3+} for Fe, and 62.6% S^{2-} vs. 37.3%. S^{6+} for S.

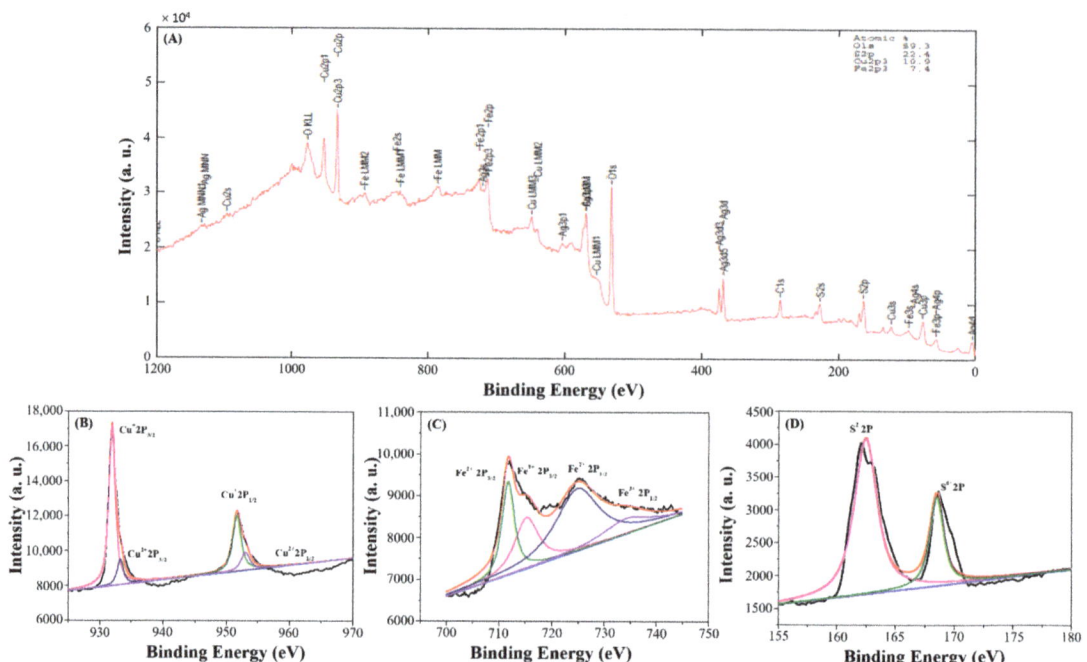

Figure 5. XPS spectra of H-CuFeS$_2$ samples: (**A**) full scan, (**B**) Cu$_{2p}$, (**C**) Fe$_{2p}$, and (**D**) S$_{2p}$.

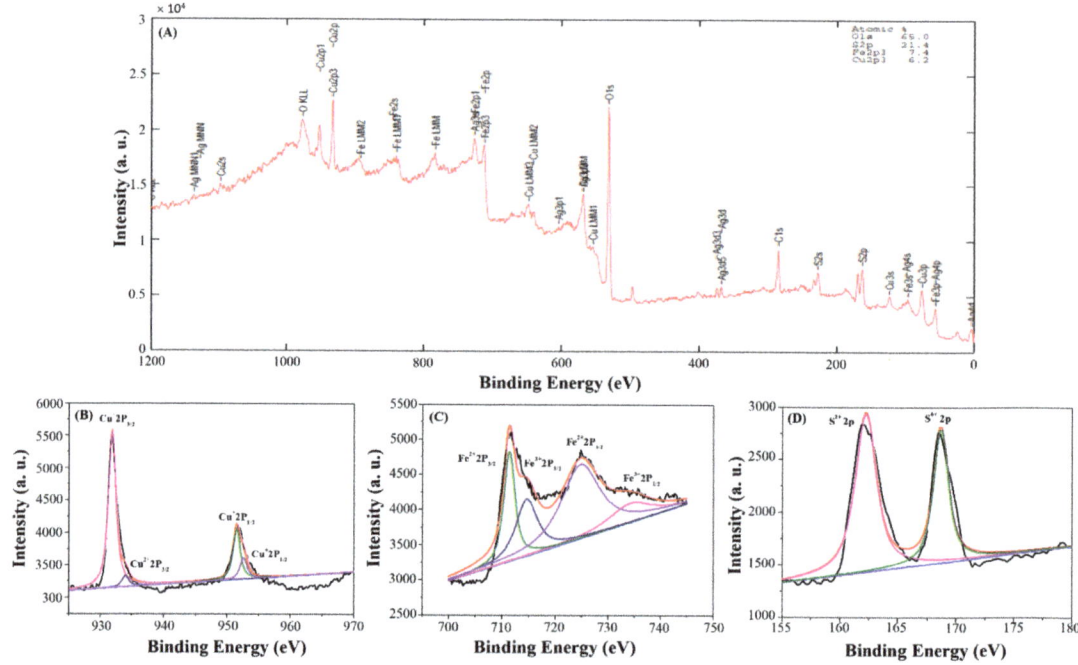

Figure 6. XPS spectra of C-CuFeS$_2$ samples: (**A**) full scan, (**B**) Cu$_{2p}$, (**C**) Fe$_{2p}$, and (**D**) S$_{2p}$.

2.2. Degradation Performance of the CuFeS₂ Samples

The degradation activity of the prepared CuFeS$_2$ samples was evaluated with RhB (20 ppm) first. According to our previous experience, the degradation efficiency decreased with an increasing dye concentration. This is because the excessive coverage of dye on the active surface of catalysts leads to a decrease in the catalytic activity. Thus, 20 ppm RhB was selected for the experiment. The variations in the RhB concentration (C/C_0), where C_0 is the initial RhB concentration, and C is the RhB concentration at time t, with the reaction time for the prepared CuFeS$_2$ samples in the presence of H$_2$O$_2$ (Fenton reaction) and Na$_2$S$_2$O$_8$ (Fenton-like reaction), were found in Figure 7. Prior to the addition of the oxidant, each catalyst (0.20 g) was introduced to the 20 ppm RhB solution for 30 min in the dark (indicated as "−30 min" in Figure 7) to reach equilibrium. The RhB concentration for H-CuFeS$_2$ samples after this equilibration time is lower than that of C-CuFeS$_2$ samples, reflecting RhB adsorption on H-CuFeS$_2$ samples. This is because smaller size of the H-CuFeS$_2$ samples had higher specific surface area than C-CuFeS$_2$ samples. Through the Fenton reaction, the degradation efficiency within 30 min was 32.3% and 26.4% for the H-CuFeS$_2$ and C-CuFeS$_2$ samples, respectively (black and blue curve). This suggests that the degradation performance of H-CuFeS$_2$ is better than that of C-CuFeS$_2$, attributable to adsorption ability of high specific surface area for the H-CuFeS$_2$ samples. The results of RhB degradation through a Fenton-like reaction by the H-CuFeS$_2$ and C-CuFeS$_2$ samples were shown in the red and pink curve. The degradation efficiency within 30 min reaction time follows this order: H-CuFeS$_2$ (93.7%) > C-CuFeS$_2$ (66.3%), indicating H-CuFeS$_2$ having higher catalytic activity to produce •SO$_4^-$ radicals. Furthermore, we found that degradation performance of •SO$_4^-$ radicals is higher than that of •OH radicals for both CuFeS$_2$ samples. This is because of the different lifetimes of radicals (•SO$_4^-$ radicals: 4 s, •OH radicals: 1 µs). Thus, the degradation system of H-CuFeS$_2$ through a Fenton-like reaction was selected for the further study.

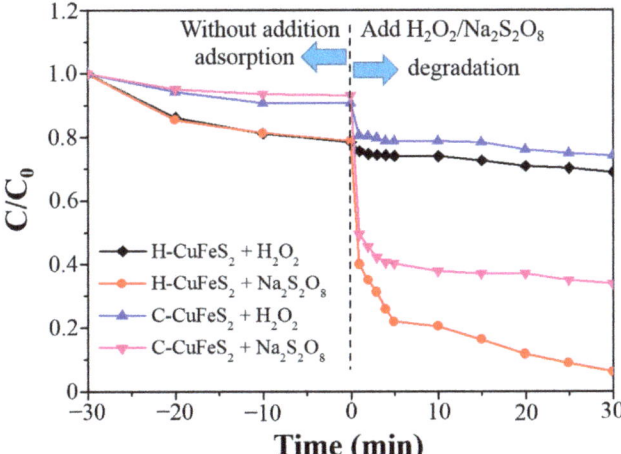

Figure 7. Fenton and Fenton-like reactions for RhB degradation at different conditions: H-CuFeS$_2$ in the presence of H$_2$O$_2$ (black), H-CuFeS$_2$ in the presence of Na$_2$S$_2$O$_8$ (red), C-CuFeS$_2$ in the presence of H$_2$O$_2$ (blue), and C-CuFeS$_2$ in the presence of Na$_2$S$_2$O$_8$ (pink).

To maximize the degradation performance of H-CuFeS$_2$, the effect from various concentrations of Na$_2$S$_2$O$_8$ was studied. As shown in Figure 8, the degradation efficiency increased with increasing Na$_2$S$_2$O$_8$ concentration. Due to low solubility of Na$_2$S$_2$O$_8$, we selected 4.0 mM of Na$_2$S$_2$O$_8$ as the optimum required concentration of Na$_2$S$_2$O$_8$. Dye adsorption on H-CuFeS$_2$ was observed in the absence of Na$_2$S$_2$O$_8$ (black cure in Figure 9). Although direct degradation of RhB by Na$_2$S$_2$O$_8$ without H-CuFeS$_2$ was noticed from the experiment due to the high oxidizing strength of $Na_2S_2O_8$ (red curve in Figure 9), its r of degradation cannot compete with H-CuFeS$_2$ samples in the presence of $Na_2S_2O_8$, wh achieved an impressive 98.8% within 10 min (blue cure in Figure 9). In addition, we also alyzed the degradation performances of Cu_2S and FeS_2 nanoparticles to investigate wh element is important for a Fenton-like reaction. As shown in Figure 10, the RhB degradati efficiency within 15 min reaches 64.1% and 89.0% for Cu_2S and FeS nanoparticles, resp tively. These results suggested that the FeS_2 nanoparticles catalyze $Na_2S_2O_8$ to produ •SO_4^- radicals better than Cu_2S nanoparticles, indicating Fe component is important th Cu component for the Fenton-like reaction.

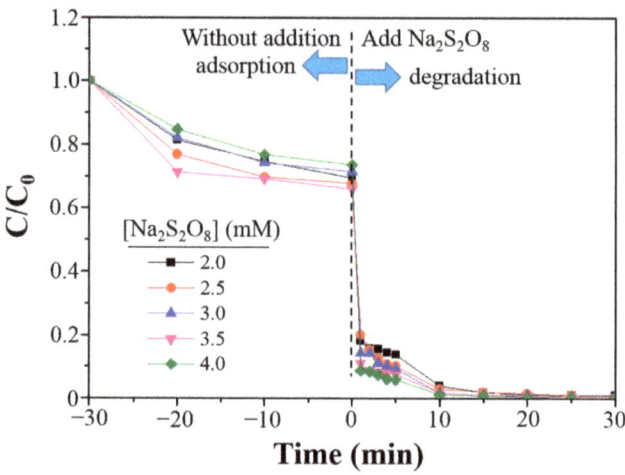

Figure 8. Fenton-like reaction for RhB degradation by H-CuFeS$_2$ samples at different concentrat of $Na_2S_2O_8$.

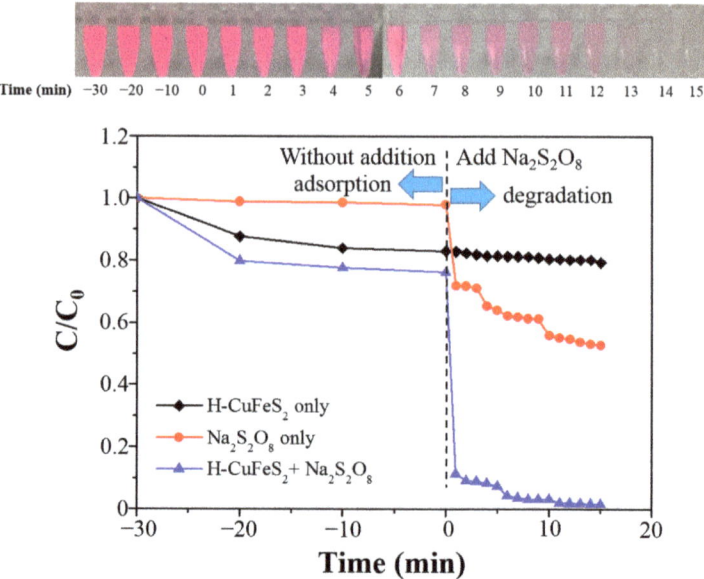

Figure 9. Fenton-like reaction for RhB degradation under different conditions: H-CuFeS$_2$ samp only (black), $Na_2S_2O_8$ only (red), and H-CuFeS$_2$ in the presence of $Na_2S_2O_8$ (blue). Top ima photographs of the RhB solution under the Fenton reaction at different reaction time.

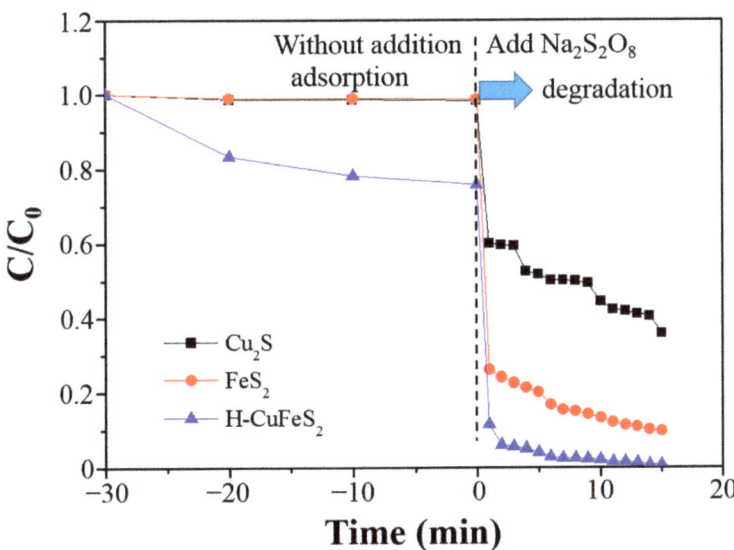

Figure 10. Fenton-like reaction for RhB degradation in the presence of $Na_2S_2O_8$ by using different catalysts: Cu_2S (black), FeS_2 (red), and H-CuFeS$_2$ (blue).

2.3. Degradation Mechanism of H-CuFeS$_2$

As a key mechanistic study, the active species involved in the degradation reaction were identified systematically using the free radical trapping experiments (Figure 11A). Methanol and NaN$_3$ were used as •OH and •SO$_4^-$ scavengers, respectively. Comparing to methanol, NaN$_3$ inhibit RhB degradation more, indicating that •SO$_4^-$ radicals are the major species involved in the Fenton-like degradation (blue curve in Figure 11A). According to the results of the scavenger test and XPS experiment, we propose a possible degradation mechanism. First, Fe^{2+}/Cu^+ ions on the CuFeS$_2$ surface catalyzed $S_2O_8^{2-}$ to produce •SO$_4^-$ radicals (Equations (1) and (2)). Due to high oxidation activity of •SO$_4^-$ radicals (E^0 = 2.5–3.1 V), they were utilized to degrade dyes and to oxidize Fe^{2+}/Cu^+ ions (Equations (3)–(5)). Then, •OH radicals also produced from the oxidation reaction between •SO$_4^-$ radicals and H$_2$O/OH$^-$ to degrade the dyes (Equations (6)–(8)). Thus, after adding methanol to the reaction mixture, RhB degradation in CuFeS$_2$ samples was slightly decreased, indicating that production of •OH radicals are considered as the indirect active species in the CuFeS$_2$ catalyzed RhB degradation (red curve in Figure 11A).

$$Fe^{2+} + S_2O_8^{2-} \rightarrow Fe^{3+} + SO_4^{\cdot -} + SO_4^{2-} \tag{1}$$

$$Cu^+ + S_2O_8^{2-} \rightarrow Cu^{2+} + SO_4^{\cdot -} + SO_4^{2-} \tag{2}$$

$$SO_4^{\cdot -} + RhB \rightarrow CO_2 + H_2O \tag{3}$$

$$SO_4^{\cdot -} + Fe^{2+} \rightarrow Fe^{3+} + SO_4^{2-} \tag{4}$$

$$SO_4^{\cdot -} + Cu^+ \rightarrow Cu^{2+} + SO_4^{2-} \tag{5}$$

$$SO_4^{\cdot -} + H_2O \rightarrow SO_4^{2-} + \cdot OH + H^+ \tag{6}$$

$$SO_4^{\cdot -} + OH^- \rightarrow SO_4^{2-} + \cdot OH \tag{7}$$

$$\cdot OH + RhB \rightarrow CO_2 + H_2O \tag{8}$$

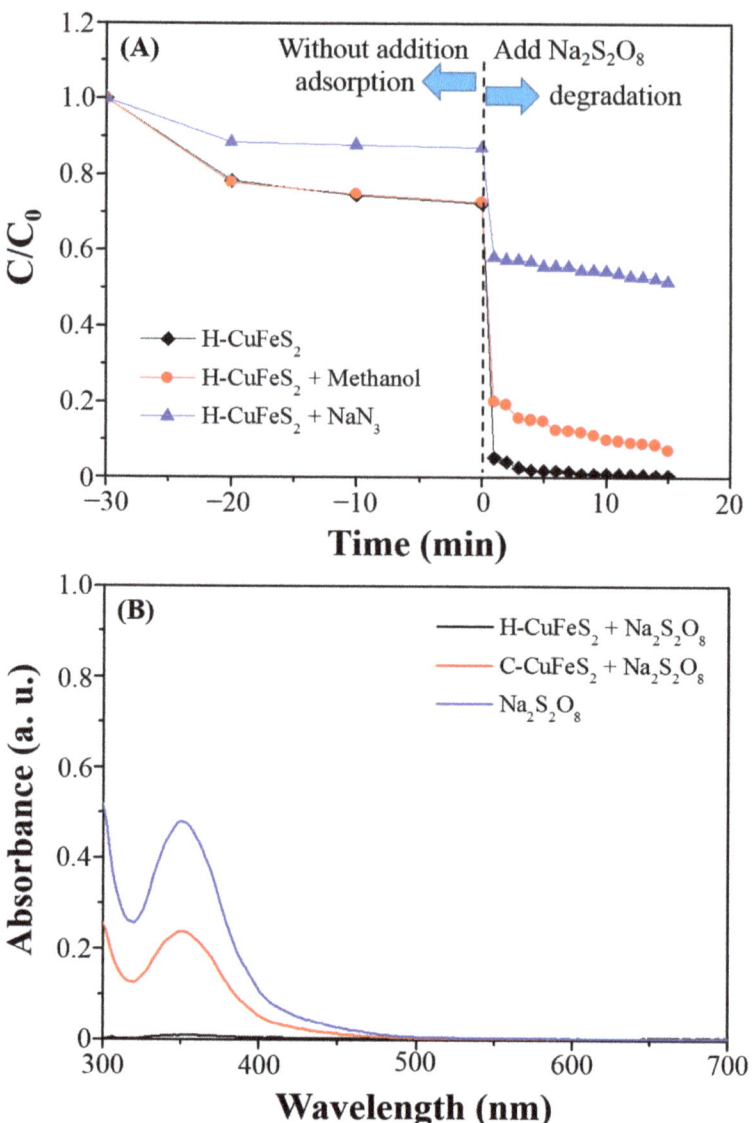

Figure 11. (A) Free radical trapping experiment and (B) absorbance spectra at different condition.

•SO$_4^-$ radical production in the Fenton-like reaction was further studied using the spectrophotometric method [50]. According to Equations (9) and (10), I$_3^-$ solution (light yellow) was found from chemical reaction between S$_2$O$_8^{2-}$ and KI. The absorbance spectra of the S$_2$O$_8^{2-}$/KI solution in the absence and presence of the prepared CuFeS$_2$ samples were evaluated. Figure 11B shows that an absorbance peak was observed at 358 nm for each sample and that the maximum absorbance was observed in the absence of the prepared CuFeS$_2$ samples (blue curve in Figure 11B). This suggests that S$_2$O$_8^{2-}$ produced the highest amount of I$_2$ compared to others, thereby leading to more chemical reactions with KI generate I$_3^-$. Due to a high specific surface area and high content of Fe^{2+} ions, H-CuFeS effectively catalyzed S$_2$O$_8^{2-}$ to produce •SO$_4^-$ radicals, as a result of a few I$_2$ production. Thus, the absorbance intensity at 358 nm of H-CuFeS$_2$/S$_2$O$_8^{2-}$/KI mixing solution (black) curve in Figure 11B) was lower than that of C-CuFeS$_2$/S$_2$O$_8^{2-}$/KI mixing solution (red curve in Figure 11B).

$$S_2O_8^{2-} + 2I^- \rightarrow 2SO_4^{2-} + I_2 \qquad (9)$$

$$I_2 + KI \rightarrow I_3^- + K^+ \qquad (10)$$

On the basis of the results described above, the degradation scheme of the H-CuFeS$_2$ samples in the Fenton-like reaction was proposed (Scheme 1). •SO$_4^-$ radicals and •OH radicals were produced from the Fenton-like reaction between S$_2$O$_8^{2-}$ and Fe^{2+}/Cu$^+$ ions on the H-CuFeS$_2$ surface to degrade RhB (Equations (1)–(8)). Then, Fe^{2+}/Cu$^+$ ions were regenerated through a series reduction of S^{2-} anions (Equations (11)–(13)). Moreover, it is also possible to produce Fe^{2+} ions by reduction reaction between Cu$^+$ and Fe^{3+} ions (Equation (14)).

$$S^{2-} + Fe^{3+}/Cu^{2+} \rightarrow Fe^{2+}/Cu^+ + S_2^{2-} \qquad (11)$$

$$S_2^{2-} + Fe^{3+}/Cu^{2+} \rightarrow Fe^{2+}/Cu^+ + S_n^{2-} \qquad (12)$$

$$S_n^{2-} + Fe^{3+}/Cu^{2+} \rightarrow Fe^{2+}/Cu^+ + SO_4^{2-} \qquad (13)$$

$$Cu^+ + Fe^{3+} \rightarrow Fe^{2+} + Cu^{2+} \qquad (14)$$

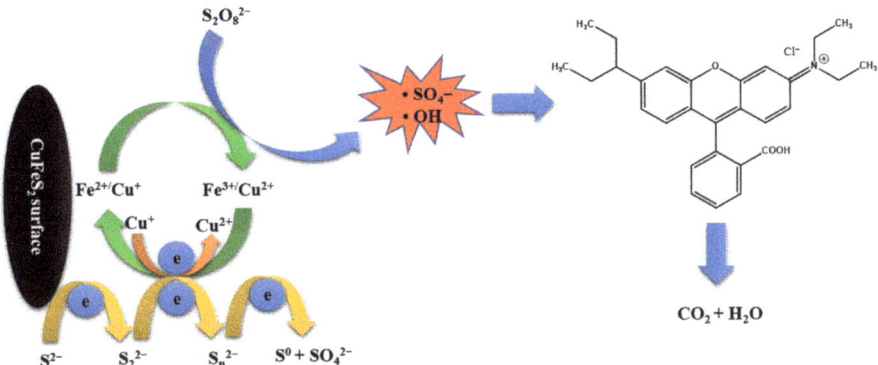

Scheme 1. Possible scheme of Fenton-like reaction for the H-CuFeS$_2$ samples.

2.4. Stability and Practical Applications of H-CuFeS$_2$

The stability of the catalyst is an essential parameter for the development of practical water treatment applications. To investigate the stability of H-CuFeS$_2$, results of pH effect, copper ions effect, and cyclic RhB degradation tests were evaluated as shown in Figures 12–14. Figure 12 showed the study of pH effect. RhB degradation by H-CuFeS$_2$ at pH 4.0 maintained a similar degradation efficiency at pH 7.0 (98.48% at pH 4.0 and 98.49% at pH 7.0, respectively), whereas that at pH 10.0 resulted in a considerable loss of efficiency (72.13% at pH 10.0). This is because most •SO$_4^-$ radicals were converted to •OH radicals at alkaline condition (Equations (6)–(8)). Thus, •OH radicals are the major active radicals involved in dye degradation at alkaline condition. In addition, inactive porphyrin ferryl complexes (FeO^{2+}) are formed as Fe^{2+} ions in the alkaline solution. As a result, a weakened degradation result at pH 10.0 was found (Figure 12).

In the study of copper ion effect as shown in Figure 13, RhB degradation efficiencies by H-CuFeS$_2$ in the presence of Cu$^+$ ioins were 88.64% at pH 4.0, 91.21% at pH 7.0, and 87.42% at pH 10.0, whereas those in the presence of Cu^{2+} ions were 93.96% at pH 4.0, 94.21% at pH 7.0, and 91.19% at pH 10.0. Comparing to that at pH 10.0 in the absence of copper ions, an obvious improvement was found. This is because •SO$_4^-$ radicals are produced in the presence of Cu$^+$ ions (Equation (2)). In addition, Fe^{2+}/Cu$^+$ ions were regenerated through reduction between S^{2-} anions and Fe^{3+}/Cu^{2+} ions (Equations (11)–(13)). As a result, an improve degradation at pH 10.0 was found.

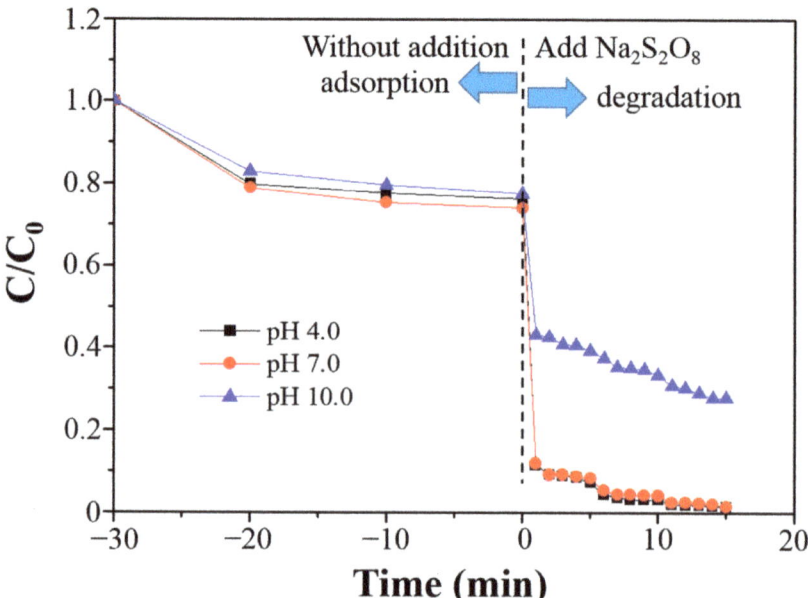

Figure 12. Fenton-like reaction for RhB degradation by the H-CuFeS$_2$ samples at different pH value system.

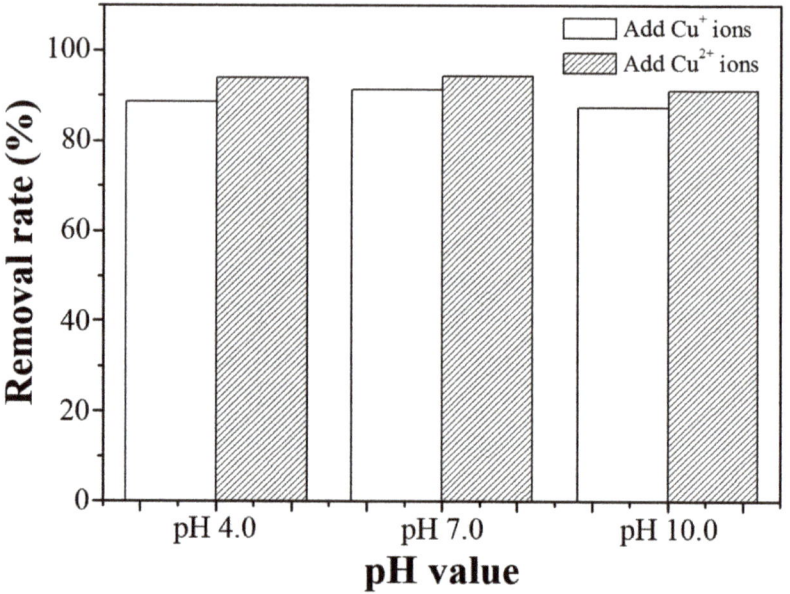

Figure 13. RhB degradation efficiency by the H-CuFeS$_2$ samples at different pH value system in presence of copper ions.

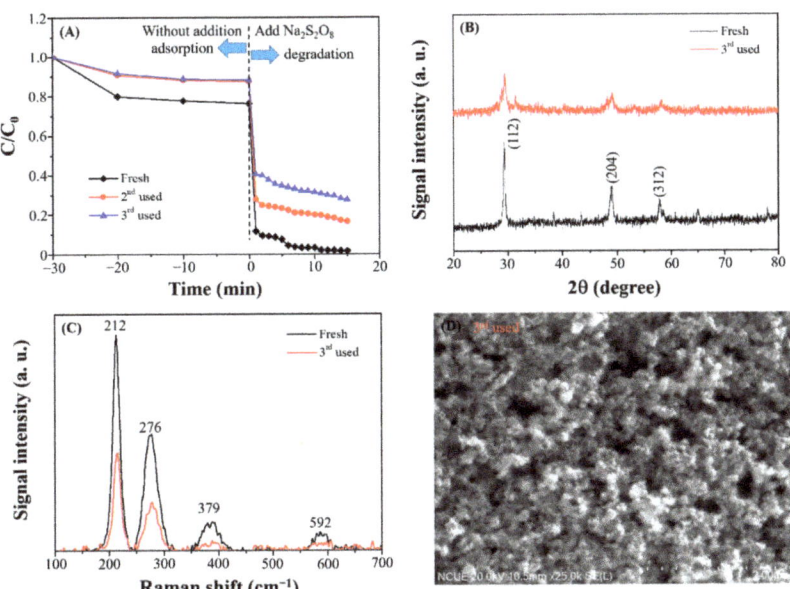

Figure 14. (**A**) Fenton-like reaction for RhB degradation by the H-CuFeS$_2$ samples for the recycling-used test, (**B**) XRD, (**C**) Raman spectra, and (**D**) SEM image of the 3rd used samples.

For recyling-used study, Figure 14A showed RhB degradation by H-CuFeS$_2$ exhibited a considerable loss of efficiency (from 98.48% to 72.46% after three cycles). Furthermore, the corresponding XRD, Raman, and SEM results (Figure 14B–D) suggest a decrease in the phase structure of the H-CuFeS$_2$ samples after the repeated reactions, indicating the destruction of the H-CufeS$_2$ sample crystalization. In addition, EDS spectrum found that the atomic ratio (Cu:Fe:S) for the third used H-CuFeS$_2$ samples was determined to be 1:1:1.9. The morphology of the third used samples still retained sphere-like structures, with the average diameter ranging 20–35 nm. Further research to improve recycling-used ability by other heterojunction, such as those doped by Ag@Ag$_3$PO$_4$ nanoparticles, is now underway in our laboratory.

To assess the practical applications of H-CuFeS$_2$ as a new water treatment option, various dyes (R6G, MB, and MO) and colorless organic compound (BPA) were tested (Figure 15A). H-CuFeS$_2$ exhibited excellent degradation efficiency toward R6G, MB, MO, and BPA, with 96.84%, 93.86%, 81.89%, and 75.24% degradation achieved within 10 min, respectively. In addition, the mineralization performance of H-CuFeS$_2$ comparing to a traditional Fenton reaction (Fe^{2+}/H$_2$O$_2$) was evaluated. From the TOC analysis (Figure 15B), mineralization efficiency for the Fe^{2+}/H$_2$O$_2$ and H-CuFeS$_2$/S$_2$O$_8^{2-}$ system was 70.0% and 80.1%, respectively, representing 10.1% improvement of RhB degradation. Finally, the prepared H-CuFeS$_2$ samples were used to degrade RhB in the environmental water samples (pond water and seawater). H-CuFeS$_2$ exhibited adequate mineralization efficiency through the Fenton-like reaction for RhB degradation. A notable difference in the mineralization efficiency for RhB was observed for the seawater samples (47.9% efficiency within 10 min) compared with pond water samples (63.8% efficiency within 10 min), probably because of the effect of higher concentration of anions or radical scavengers in the seawater sample that reduced the degradation activity of H-CuFeS$_2$. Nevertheless, the studies on the environmental water samples strongly support the benefits of this newly developed H-CuFeS$_2$-based Fenton-like water treatment option.

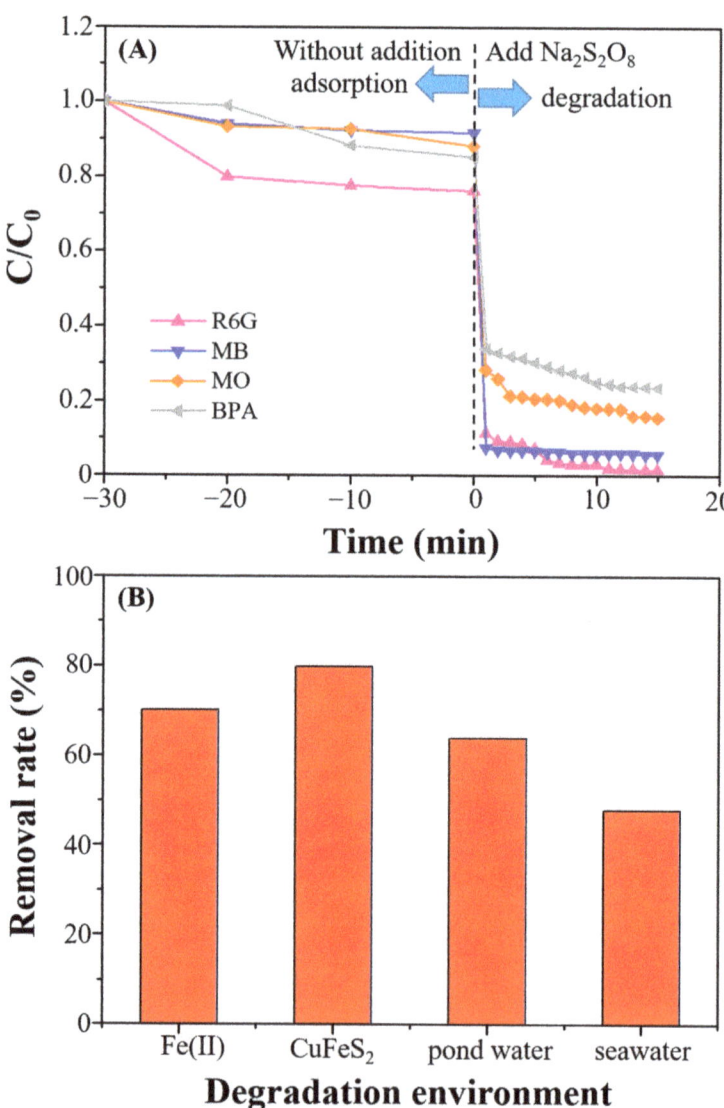

Figure 15. Fenton-like reaction of (**A**) different dyestuff by the H-CuFeS$_2$ samples, (**B**) TOC analysis of different degradation systems by using Fe(II) and the H-CuFeS$_2$ samples in the different environmental water samples.

3. Materials and Methods

3.1. Preparation of CuFeS$_2$

All chemicals were purchased from Sigma Aldrich (St. Louis, MO, USA) and were of analytical grade and used without further purification. In this study, hydrothermal (H) and co-precipitated method (C) were used to prepare CuFeS$_2$ samples, representing H-CuFeS$_2$ and C-CuFeS$_2$, respectively. For hydrothermal procedure, 0.989 g of CuCl, and 2.703 g of FeCl$_3$·6H$_2$O were added to 57 mL of deionized water, with stirring for 10 min. Then, 8 mL of Na$_2$S·9H$_2$O (0.02 mol) was added dropwisely into the above green mixture. After stirring for 30 min, the black mixture was transferred into a Teflon-lined stainless-ste autoclave. The autoclave was sealed and heated in an electric oven at 200 °C for 10 h. After the autoclave naturally cooled to room temperature, the precipitates were centrifuged (5000 rpm, 15 min) and washed three times with ethanol and deionized water, and then dried in vacuum at 60 °C overnight. In addition, Cu_2S and FeS_2 nanoparticles were prepared following similar method without adding $FeCl_3 \cdot 6H_2O$ and CuCl precursor, respectively.

For the co-precipitated method, 4.95 mg of CuCl, and 0.0135 g of $FeCl_3 \cdot 6H_2O$ were added to 20 mL of deionized water, with stirring at 70 °C for 10 min. Then, 1 mL of NH_4OH (30%) and 1 mL of $N_2H_4 \cdot H_2O$ (64–65%) were added dropwise into the above mixture with stirring at 70 °C for 3 h. After that, 0.024 g of $Na_2S \cdot 9H_2O$ was added into the above brown mixture with stirring at 70 °C for 3 h. Finally, the black precipitates were centrifuged (5000 rpm, 15 min) and washed three times with ethanol and deionized water, and then dried in vacuum at 60 °C overnight.

3.2. Characterization of $CuFeS_2$

The morphological and compositional characteristics of all as-prepared samples were observed with scanning electron microscopy (SEM) on a HITACHI S-4300 (Hitachi, Tokyo, Japan) and transmission electron microscopy (TEM) on a 1200EX II (JEOL, Tokyo, Japan) equipped with a QUANTAX Annular XFlash QUAD FQ5060 (Bruker Nano, Berlin, Germany). The crystallographic texture of the samples was measured by powder X-ray diffraction (XRD) on SMART APEX II (Bruker AXS, Billerica, MA, USA) using Cu Kα radiation ($\lambda = 1.5406$ Å). Raman spectra were collected at room temperature using a confocal micro-Raman system (Thermo Scientific Inc., New York, NY, USA). A 532 nm laser line was used as the photoexcitation source with a laser power of 2 mW focused on the sample for 10 s. The binding energy of elements was determined through X-ray photoelectron spectroscopy (XPS) on a VG ESCA210 (VG Scientific, West Sussex, UK).

3.3. Degradation Procedure

RhB degradation was used to assess the degradation activity of the prepared samples. For the Fenton-like reaction, 20 mg of the prepared catalyst samples was added into the RhB solution (20 ppm, 50 mL), and the solution was stirred in the dark for 30 min. At 10 min before adding $Na_2S_2O_8$, the absorbance at its characteristic absorption peak of 550 nm was measured to check the adsorption ability of the prepared samples. Subsequently, 100 μL of $Na_2S_2O_8$ (2 M) was added to dye solution. After a given time interval, 1 mL of suspension was sampled with a plastic pipette and this aliquot was quenched immediately by adding 10 μL NaN_3 (1 M) and filtered by a 0.22-μm syringe filter organic membrane to remove catalyst particles. The concentration of RhB was measured using a Synergy H1 hybrid multimode microplate reader (BioTek Instruments, Winooski, VT, USA) at its characteristic absorption peak of 550 nm. Similar processes were performed for other catalysts (Cu_2S and FeS_2), dyestuffs (R6G, MB, and MO), and organic pollutant (BPA). After the experiment, TOC concentration was determined on an Elementar Acquray TOC analyzer (Elementar Analysensysteme GmbH, Langenselbold, Germany) to evaluate the extent of mineralization.

3.4. Free Radical Trapping Experiment

To investigate the active species generated during RhB degradation over $H-CuFeS_2$, the trapping experiment was conducted using NaN_3 and methanol (each 0.1 M) as the capturing agent for •SO_4^- radicals and •OH radicals, respectively. The implemented trapping experimental procedure was identical to the steps mentioned in the degradation section with an additional step of adding the capturing agent at each run.

4. Conclusions

The prepared $H-CuFeS_2$ samples showed higher RhB degradation efficiency through the Fenton-like reaction than the prepared $C-CuFeS_2$, FeS_2, Cu_2S nanoparticles, and previously reported samples (Table 1). This high enhancement in the degradation efficiency

(98.8% RhB degradation within 10 min) was attributed to the prepared H-CuFeS$_2$ sample possessed smaller size and higher surface area. Based on the results of scavenger test and radicals' quantitation experiments, H-CuFeS$_2$ catalyzed Na$_2$S$_2$O$_8$ to produce •SO$_4^-$ radicals and •OH radicals for the organics degradation. As we know, the three limiting factors to address prior to industrial application were viable methods of catalyst preparation, the catalyst durability and universality under operating conditions. The prepared H-CuFeS$_2$ samples possessed several attractive features. First, the prepared H-CuFeS$_2$ samples in the presence of Na$_2$S$_2$O$_8$ had 98.8% RhB degradation performance within 10 min. In addition, various organics (R6G, MB, MO, and BPA) with 75.24–96.84% degradation efficiency could be achieved. However, the repeated use of H-CuFeS$_2$ showed performance deterioration due to the change in the crystal phase of used H-CuFeS$_2$. Further research on the high recycling-used ability of other heterojunction CuFeS$_2$ composites, such as those doped by Ag@Ag$_3$PO$_4$ nanoparticles, is now underway in our laboratory. Finally, the prepared H-CuFeS$_2$ samples were used to degrade RhB with 10.1% mineralization improvement comparing to traditional Fenton reaction (Fe^{2+}/H$_2$O$_2$). It is also easy to recover H-CuFeS$_2$ catalyst comparing to Fe^{2+} ions. In addition, H-CuFeS$_2$ catalyst deposited on a cellulose based substrate is ongoing in our lab. The difficult separation and recycle of powder catalyst may result in high cost and secondary pollution, therefore, the powder form catalyst greatly limited the commercial industrial application. More importantly, H-CuFeS$_2$ deposited on cellulose is very suitable for the dynamic-flow water treatment system. We will propose a new adsorption-degradation strategy for the pollutant removal in industrial level application in the future.

Table 1. Comparison of degradation performance using the (photo-) Fenton-like reaction.

Samples	Preparation	Degradation Performance	Target	Ref.
CuO/MSS	Hydrothermal method	90% degradation (0.15 g catalyst/50 ppm BPA) within 45 min	BPA	[20]
MMSS	Surface etching method	90% degradation (0.1 g catalyst/50 ppm MB) within 60 min	MB	[21]
Natural pyrite	Mined from Anhui, China	90% degradation (0.1 g catalyst/100 ppm MB) within 120 min	MB	[26]
CDs/Fe$_3$O$_4$@CS	Solvothermal method	96% degradation (0.3 g catalyst/50 µM Ibuprofen (IBP)) within 2 h (350-W Xe lamp)	IBP	[23]
TiO$_2$ nanotubes arrays	Anodization	94.6% degradation (1 ppm BPA) within 30 min (300-W Xe lamp)	BPA	[51]
Fe^0@Fe$_3$O$_4$ nanowires	Reduction method	100% degradation (2.5 mg catalyst/0.5 ppm Atrazine (ATZ)) within 6 min	ATZ	[24]
FeOCl nanosheets	Pyrolysis method	86.5% degradation (0.05 g catalyst/10 µM Phenacetin (PCNT)) within 30 min	PCNT	[25]
CuFeS$_2$	Hydrothermal method	98.8% degradation (0.02 g catalyst/20 ppm RhB) within 10 min	RhB, R6G, MB, MO, BPA	This study

In summary, this study discovered the hydrothermal synthesis of CuFeS$_2$ samples and successfully demonstrated the application of the Fenton-like reaction in the environmental water samples. The current findings can be used to the application of AOPs in wastewater treatment in the future.

Author Contributions: Conceptualization, P.-Y.W. and Y.-W.L.; methodology, Y.-W.L.; software, Y.-W.L.; validation, P.-Y.W., T.-Y.L. and T.W.; formal analysis, P.-Y.W., T.-Y.L. and Y.-W.L.; investigation, P.-Y.W.; resources, Y.-W.L.; data curation, P.-Y.W. and T.-Y.L.; writing—original draft preparation, Y.-W.L.; writing—review and editing, T.W. and Y.-W.L.; visualization, Y.-W.L.; supervision, Y.-W.L.; project administration, Y.-W.L.; funding acquisition, Y.-W.L. All authors have read and agreed to the published version of the manuscript.

Funding: This study was supported by the Ministry of Science and Technology of Taiwan under contract (MOST 110-2113-M-018-001).

Conflicts of Interest: The authors declare no conflict of interest.

References

Su, Y. Revisiting carbon, nitrogen, and phosphorus metabolisms in microalgae for wastewater treatment. *Sci. Total Environ.* **2021**, *762*, 144590. [CrossRef] [PubMed]

Saeed, M.; Khan, I.; Adeel, M.; Akram, N.; Muneer, M. Synthesis of CoO-ZnO photocatalyst for enhanced visible-light assisted photodegradation of methylene blue. *New J. Chem.* **2022**. [CrossRef]

Nipa, S.T.; Akter, R.; Raihan, A.; Rasul, S.B.; Som, U.; Ahmed, S.; Alam, J.; Khan, M.R.; Enzo, S.; Rahman, W. State-of-the-art biosynthesis of tin oxide nanoparticles by chemical precipitation method towards photocatalytic application. *Environ. Sci. Pollut. Res.* **2022**. [CrossRef] [PubMed]

Varsha, M.; Kumar, P.S.; Rathi, B.S. A review on recent trends in the removal of emerging contaminants from aquatic environment using low-cost adsorbents. *Chemosphere* **2022**, *287*, 132270. [CrossRef]

Bilińska, L.; Gmurek, M. Novel trends in AOPs for textile wastewater treatment. Enhanced dye by-products removal by catalytic and synergistic actions. *Water Resour. Ind.* **2021**, *26*, 100160. [CrossRef]

Qasem, N.A.; Mohammed, R.H.; Lawal, D.U. Removal of heavy metal ions from wastewater: A comprehensive and critical review. *NPJ Clean Water* **2021**, *4*, 36. [CrossRef]

Kamran, U.; Bhatti, H.N.; Noreen, S.; Tahir, M.A.; Park, S.-J. Chemically modified sugarcane bagasse-based biocomposites for efficient removal of acid red 1 dye: Kinetics, isotherms, thermodynamics, and desorption studies. *Chemosphere* **2022**, *291*, 132796. [CrossRef]

Kamran, U.; Bhatti, H.N.; Iqbal, M.; Jamil, S.; Zahid, M. Biogenic synthesis, characterization and investigation of photocatalytic and antimicrobial activity of manganese nanoparticles synthesized from Cinnamomum verum bark extract. *J. Mol. Struct.* **2019**, *1179*, 532–539. [CrossRef]

Li, J.; He, C.; Cao, X.; Sui, H.; Li, X.; He, L. Low temperature thermal desorption-chemical oxidation hybrid process for the remediation of organic contaminated model soil: A case study. *J. Contam. Hydrol.* **2021**, *243*, 103908. [CrossRef]

Huang, R.; Zhu, Y.; Curnan, M.T.; Zhang, Y.; Han, J.W.; Chen, Y.; Huang, S.; Lin, Z. Tuning reaction pathways of peroxymonosulfate-based advanced oxidation process via defect engineering. *Cell Rep. Phys. Sci.* **2021**, *2*, 100550. [CrossRef]

Sun, Y.; Li, D.; Zhou, S.; Shah, K.J.; Xiao, X. Research Progress of Advanced Oxidation Water Treatment Technology. *Adv. Waterwater Treat. II* **2021**, *102*, 1–47.

Walton, J.; Labine, P.; Reidies, A. The chemistry of permanganate in degradative oxidations. In *Chemical Oxidation*; Eckenfelder, W.W., Bowers, A.R., Roth, J.A., Eds.; Technomic Publishing Co., Inc.: Lancaster, Basel, 1991; Volume 1, pp. 205–219.

Luo, H.; Zeng, Y.; He, D.; Pan, X. Application of iron-based materials in heterogeneous advanced oxidation processes for wastewater treatment: A review. *Chem. Eng. J.* **2021**, *407*, 127191. [CrossRef]

Sgroi, M.; Anumol, T.; Vagliasindi, F.G.; Snyder, S.A.; Roccaro, P. Comparison of the new $Cl_2/O_3/UV$ process with different ozone-and UV-based AOPs for wastewater treatment at pilot scale: Removal of pharmaceuticals and changes in fluorescing organic matter. *Sci. Total Environ.* **2021**, *765*, 142720. [CrossRef] [PubMed]

Deniere, E.; Alagappan, R.P.; Van Langenhove, H.; Van Hulle, S.; Demeestere, K. The ozone-activated peroxymonosulfate process (O_3/PMS) for removal of trace organic contaminants in natural and wastewater: Effect of the (in) organic matrix composition. *Chem. Eng. J.* **2022**, *430*, 133000. [CrossRef]

Ganiyu, S.O.; Zhou, M.; Martinez-Huitle, C.A. Heterogeneous electro-Fenton and photoelectro-Fenton processes: A critical review of fundamental principles and application for water/wastewater treatment. *Appl. Catal. B Environ.* **2018**, *235*, 103–129. [CrossRef]

Labiadh, L.; Ammar, S.; Kamali, A.R. Oxidation/mineralization of AO7 by electro-Fenton process using chalcopyrite as the heterogeneous source of iron and copper catalysts with enhanced degradation activity and reusability. *J. Electroanal. Chem.* **2019**, *853*, 113532. [CrossRef]

Ribeiro, J.P.; Nunes, M.I. Recent trends and developments in Fenton processes for industrial wastewater treatment–A critical review. *Environ. Res.* **2021**, *197*, 110957. [CrossRef]

Mahtab, M.S.; Farooqi, I.H.; Khursheed, A. Sustainable approaches to the Fenton process for wastewater treatment: A review. *Mater. Today* **2021**, *47*, 1480–1484. [CrossRef]

20. Liang, S.; Ziyu, Z.; Fulong, W.; Maojuan, B.; Xiaoyan, D.; Lingyun, W. Activation of persulfate by mesoporous silica spheres-doped CuO for bisphenol A removal. *Environ. Res.* **2022**, *205*, 112529. [CrossRef]
21. Liang, S.; Ziyu, Z.; Han, J.; Xiaoyan, D. Facile synthesis of magnetic mesoporous silica spheres for efficient removal of methylene blue via catalytic persulfate activation. *Sep. Purif. Technol.* **2021**, *256*, 117801. [CrossRef]
22. Kiejza, D.; Kotowska, U.; Polińska, W.; Karpińska, J. Peracids-New oxidants in advanced oxidation processes: The Use peracetic acid, peroxymonosulfate, and persulfate salts in the removal of organic micropollutants of emerging concern—A review. *Sci. Total Environ.* **2021**, *790*, 148195. [CrossRef] [PubMed]
23. Zhang, B.-T.; Wang, Q.; Zhang, Y.; Teng, Y.; Fan, M. Degradation of ibuprofen in the carbon dots/Fe$_3$O$_4$@carbon sphere pomegranate-like composites activated persulfate system. *Sep. Purif. Technol.* **2020**, *242*, 116820. [CrossRef]
24. Feng, Y.; Zhong, J.; Zhang, L.; Fan, Y.; Yang, Z.; Shih, K.; Li, H.; Wu, D.; Yan, B. Activation of peroxymonosulfate by Fe^0@Fe$_3$ core-shell nanowires for sulfate radical generation: Electron transfer and transformation products. *Sep. Purif. Technol.* **20**, *247*, 116942. [CrossRef]
25. Tan, C.; Xu, Q.; Sheng, T.; Cui, X.; Wu, Z.; Gao, H.; Li, H. Reactive oxygen species generation in FeOCl nanosheets activated peroxymonosulfate system: Radicals and non-radical pathways. *J. Hazard. Mater.* **2020**, *398*, 123084. [CrossRef]
26. Sun, L.; Hu, D.; Zhang, Z.; Deng, X. Oxidative degradation of methylene blue via PDS-based advanced oxidation process using natural pyrite. *Int. J. Environ. Res. Public Health* **2019**, *16*, 4773. [CrossRef]
27. Zhu, Y.; Liu, Y.; Li, P.; Zhang, Y.; Wang, G.; Zhang, Y. A comparative study of peroxydisulfate and peroxymonosulfate activation by a transition metal–H$_2$O$_2$ system. *Environ. Sci. Pollut. Res.* **2021**, *28*, 47342–47353. [CrossRef]
28. Ghanbari, F.; Khatebasreh, M.; Mahdavianpour, M.; Mashayekh-Salehi, A.; Aghayani, E.; Lin, K.-Y.A.; Noredinvand, B. Evaluation of peroxymonosulfate/O$_3$/UV process on a real polluted water with landfill leachate: Feasibility and comparative study. *Korean J. Chem. Eng.* **2021**, *38*, 1416–1424. [CrossRef]
29. Dung, N.T.; Thao, V.D.; Huy, N.N. Decomposition of glyphosate in water by peroxymonosulfate activated with CuCoFe-LD material. *Vietnam J. Chem.* **2021**, *59*, 813–822.
30. Wang, W.; Chen, M.; Wang, D.; Yan, M.; Liu, Z. Different activation methods in sulfate radical-based oxidation for organic pollutants degradation: Catalytic mechanism and toxicity assessment of degradation intermediates. *Sci. Total Environ.* **20**, *772*, 145522. [CrossRef]
31. Xu, X.; Tang, D.; Cai, J.; Xi, B.; Zhang, Y.; Pi, L.; Mao, X. Heterogeneous activation of peroxymonocarbonate by chalcopyrite (CuFeS$_2$) for efficient degradation of 2, 4-dichlorophenol in simulated groundwater. *Appl. Catal. B Environ.* **2019**, *251*, 273–2 [CrossRef]
32. Karim, A.V.; Jiao, Y.; Zhou, M.; Nidheesh, P. Iron-based persulfate activation process for environmental decontamination in water and soil. *Chemosphere* **2021**, *265*, 129057. [CrossRef] [PubMed]
33. Hou, K.; Pi, Z.; Yao, F.; Wu, B.; He, L.; Li, X.; Wang, D.; Dong, H.; Yang, Q. A critical review on the mechanisms of persulfate activation by iron-based materials: Clarifying some ambiguity and controversies. *Chem. Eng. J.* **2021**, *407*, 127078. [CrossRef]
34. Zheng, X.; Niu, X.; Zhang, D.; Lv, M.; Ye, X.; Ma, J.; Lin, Z.; Fu, M. Metal-based catalysts for persulfate and peroxymonosulfate activation in heterogeneous ways: A review. *Chem. Eng. J.* **2022**, *429*, 132323. [CrossRef]
35. Li, B.; Wang, Y.-F.; Zhang, L.; Xu, H.-Y. Enhancement strategies for efficient activation of persulfate by heterogeneous cobalt containing catalysts: A review. *Chemosphere* **2021**, *291*, 132954. [CrossRef]
36. Fayyaz, A.; Saravanakumar, K.; Talukdar, K.; Kim, Y.; Yoon, Y.; Park, C.M. Catalytic oxidation of naproxen in cobalt spinel ferrite decorated Ti$_3$C$_2$T$_x$ MXene activated persulfate system: Mechanisms and pathways. *Chem. Eng. J.* **2021**, *407*, 127842. [CrossRef]
37. Nie, W.; Mao, Q.; Ding, Y.; Hu, Y.; Tang, H. Highly efficient catalysis of chalcopyrite with surface bonded ferrous species activation of peroxymonosulfate toward degradation of bisphenol A: A mechanism study. *J. Hazard. Mater.* **2019**, *364*, 59– [CrossRef]
38. Huang, X.; Zhu, T.; Duan, W.; Liang, S.; Li, G.; Xiao, W. Comparative studies on catalytic mechanisms for natural chalcopyrite induced Fenton oxidation: Effect of chalcopyrite type. *J. Hazard. Mater.* **2020**, *381*, 120998. [CrossRef]
39. Li, Y.; Dong, H.; Li, L.; Tang, L.; Tian, R.; Li, R.; Chen, J.; Xie, Q.; Jin, Z.; Xiao, J. Recent advances in wastewater treatment through transition metal sulfides-based advanced oxidation processes. *Water Res.* **2021**, *192*, 116850. [CrossRef]
40. Xia, Q.; Zhang, D.; Yao, Z.; Jiang, Z. Investigation of Cu heteroatoms and Cu clusters in Fe-Cu alloy and their special effects mechanisms on the Fenton-like catalytic activity and reusability. *Appl. Catal. B Environ.* **2021**, *299*, 120662. [CrossRef]
41. Xia, Q.; Zhang, D.; Yao, Z.; Jiang, Z. Revealing the enhancing mechanisms of Fe–Cu bimetallic catalysts for the Fenton-like degradation of phenol. *Chemosphere* **2022**, *289*, 133195. [CrossRef]
42. Da Silveira Salla, J.; da Boit Martinello, K.; Dotto, G.L.; García-Díaz, E.; Javed, H.; Alvarez, P.J.; Foletto, E.L. Synthesis citrate–modified CuFeS$_2$ catalyst with significant effect on the photo–Fenton degradation efficiency of bisphenol a under visible light and near–neutral pH. *Colloid Surf. A Physicochem. Eng. Asp.* **2020**, *595*, 124679. [CrossRef]
43. Ltaïef, A.H.; Pastrana-Martínez, L.M.; Ammar, S.; Gadri, A.; Faria, J.L.; Silva, A.M. Mined pyrite and chalcopyrite as catalysts for spontaneous acidic pH adjustment in Fenton and LED photo-Fenton-like processes. *J. Chem. Technol. Biotechnol.* **20**, *93*, 1137–1146. [CrossRef]
44. Chang, S.-A.; Wen, P.-Y.; Wu, T.; Lin, Y.-W. Microwave-Assisted Synthesis of Chalcopyrite/Silver Phosphate Composites with Enhanced Degradation of Rhodamine B under Photo-Fenton Process. *Nanomaterials* **2020**, *10*, 2300. [CrossRef] [PubMed]

Kamran, U.; Park, S.-J. Hybrid biochar supported transition metal doped MnO_2 composites: Efficient contenders for lithium adsorption and recovery from aqueous solutions. *Desalination* **2022**, *522*, 115387. [CrossRef]

Kamran, U.; Park, S.-J. Acetic acid-mediated cellulose-based carbons: Influence of activation conditions on textural features and carbon dioxide uptakes. *J. Colloid Interface Sci.* **2021**, *594*, 745–758. [CrossRef] [PubMed]

Kamran, U.; Park, S.-J. Tuning ratios of KOH and NaOH on acetic acid-mediated chitosan-based porous carbons for improving their textural features and CO_2 uptakes. *J. CO2 Util.* **2020**, *40*, 101212. [CrossRef]

Ghahremaninezhad, A.; Dixon, D.; Asselin, E. Electrochemical and XPS analysis of chalcopyrite ($CuFeS_2$) dissolution in sulfuric acid solution. *Electrochim. Acta* **2013**, *87*, 97–112. [CrossRef]

Jiang, L.; Luo, Z.; Li, Y.; Wang, W.; Li, J.; Li, J.; Ao, Y.; He, J.; Sharma, V.K.; Wang, J. Morphology-and phase-controlled synthesis of visible-light-activated S-doped TiO_2 with tunable S^{4+}/S^{6+} ratio. *Chem. Eng. J.* **2020**, *402*, 125549. [CrossRef]

Liang, C.; Huang, C.-F.; Mohanty, N.; Kurakalva, R.M. A rapid spectrophotometric determination of persulfate anion in ISCO. *Chemosphere* **2008**, *73*, 1540–1543. [CrossRef]

Jia, J.; Liu, D.; Wang, S.; Li, H.; Ni, J.; Li, X.; Tian, J.; Wang, Q. Visible-light-induced activation of peroxymonosulfate by TiO_2 nano-tubes arrays for enhanced degradation of bisphenol A. *Sep. Purif. Technol.* **2020**, *253*, 117510. [CrossRef]

MDPI
St. Alban-Anlage 66
4052 Basel
Switzerland
Tel. +41 61 683 77 34
Fax +41 61 302 89 18
www.mdpi.com

Catalysts Editorial Office
E-mail: catalysts@mdpi.com
www.mdpi.com/journal/catalysts

www.ingramcontent.com/pod-product-compliance
Lightning Source LLC
LaVergne TN
LVHW070421100526
838202LV00014B/1500